ESSENTIALS OF CHEMICAL BIOLOGY

Leandro Oviedo

TAP

Toronto Academic Press

ESSENTIALS OF CHEMICAL BIOLOGY

Leandro Oviedo

Toronto Academic Press
4164 Lakeshore Road
Burlington ON L7L 1A4
Canada
www.tap-books.com

Email: orders@arclereducation.com

ISBN: 978-1-77956-325-5

ABOUT THE AUTHOR

Leandro Rodrigues Oviedo is a chemical engineer with a solid background in machine/deep learning, chemical kinetics, reactor design, and water/wastewater treatment (especially Advanced Oxidation Processes - AOPs). Master in Nanosciences (Nanoscience Post-Graduation Program at Franciscan University - UFN, Brazil), with expertise in nanozeolite synthesis from solid (agro)industrial wastes for adsorption and photocatalytic degradation of organic pollutants.

Table of Contents

List of Figures

points being denoted as the letters on the right side of the phylogeny. Numbers above the branches indicate the actual or estimated number of nucleotide substitutions that occurred along each branch, respectively. Actual substitutions were determined by sequencing 1091 base pairs of the ancestral viruses at each branch point in the tree

Figure 10.3. Methods for the randomization of DNA sequences using PCR. Each line in left column represents DNA duplex and each arrow a primer; the 5'-sense primer initiates strand copying using the complementary strand as a template, and the 5'-complementary primer initiates strand copying of the complementary strand using the sense strand as a template. Random methods such as error-prone PCR introduce random changes at positions throughout the gene sequence. Directed methods will introduce changes at a specific position(s) defined by lesions in the primers themselves. Recombination methods such as family shuffling start from a library of homologous genes from related species that can be artificially fragmented by DNAse I treatment and recombined by PCR (without added primers) leading to chimeric genes that code for chimeric proteins. These methods confer high levels of molecular diversity upon which to exercise selection screening for novel or improved characteristics

Figure 10.4. Strategy for evolving an enantioselective lipase. Molecular diversity is introduced by random mutagenesis of the lipase gene and site-specific random or saturation mutagenesis. Mutant genes resulting are segregated and separately expressed in individual bacterial clones. Protein products are secreted into the extracellular medium and a spectrophotometric assay used to assay for nitrophenolate production from (R) and (S) substrates by action of active lipase enzymes. Relative quantities of (R) and (S) product are quantified by gas chromatography and where enantioselectivity is also discovered then the corresponding gene is harvested for another round of mutagenesis and selection

List of Tables

PREFACE

Chemistry, biology, and physics are the essential pillars sustaining the development of the natural sciences. The tremendous progress in our understanding of biological systems is largely due to the skillful application of the principles and techniques of organic chemistry, with synthetic organic chemistry playing a key role in biological discovery. This progression illustrates how biology has transitioned from a descriptive (phenomenological) level to a molecular (biochemical) level, giving rise to new disciplines such as structural biology and molecular biology, which are now integral to the natural sciences.

Living organisms produce and release chemical compounds into their environment, significantly affecting other organisms and determining the existence of chemical interactions between them. Essentially, all organisms generate chemical signals and respond to the chemical signals of others. These signals are compounds produced through secondary metabolic pathways, which are closely linked to primary metabolic pathways and metabolites (carbohydrates, lipids, proteins, and nucleic acids).

In biology, the analysis of these interactions can be conducted "top-down," starting with a cell, tissue, limb, or organism and working down to the molecular level. Conversely, in chemistry, a "bottom-up" approach begins with molecules and native macromolecules, building up to the totality of the cell and its higher organizations, emerging through modular motifs and supramodular functional units.

Since 1839, it has been recognized that the cell is the simplest unit in living organisms. The cell is a protected region where diverse small molecules and macromolecular clusters interact harmoniously through self-assembly. Much of the cell's content consists of an aqueous solution containing small molecules (e.g., simple sugars, amino acids, vitamins) and ions (e.g., sodium, chloride, calcium ions). To study living systems and biochemical processes further, tools were needed to disrupt these systems using small molecules, providing deeper and more detailed information on the operation of living systems.

Following this trend, a new discipline, chemical biology, emerged at the interface of synthetic organic chemistry and molecular, structural, and cellular biology. The main goal of chemical biology is to elucidate the fundamental principles of the chemistry of life and to utilize our understanding of the behaviors of living organisms to study the interactions between biological macromolecules (endogenous) and small organic molecules (exogenous). This approach goes beyond understanding biological processes at the molecular level.

For chemists, chemical biology provides a field to explore new chemistry with biological relevance. For biologists, it offers a way to apply chemical principles to fundamental biological questions. The value of such studies has become evident with the global spread of COVID-19 (SARS-CoV-2). This pandemic forced scientists to consider the regulatory processes involved in infection and to develop a novel chemistry for creating vaccines and treatments that modulate these processes. Additionally, COVID-19 has taught us that the chemistry of biological systems is influenced by mutations and evolution and is constantly changing, leading to biological organisms with enhanced survival and adaptation mechanisms.

This book covers the fundamental aspects of chemical biology, encompassing a wide range of topics across 10 detailed chapters. Each chapter is designed to provide an in-depth understanding of the principles and techniques that are pivotal to the field.

Chapter 1 lays the groundwork by exploring the structures of biological macromolecules, including proteins, nucleic acids, and carbohydrates as well as lipid assemblies. It discusses the intricate details of their three-dimensional structures, the forces that stabilize them, and their functional implications in biological systems.

Chapter 2 focuses on the methods of synthesizing biological molecules. It covers both chemical synthesis (e.g., solid-phase peptide synthesis and oligonucleotide synthesis) and biological synthesis methods such as recombinant DNA technology and protein expression systems.

Chapter 3 examines how molecular biology techniques are utilized in chemical biology. Techniques such as gene cloning, PCR, and sequencing are discussed, highlighting their applications in manipulating and studying biological molecules.

Chapter 4 introduces electronic and vibrational spectroscopy methods used to analyze biological molecules. It explains the principles behind the techniques such as UV-Vis spectroscopy, fluorescence spectroscopy, and infrared spectroscopy, and their applications in determining molecular structure and dynamics.

Chapter 5 examines chemical biology within cellular contexts. It covers topics such as cell signaling pathways, the chemical modification of biomolecules within cells, and the use of chemical probes to study cellular processes.

The book then explores diffraction and microscopy techniques essential for visualizing biological structures at high resolution in Chapter 6. It discusses X-ray crystallography, NMR spectroscopy, and various forms of microscopy including electron microscopy and fluorescence microscopy.

In Chapter 7, the focus is on the principles of molecular recognition and binding. It discusses how molecules interact with each other through non-covalent interactions and the techniques used to study these interactions, such as surface plasmon resonance and isothermal titration calorimetry.

Chapter 8 provides an understanding of the kinetic and catalytic properties of biological molecules. It covers enzyme kinetics, the mechanisms of enzyme action, and the methods used to study catalytic processes.

Mass spectrometry and its applications in proteomics are the focus of Chapter 9. The chapter explains the principles of mass spectrometry, the different types of mass spectrometers, and how these techniques are used to analyze proteins and other biomolecules.

The final Chapter 10 explores the concepts of molecular selection and evolution. It discusses methods such as directed evolution and phage display, which are used to evolve proteins and nucleic acids with desired properties, and their applications in chemical biology and biotechnology.

This book aims to provide a thorough understanding of its topic, making the book an invaluable resource for students and professionals in the field of chemical biology.

Audience

The audience of this book includes students, researchers, and professionals in the fields of chemistry and biology who seek a comprehensive understanding of the intersection between these disciplines. It caters to individuals interested in exploring the fundamental principles and techniques of chemical biology, from synthetic organic chemistry to molecular, structural, and cellular biology. The book is designed to be accessible to those with varying levels of expertise, providing a solid foundation for beginners while offering insights and advanced concepts for experienced practitioners. It serves as a valuable resource for anyone looking to deepen their knowledge of the chemical processes underlying biological systems.

Learning Features

- Start-of-chapter learning objectives
- Key terms in each chapter
- Summary Points at the end of each chapter
- Review Questions at the end of each chapter
- Case Studies and Activities in each chapter for solving real problems
- References at the end of each chapter

CHAPTER 1

The Structures of Biological Macromolecules and Lipid Assemblies

LEARNING OBJECTIVES

After studying this chapter, you will be able to:

- Understand the four levels of protein structure.
- Describe the carbohydrate structure.
- Describe the basic structure of nucleic acids.
- Understand the macromolecular lipid assemblies.

KEY TERMS FROM THIS CHAPTER

Amino acid	Carbohydrates	Cells
Central dogma	Chemical biology	Cytosol
DNA	Fibrous proteins	Globular proteins
Nucleic acids	Nucleus	Organisms
Organs	Peptide bonds	Polypeptide chains
Proteins	RNA	

INTRODUCTION

Every living organism is made up of individual cells, which can differ greatly in size, shape, and appearance. In complex multicellular organisms, numerous cells are arranged into various functional organs that work together to carry out a specific task (Figure 1.1).

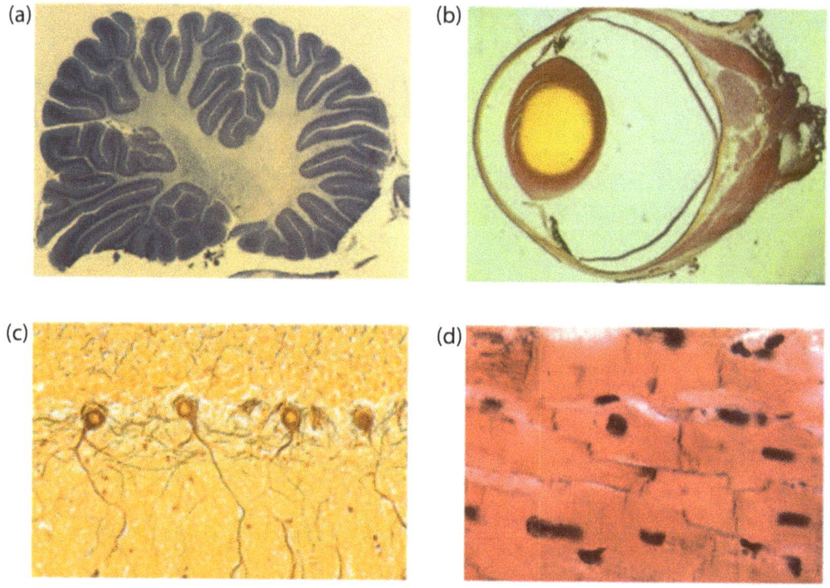

Figure 1.1. *Organs and cells.*

Note: (a) cross-section of mammalian brain showing the complex surface folds. There are an incalculable number of cells that make up the mammalian brain; (b) cross section of mammalian eyeball in which the lens is made of proteins controlled in function by peripheral muscles. There is an enormous morphological and functional diversity between cells required for muscle control, light reception, and signal transduction along the optic nerve; (c) cross-section of mammalian neurological tissue illustrating the neuron cell bodies with complex axonal/dendritic processes surrounded by support cells all of a wide range of size, shape, structure and function; (d) cross section of mammalian heart tissue showing clusters of muscle fibers (single cell myocytes) that make up the heart wall. Myocytes are multinucleate with a very different shape, composition and function to neurological cells.

Source: Andrew Miller and Julian Tanner. Essentials of Chemical Biology: Structure and Dynamics of Biological Macromolecules. Wiley. ISBN 978–0-470–84530–1. https://nattabiochem.wordpress.com/wp-content/uploads/2009/08/essentials_chemical_biology.pdf.

Regardless of their physical makeup, all living organisms' cells are composed of lipid assemblies, proteins, carbohydrates, and nucleic acids, which collectively determine a cell's structure and functionality. This collective composition determines a cell's structure and functionality. Knowing and comprehending the chemistry of these biological macromolecules entails understanding the fundamental building blocks of living things, including cells. Functionally speaking, macromolecular lipid assemblies enable compartmentalization through membrane barriers that separate the intracellular environment into distinct organelles or functional zones in addition to defining the "outer limits" of individual cells (Figure 1.2).

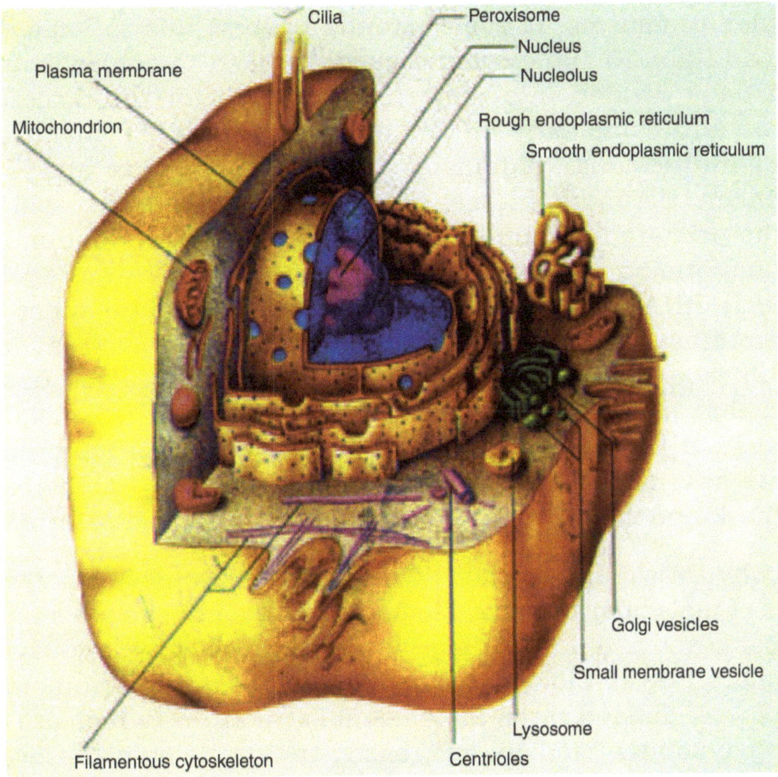

Figure 1.2. *General structure of a cell, with the main compartments (organelles) into which the interior is partitioned. All cells of all organisms are constructed from the main biological macromolecules proteins, carbohydrates, and nucleic acids; together with macromolecular lipid structures that comprise the membranes.*

Source: Andrew Miller and Julian Tanner. Essentials of Chemical Biology: Structure and Dynamics of Biological Macromolecules. Wiley. ISBN 978-0-470-84530-1. https://nattabiochem.wordpress.com/wp-content/uploads/2009/08/essentials_chemical_biology.pdf.

Proteins play a supporting and scaffolding role in both organelles and the cytosol, the fluid bulk of the cell, because membrane barriers are fluidic and lack rigidity. Additionally, proteins in the nucleus perform the role of nucleic acid packaging, which limits and confines enormous amounts of nucleic acids inside the nuclear volume. Additionally, proteins fulfill a variety of specific purposes in every part of a cell, including the cytosol or organelle volumes, the nucleus, the outer membranes (such as pores or receptors), the organelle membranes (such as selective transporters, redox acceptors, or energy transducers), and the cytosol or organelle volumes (such as enzyme catalysts, molecular chaperones, or "communication and control" centers). Both chemical biologists and biochemists find proteins to be endlessly fascinating due to their remarkable diversity of functions and "workhorse"-like nature in biology.

Deoxyribonucleic acid (DNA) and ribonucleic acid (RNA) are the two primary classes of nucleic acids. DNA is primarily found in the nucleus and contains genetic information that determines the make-up, organization, and structure of cells, as well as the multicellular structure of complex organisms. It even extends beyond this to impact the behavior of the organisms. In addition to many other divisions linked to control over gene use, DNA molecules are partially divided into genes, which code for the structures of proteins. Indeed, the degree and complexity of this control may be the main factor influencing the complexity of multicellular organisms; the more complex and sophisticated the multicellular organism, the more extensive and sophisticated the level of control is. On the other hand, the primary function of RNA is to transfer information from the nucleus to the cytosol. Transforming genetic information from the DNA storage form into actual protein structures is RNA's primary function. The fundamental tenet of biology—that genes code for proteins—is made possible by RNA. Lastly, carbohydrates are known to adorn certain intracellular proteins and bind to proteins in the outer membrane to form a glycocalyx that covers the surface of many cells and is crucial for intercellular communication, even if they are not stored in complex forms for primary metabolism. Huge carbohydrate assemblies also provide the exoskeleton framework to which cells are attached in the kingdoms of plants and insects, giving both form and function.

Proteins, carbohydrates, and nucleic acids are all examples of polymers constructed from common sets of molecular building blocks. Lipid assemblies are similarly constructed from a standard basis set of lipid building blocks connected by non-covalent bonds. The common feature shared by all biological macromolecules and macromolecular assemblies is their adoption of well-defined three-dimensional structures, which are essential to their functions, including dynamics, binding, and reactivity. Surprisingly, these three-dimensional structures are not only essential to function but also the product of stereoelectronic characteristics that are present in every class of polymer or macromolecular assembly combined with weak, non-covalent forces of association. Although structure does not always predict function, it is difficult to understand function in the absence of structure. Therefore, before diving into any further sections of this fascinating subject, readers interested in chemical biology should have a basic understanding of the structures of proteins, carbohydrates, nucleic acids, and lipid assemblies.

1.1. PROTEIN STRUCTURE

Protein structures are created when amino acids condense to form peptide bonds (Branden, 1990; Tooze, 1991). A protein's primary structure refers to the arrangement of its amino acids. Peptide bond dihedral angles determine the secondary structure, and protein chain folding in space determines the tertiary structure. Quaternary structure is the result of folded polypeptide molecules attaching themselves to intricately functional proteins (Kyte, 1995).

The polypeptide chain's amino acid sequence is referred to as a protein's primary structure (Sanger, 1952). The principal structure is maintained by peptide bonds formed during the process of protein biosynthesis. Based on the molecular shape, proteins can be classified into two types.

1. **Fibrous Proteins:** When the polypeptide chains run parallel and are held together by hydrogen and disulfide bonds, then the fiber-like structure is formed. Such proteins are generally insoluble in water. These are water-insoluble proteins. Example – keratin (present in hair, wool, and silk) and myosin (present in muscles), etc.
2. **Globular Proteins:** This structure results when the chains of polypeptides coil around to give a spherical shape. These are usually soluble in water. Example – Insulin and albumins are common examples of globular proteins.

1.1.1. Primary Structure

The basic level of the hierarchy is called the primary structure of a protein, and it is the specific linear sequence of amino acids that makes up a single polypeptide chain.

The 20 naturally occurring l-α-amino acids, which are shown in Table 1.1 and Figure 1.3, are the main building blocks of proteins, which are (bio)polymers. These basic 20 α-amino acid building blocks make up nearly all known protein structures. The side-chain characteristics of these building blocks categorize them into two primary groups: hydrophobic (apolar) and hydrophilic (polar) (Table 1.1). Having two sets of α-amino-acid building blocks that are opposite to each other is essential for protein architecture. A peptide link connects the individual α-amino acid building blocks (Figure 1.4). An oligopeptide is created when a few (2–20) amino acids are linked together by peptide bonds to create an unbranched chain (Figure 1.5). Peptide linkages, on the other hand, can connect any length of 20–2,000 amino acid residues to create substantial, unbranched polymeric chains of l-α-amino acids. We call these polypeptides. The repeat unit (–N–C_α–C(O)–)n, which ignores the α-amino acid sidechains within each polypeptide chain, is referred to as the main chain or backbone, and each constituent that is linked to an α-amino acid building block is referred to as an amino-acid residue. The term amino-acid sequence refers to the arrangement of amino-acid residues in a specific order, starting from the free, uncombined α-amino-terminal end (N-terminus) and ending at the free, uncombined α-carboxyl terminus (C-terminus).

Table 1.1. Structures of all Naturally Occurring L-α-Amino Acids That are Found in All Proteins of all Organisms

pK_a 7.8 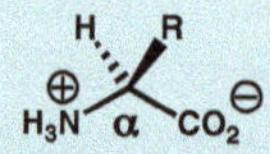 pK_a 3.6

R	Name	Abbrev.	R	Name	Abbrev.	pK_a
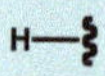	Glycine	Gly (G)	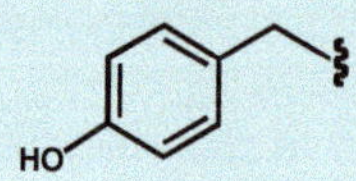	Tyrosine	Tyr (Y)	9.7
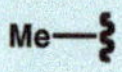	Alanine	Ala (A)				
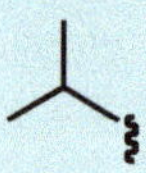	Valine	Val (V)	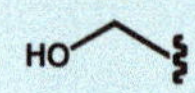	Serine	Ser (S)	15
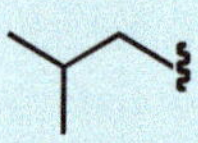	Leucine	Leu (L)	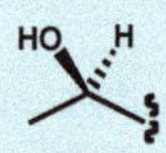	Threonine	Thr (T)	15
			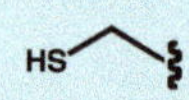	Cysteine	Cys (C)	9.1
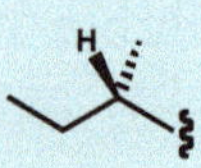	Isoleucine	Ile (I)	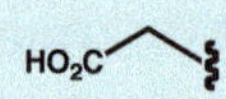	Aspartic Acid	Asp (D)	4.0
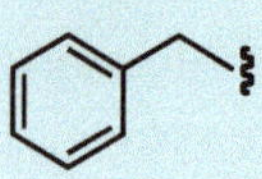	Phenylalanine	Phe (F)	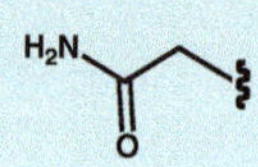	Asparagine	Asn (N)	
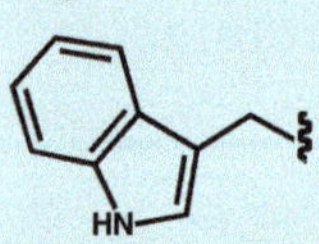	Tryptophan	Trp (W)	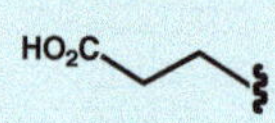	Glutamic Acid	Glu (E)	4.5
			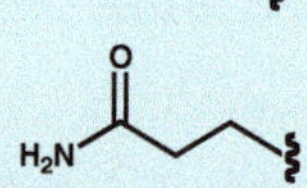	Glutamine	Gln (Q)	
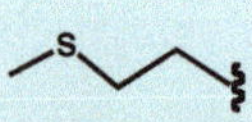	Methionine	Met (M)	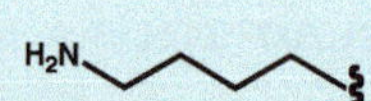	Lysine	Lys (K)	10.4
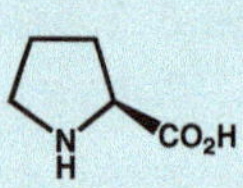	Proline	Pro (P)	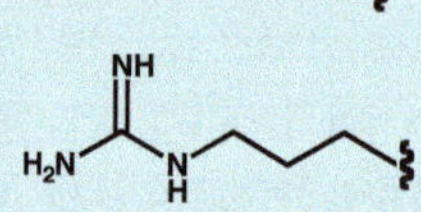	Arginine	Arg (R)	12
			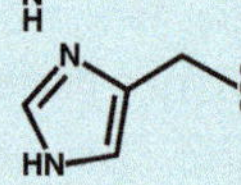	Histidine	His (H)	6.0

Source: Andrew Miller and Julian Tanner. Essentials of Chemical Biology: Structure and Dynamics of Biological Macromolecules. Wiley. ISBN 978-0-470-84530-1. https://nattabiochem.wordpress.com/wp-content/uploads/2009/08/essentials_chemical_biology.pdf.

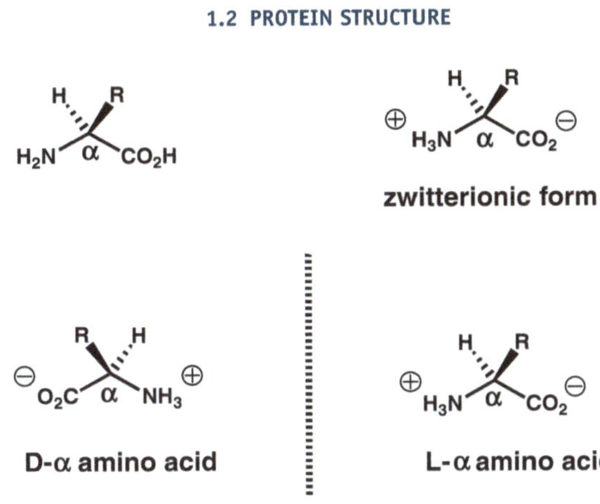

Figure 1.3. *Structures of α-amino acids the monomeric building blocks of proteins.*

Source: Andrew Miller and Julian Tanner. Essentials of Chemical Biology: Structure and Dynamics of Biological Macromolecules. Wiley. ISBN 978–0-470–84530–1. https://nattabiochem.wordpress.com/wp-content/uploads/2009/08/essentials_chemical_biology.pdf.

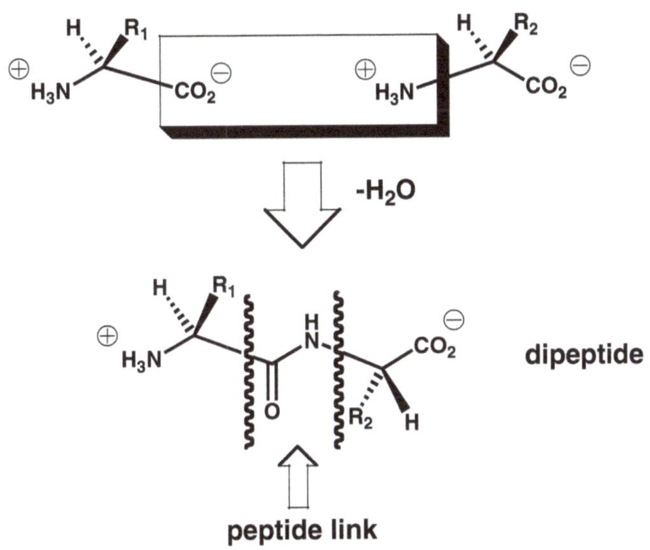

Figure 1.4. *Schematic of peptide link formation, central virtual bond of polypeptide and protein backbones.*

Source: Andrew Miller and Julian Tanner. Essentials of Chemical Biology: Structure and Dynamics of Biological Macromolecules. Wiley. ISBN 978–0-470–84530–1. https://nattabiochem.wordpress.com/wp-content/uploads/2009/08/essentials_chemical_biology.pdf.

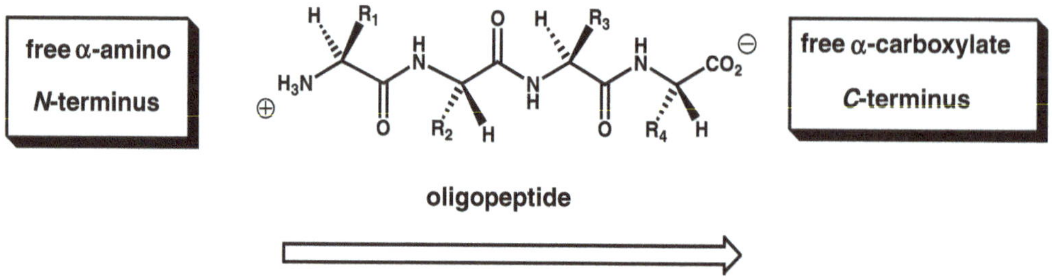

Figure 1.5. *General structure of tetrapeptide; by convention the highest priority end is the free N-terminus and the lowest priority the free C-terminus giving the backbone a directionality illustrated by the arrow. This convention applies to all peptides, polypeptides and proteins.*

Source: Andrew Miller and Julian Tanner. Essentials of Chemical Biology: Structure and Dynamics of Biological Macromolecules. Wiley. ISBN 978-0-470-84530-1. https://nattabiochem.wordpress.com/wp-content/uploads/2009/08/essentials_chemical_biology.pdf.

Each peptide link is, quite obviously, a basic secondary amide functional group with a few peculiar characteristics. The peptide link's N, H, C, and O atoms, along with each pair of neighboring α-carbon atoms, actually form a rigid, coplanar unit that functions nearly like a single bond because of the nitrogen atom lone pair resonance-induced restricted rotation about the N–C(O) bond and the accumulation of N–C(O) double bond character (Figure 1.6). This is the reason why the peptide link and adjacent C_α atoms are sometimes called a virtual bond. It is possible to state that each amino acid residue in a polypeptide chain has a C_α atom that concurrently belongs to two of these virtual bonds (Figure 1.7). The conformational angles φ and ψ, which are the main chain dihedral angles subtended about the N(H)–C_α and C_α–C(O) ⊠ bonds of each amino acid residue, respectively, are then used to define the spatial relationship between each C_α-linked pair of virtual bonds (Figures 1.7 and 1.8). Due to steric congestion between the sidechains of adjacent amino-acid residues, only specific combinations of φ and ψ are permitted at this time (Figure 1.8). As a result, the overall conformation of a particular polypeptide chain is also highly constrained, which directly affects the three-dimensional protein structures. The conformational freedom of a polypeptide and, consequently, the range of potential three-dimensional structures that any given polypeptide polymer may form are significantly limited by conformational constraints imposed by the lack of free rotation in the peptide link and the characteristics of each peptide-linked amino-acid residue. In actuality, a protein's primary structure, or amino acid sequence, both determines and influences its three-dimensional structure. Stated differently, the primary structure of a protein contains all the "information" required for its three-dimensional structure. This is the fundamental idea behind biological self-assembly and the reason proteins can serve as such great "workbenches" or platforms for the elaboration of specific roles and the progression of living things.

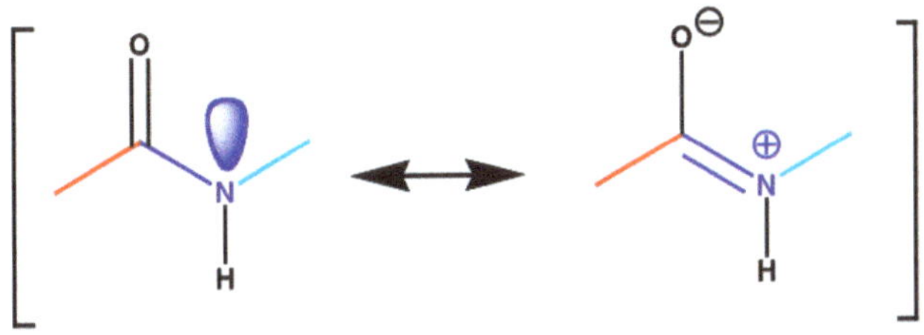

Figure 1.6. *Peptide link resonance structures illustrating partial double character in the C(O) N bond (blue) sufficient to prevent free rotation thereby restricting conformational freedom of peptide or polypeptide backbones.*

Source: Andrew Miller and Julian Tanner. Essentials of Chemical Biology: Structure and Dynamics of Biological Macromolecules. Wiley. ISBN 978-0-470-84530-1. https://nattabiochem.wordpress.com/wp-content/uploads/2009/08/essentials_chemical_biology.pdf.

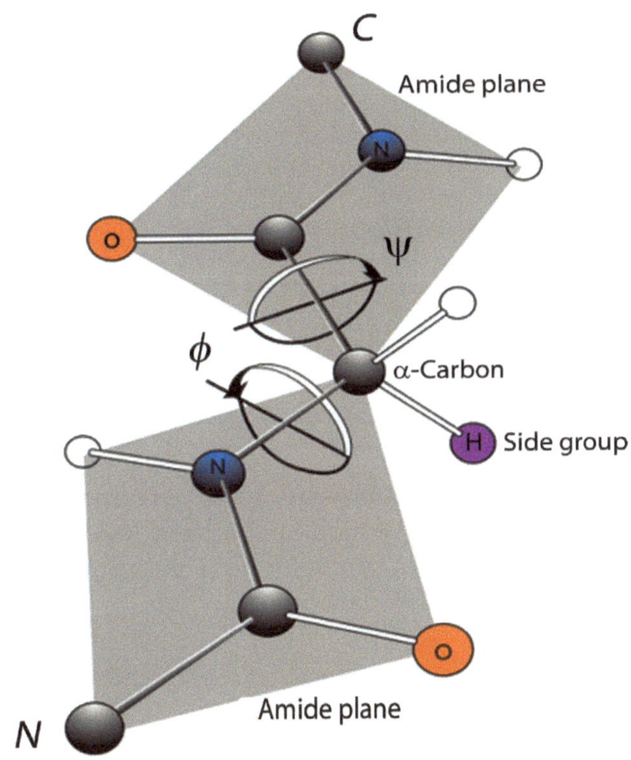

Figure 1.7. *Peptide link C, O, N and H atoms act as a rigid coplanar units equivalent to a single bond (virtual bond) so that consecutive peptide links act as rigid coplanar units that pivot around individual C α-atoms.*

Source: Andrew Miller and Julian Tanner. Essentials of Chemical Biology: Structure and Dynamics of Biological Macromolecules. Wiley. ISBN 978-0-470-84530-1. https://nattabiochem.wordpress.com/wp-content/uploads/2009/08/essentials_chemical_biology.pdf.

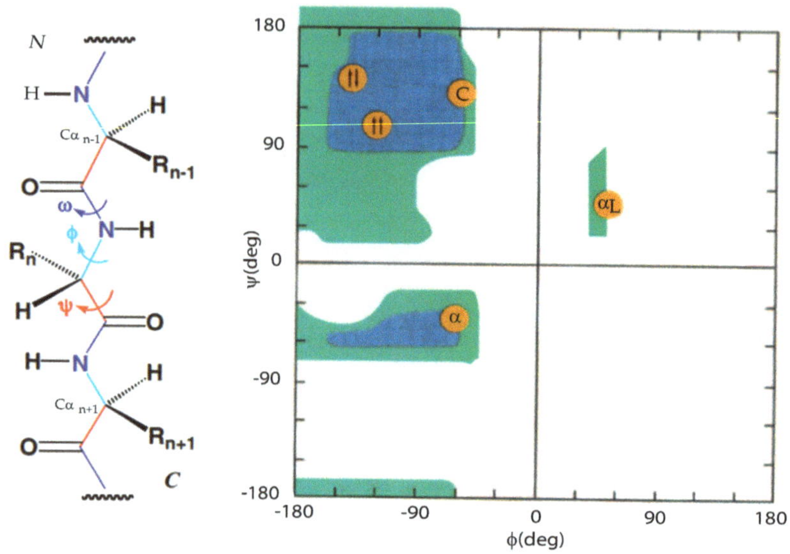

Figure 1.8. *Amino acid residue side-chain interactions further restrict free rotation in peptide or polypeptide backbone. Rotational possibilities are defined by allowed values of dihedral angle φ subtended about N–C_α bond and ψ subtended about C_α–C(O) bond (left). Theoretically allowed angles are shown in Ramachandran plot (right) together with positions of actual angles found in real protein secondary structures; α: right-handed α-helix; α_L: left-handed α-helix; ↑↑: parallel β-sheet; ↑↓: anti-parallel β-sheet; C: collagen, P_{II} helix.*

Source: Andrew Miller and Julian Tanner. Essentials of Chemical Biology: Structure and Dynamics of Biological Macromolecules. Wiley. ISBN 978-0-470-84530-1. https://nattabiochem.wordpress.com/wp-content/uploads/2009/08/essentials_chemical_biology.pdf.

1.1.2. Secondary Structure

The regular folding of regions into particular structural patterns within a single polypeptide chain constitutes secondary structure, which is the next level up from primary structure (Sun et al., 2004). Secondary structures typically hold hydrogen bonds together between the peptide bond amide hydrogen and carbonyl oxygen.

1.1.2.1. Repetitive Secondary Structure

The first significant steps toward a functional three-dimensional structure are represented by the secondary structure if the primary structure is an amino acid residue sequence. Polypeptides can form secondary structures in solution, which are essentially transient three-dimensional structural elements that can interlock or dock together for stability. The right-handed α helix (α_R) and the right-handed 3_{10} helix are two exceptionally beautiful helical structures that polypeptides can form (Schulz GE and Schirmer RH 1979). The term "right-handed" describes how the main chain of a polypeptide follows the path of a right-handed corkscrew (it should be noted that while the left-handed α helix, or α_L, is theoretically possible, it has not yet been discovered in any natural protein). Remarkably strong, resilient, and regular, the α_R helix can be a structural feature (Figures 1.9 and

1.10). α_R helices are generally composed of up to 35 amino acid residues and exhibit high levels of stereoregularity. The conformational angles of φ and ψ for each amino acid residue in the α_R helix are consistently around –60° (Figure 1.11). A regular web of non-covalent hydrogen bonds between the C=O peptide bond and the N–H groups of nearby amino acid residues holds helices together (Figure 1.12). As a result, hydrogen bonds are formed between the N–H groups (hydrogen bond donors) of the (n + 4)th residues and the C=O group lone pairs (hydrogen bond acceptors) of the nth residues. There are three to six amino acid residues per turn. 13 atoms make up the closed loop made of one of these hydrogen bonds and the polypeptide main chain segment that lies in between (Figure 1.12). Thus, it has also been decided to refer to the α_R helix as a 3.6_{13} helix. However, with only three amino-acid residues per turn and 10 atoms involved in the intervening stretch of the polypeptide main chain, the 3_{10} helix (also known as the α_{II} helix) is essentially a smaller and slightly distorted version of the α_R helix (Figures 1.13 and 1.14). Thus, hydrogen bonds are created between the N–H groups (hydrogen bond donors) of the (n + 3)th residues and the C=O group lone pairs of the nth residues; the conformational angles of φ and ψ are roughly –60 and –30°, respectively (Figure 1.15).

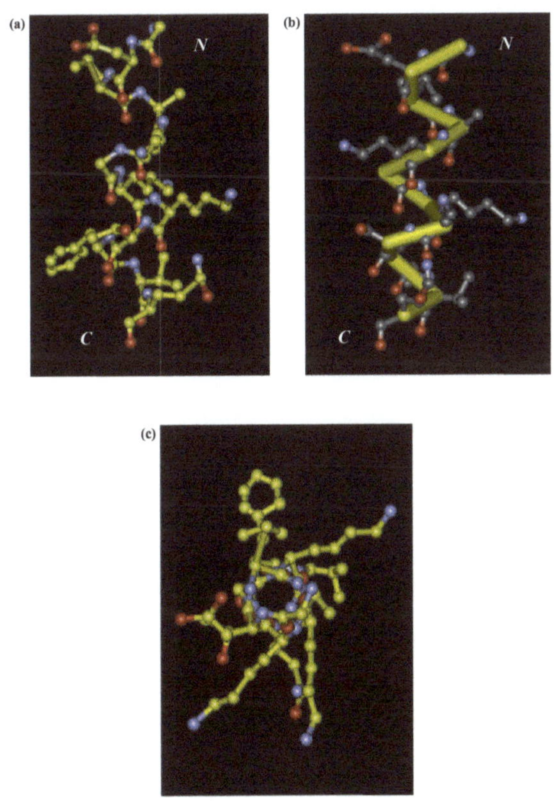

Figure 1.9. *Various depictions of an α-helix from triose phosphate isomerase (chicken muscle) (pdb: 1tim).*

Note: (a) ball and stick representation (side view) of atoms and bonds shown with carbon (yellow), nitrogen (blue) and oxygen (red); (b) CA stick display of α-carbon backbone, atoms and bonds of amino acid side-chains are rendered in ball and stick representations with carbon (gray), nitrogen (blue) and oxygen (red); (c) ball and stick representation (top view) of atoms and bonds with labels as per (a).

Source: Andrew Miller & Julian Tanner. Essentials of Chemical Biology: Structure and Dynamics of Biological Macromolecules. Wiley. ISBN 978-0-470-84530-1. https://nattabiochem.wordpress.com/wp-content/uploads/2009/08/essentials_chemical_biology.pdf.

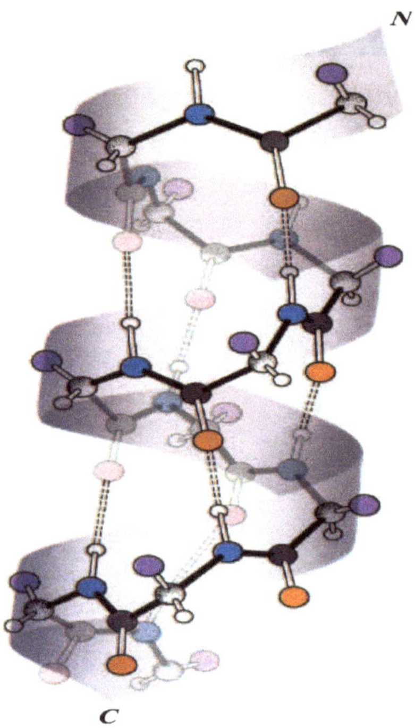

Figure 1.10. *Cartoon rendition of α-helix in which right-hand helix path is illustrated as a ribbon over which a ball and stick representation of the α-carbon backbone is drawn using the code hydrogen (white), carbon (gray), nitrogen (blue) oxygen (red) and side chain atom (purple), in order to illustrate general hydrogen bonding patterns in the helix.*

Source: Andrew Miller and Julian Tanner. Essentials of Chemical Biology: Structure and Dynamics of Biological Macromolecules. Wiley. ISBN 978-0-470-84530-1. https://nattabiochem.wordpress.com/wp-content/uploads/2009/08/essentials_chemical_biology.pdf.

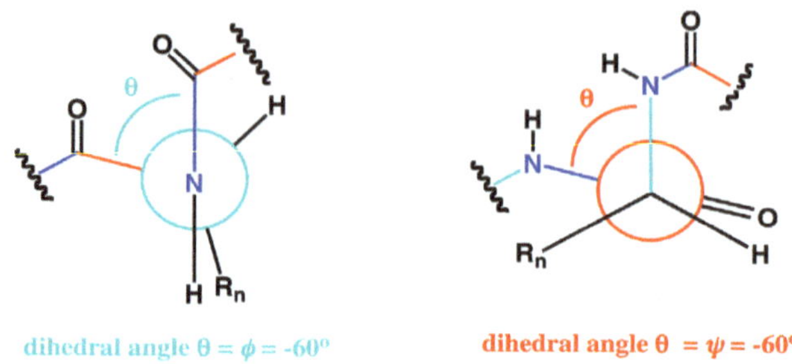

Figure 1.11. *Newman projections involving the N–C_α bond and C_α–C(O) bond of the α-helix to demonstrate the consequences of highly regular dihedral angles φ and ψ respectively.*

Source: Andrew Miller and Julian Tanner. Essentials of Chemical Biology: Structure and Dynamics of Biological Macromolecules. Wiley. ISBN 978-0-470-84530-1. https://nattabiochem.wordpress.com/wp-content/uploads/2009/08/essentials_chemical_biology.pdf.

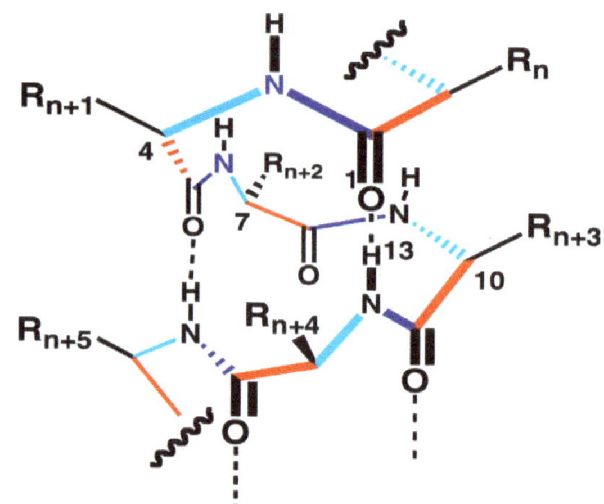

Figure 1.12. *Stereo-defined structure of first turn of an α-helix to demonstrate the atom separation between N–H hydrogen bond donors and C=O hydrogen bond acceptors. The C=O acceptor of each n-th residue forms a hydrogen bond link with the N–H bond donor of the (n + 4)-th residue defining an atom separation of 13 between acceptor O-atom and donor H-atom.*

Source: Andrew Miller and Julian Tanner. Essentials of Chemical Biology: Structure and Dynamics of Biological Macromolecules. Wiley. ISBN 978-0-470-84530-1. https://nattabiochem.wordpress.com/wp-content/uploads/2009/08/essentials_chemical_biology.pdf.

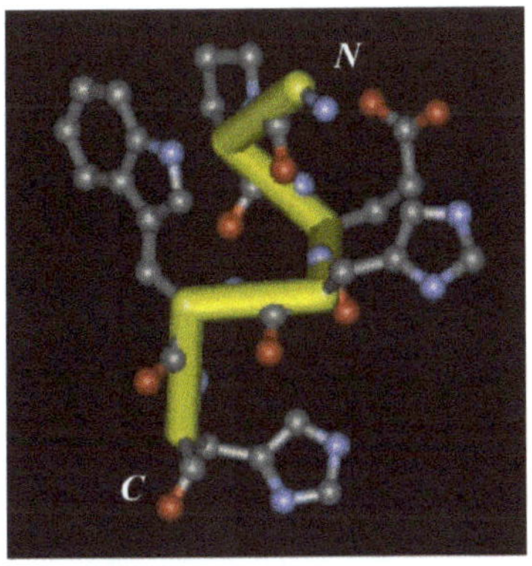

Figure 1.13. *Depiction of 3_{10}-helix (turn) from triose phosphate isomerase (chicken muscle) (pdb: 1tim). CA stick display of α-carbon backbone, atoms and bonds of amino acid side chains are rendered in ball and stick representations with carbon (gray), nitrogen (blue) and oxygen (red).*

Source: Andrew Miller and Julian Tanner. Essentials of Chemical Biology: Structure and Dynamics of Biological Macromolecules. Wiley. ISBN 978-0-470-84530-1. https://nattabiochem.wordpress.com/wp-content/uploads/2009/08/essentials_chemical_biology.pdf.

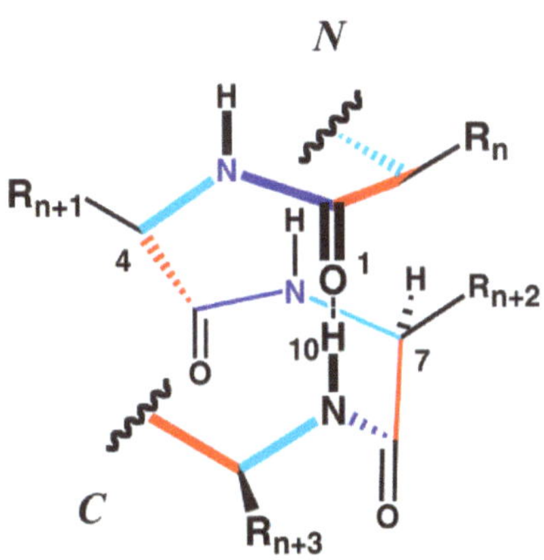

Figure 1.14. *Stereo-defined structure of the first turn of a 3_{10}-helix to demonstrate the atom separation between N H hydrogen bond donors and C=O hydrogen bond acceptors. The C=O acceptor of each n-th residue forms a hydrogen bond link with the N–H bond donor of the (n + 3)-th residue defining an atom separation of 10 between acceptor O-atom and donor H-atom.*

Source: Andrew Miller and Julian Tanner. Essentials of Chemical Biology: Structure and Dynamics of Biological Macromolecules. Wiley. ISBN 978-0-470-84530-1. https://nattabiochem.wordpress.com/wp-content/uploads/2009/08/essentials_chemical_biology.pdf.

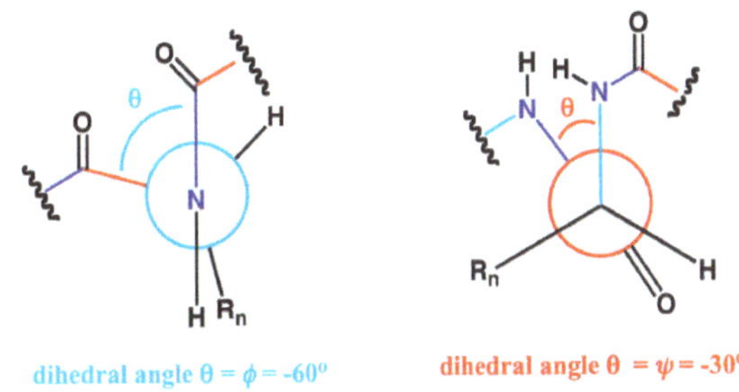

Figure 1.15. *Newman projections involving the N–C_α bond and C_α–C(O) bond of the 3_{10}-helix to demonstrate the consequences of highly regular dihedral angles φ and ψ respectively.*

Source: Andrew Miller and Julian Tanner. Essentials of Chemical Biology: Structure and Dynamics of Biological Macromolecules. Wiley. ISBN 978-0-470-84530-1. https://nattabiochem.wordpress.com/wp-content/uploads/2009/08/essentials_chemical_biology.pdf.

The primary substitute for helices in structures is sheetlike. The behavior of polypeptide chains in their fully extended β-strand conformations serves as the model for these structures. A β strand appears "pleated," with side chains of amino acid residues alternating above and below and peptide bonds oriented perpendicular to the main chain (Figure 1.16). Figure 1.17 shows that while both φ and ψ conformational angles are close to 180°, they are usually within the range of −120 and −150° and +120 and +150°, respectively. While each β-strand conformation is unstable on its own, non-covalent hydrogen bonding between strands can stabilize them and produce a β sheet (Figures 1.16 and 1.18). Depending on whether the β strands are oriented in the same direction or opposite to one another, these β sheets can be either antiparallel (β) or parallel (β') (Figures 1.19 and 1.20). The same functional groups that form helices also form the hydrogen bonds that join β strands together. Hydrogen bonds are sporadically spaced closely together and widely apart in antiparallel (β) β sheets, but uniformly spaced throughout in parallel (β') β sheets (Figure 1.20).

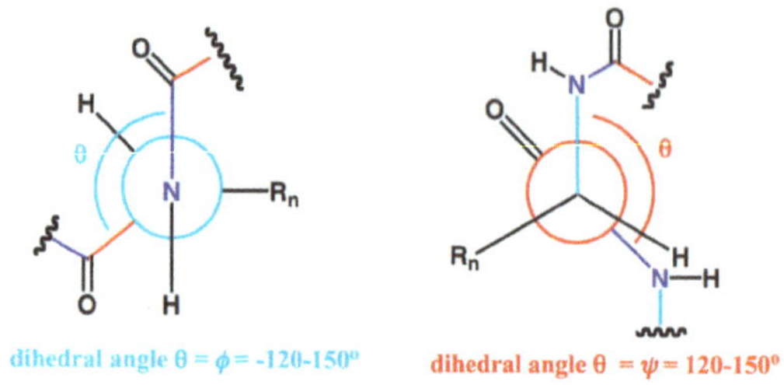

Figure 1.17. *Newman projections involving the N–C_α bond and C_α–C(O) bond of a β-strand to demonstrate the consequences of highly regular dihedral angles φ and ψ respectively and extending the conformation.*

Source: Andrew Miller and Julian Tanner. Essentials of Chemical Biology: Structure and Dynamics of Biological Macromolecules. Wiley. ISBN 978-0-470-84530-1. https://nattabiochem.wordpress.com/wp-content/uploads/2009/08/essentials_chemical_biology.pdf.

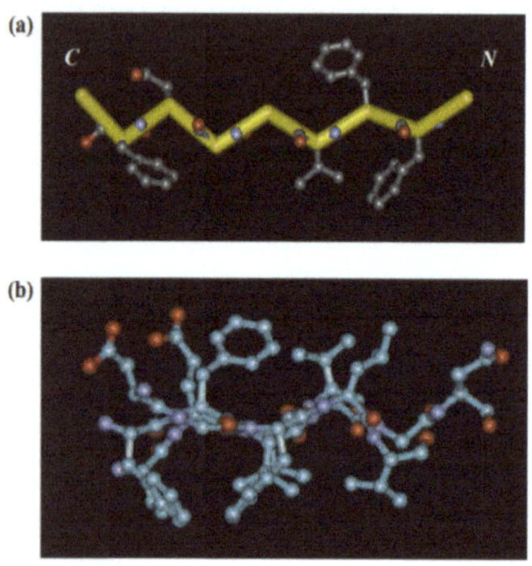

Figure 1.16. *Depiction of β-strand and β-sheet from triose phosphate isomerase (chicken muscle) (pdb: 1tim).*

Note: (a) CA stick Display of α-carbon backbone (side view), atoms and bonds of amino acid side-chains are rendered in ball and stick representation with carbon (gray), nitrogen (blue) and oxygen (red). β-strand is shown to illustrate "zig-zag" extended conformation; (b) Ball and stick representation of β-sheet (side view) is shown with carbon (gray), nitrogen (blue) and oxygen (red) to illustrate "zig-zag" pleating and to show regular arrangement of amino acid residue side chains in close juxtaposition.

Source: Andrew Miller and Julian Tanner. Essentials of Chemical Biology: Structure and Dynamics of Biological Macromolecules. Wiley. ISBN 978-0-470-84530-1. https://nattabiochem.wordpress.com/wp-content/uploads/2009/08/essentials_chemical_biology.pdf.

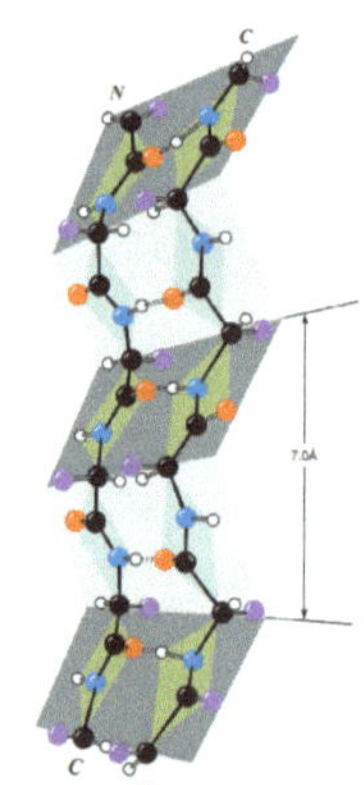

Figure 1.18. *Cartoon rendition of a β-sheet in which pleating is illustrated as a sequence of intersecting planes over which a ball and stick representation of the α-carbon backbone is drawn using the code hydrogen (white), carbon (black), nitrogen (blue) oxygen (red) and side chain atom (purple), in order to illustrate general positioning of side chains above and below the sheet.*

Source: Andrew Miller and Julian Tanner. Essentials of Chemical Biology: Structure and Dynamics of Biological Macromolecules. Wiley. ISBN 978-0-470-84530-1. https://nattabiochem.wordpress.com/wp-content/uploads/2009/08/essentials_chemical_biology.pdf.

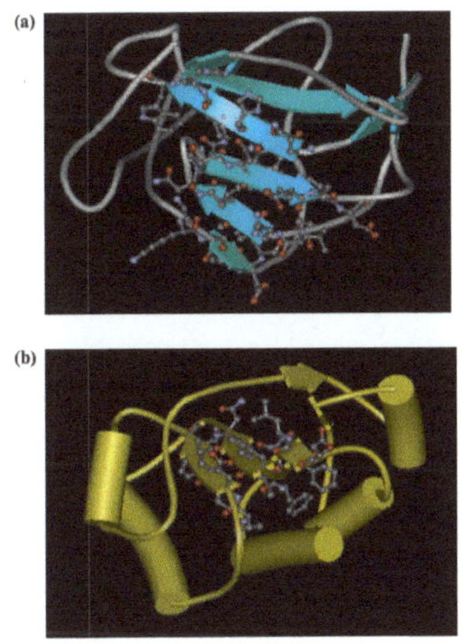

Figure 1.19. *Depiction of β-sheet structures from indicated proteins.*

Note: (a) Schematic Display Structure of anti-parallel β-sheet segment of carbonic anhydrase I (human erythrocyte) (pdb: 2cab), atoms and bonds of amino acid side-chains are rendered in ball and stick representation with carbon (gray), nitrogen (blue) and oxygen (red); (b) Schematic Display Structure of parallel β-sheet segment of triose phosphate isomerase (chicken muscle) (pdb: 1tim), atoms and bonds of side-chains are rendered as in (a).

Source: Andrew Miller and Julian Tanner. Essentials of Chemical Biology: Structure and Dynamics of Biological Macromolecules. Wiley. ISBN 978–0-470–84530–1. https://nattabiochem.wordpress.com/wp-content/uploads/2009/08/essentials_chemical_biology.pdf.

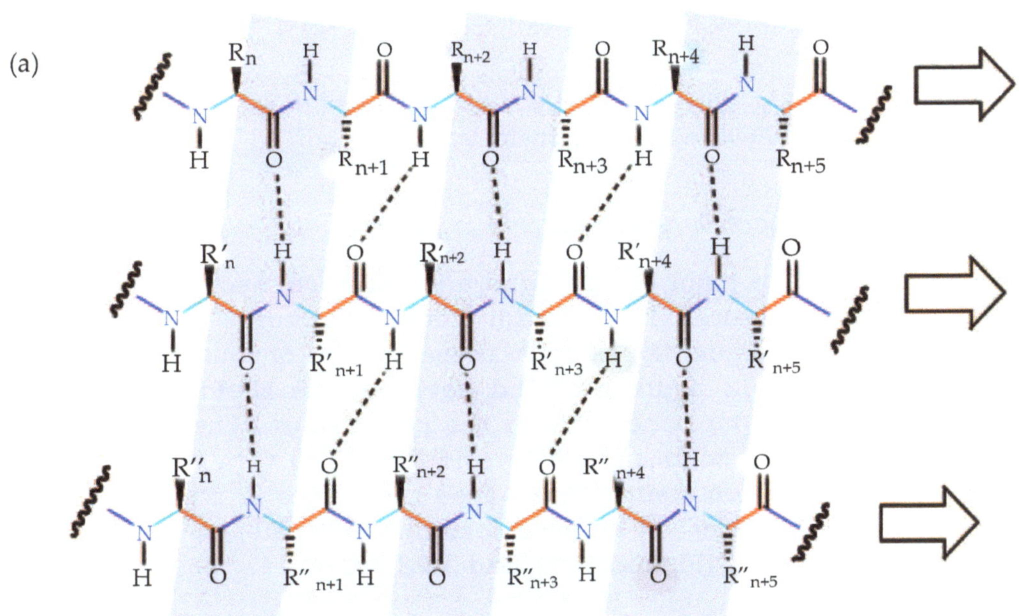

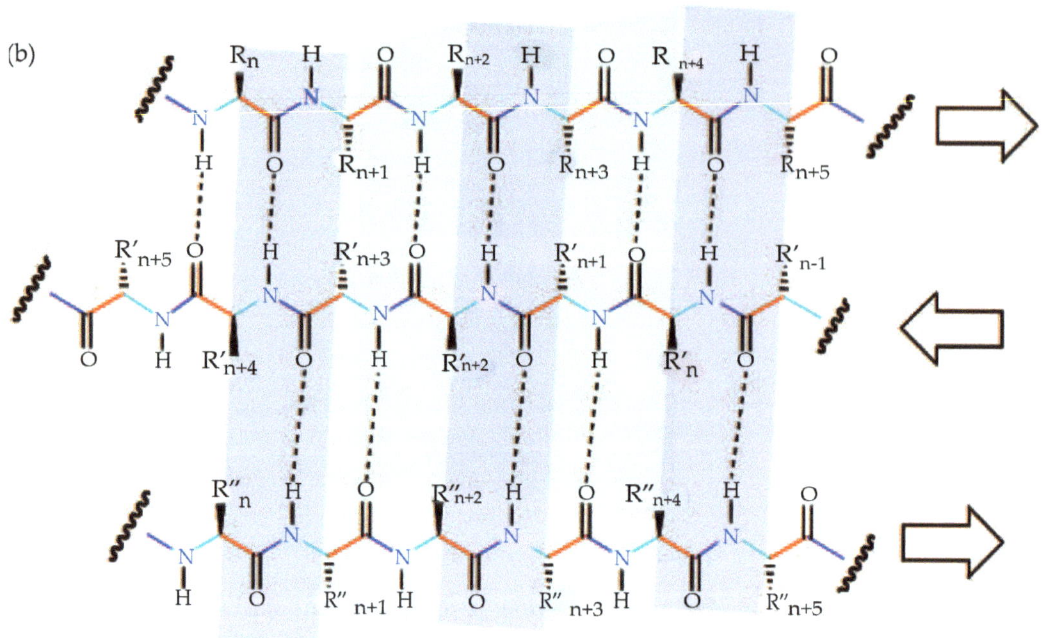

Figure 1.20. *Stereo-defined structures of β-sheets.*

Note: (a) Three stranded parallel β-sheet structure showing hydrogen bonding relationship between parallel β-strands. The N–H donor of each peptide link is able to form a hydrogen bond with the C=O acceptor of a peptide link in a parallel β-strand. Shading is used to demonstrate pleating and emphasize amino acid residue side-chain orientations with respect to the sheet and with respect to each other; (b) three stranded antiparallel β-sheet structure as for (a) except that hydrogen bonding occurs between peptide links in neighboring antiparallel β-strands.

Source: Andrew Miller and Julian Tanner. Essentials of Chemical Biology: Structure and Dynamics of Biological Macromolecules. Wiley. ISBN 978-0-470-84530-1. https://nattabiochem.wordpress.com/wp-content/uploads/2009/08/essentials_chemical_biology.pdf.

1.1.2.2. Non-Repetitive Secondary Structure

In a particular polypeptide main chain, turns and loops connect and/or hold together helices and sheetlike structures. In the main chain, tight turns—also referred to as β bends or β turns—occur frequently. Usually, these consist of four units of amino acids separated by a non-covalent hydrogen bond between the N–H group of the (n + 3) th residue and the C=O group lone pairs of the nth residue. The conformation of the peptide link between the second and third residues of the turn is the primary source of difference between the two main classes, type I and type II, which are composed of at least six possible variants based on variations in the possible φ and ψ angles of the amino-acid residues involved (Figures 1.21 and 1.22).

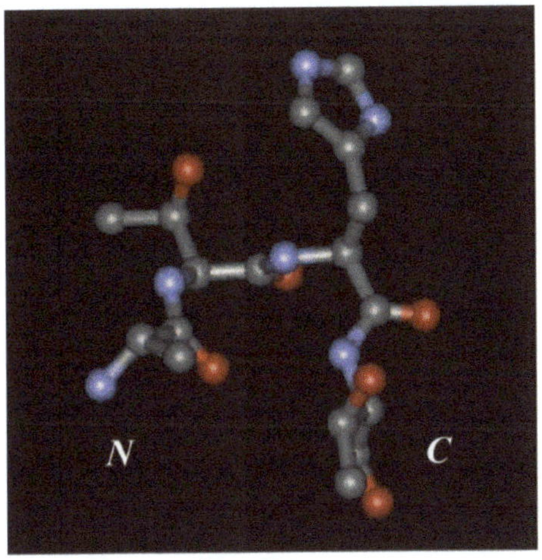

Figure 1.21. *Depiction of β-turn (type I) from carbonic anhydrase I (human erythrocyte) (pdb: 2cab). Ball and stick representation of atoms and bonds with carbon (gray), nitrogen (blue) and oxygen (red).*

Source: Andrew Miller and Julian Tanner. Essentials of Chemical Biology: Structure and Dynamics of Biological Macromolecules. Wiley. ISBN 978–0-470–84530–1. https://nattabiochem.wordpress.com/wp-content/uploads/2009/08/essentials_chemical_biology.pdf.

Although sidechain interactions between amino acid residues give stability instead of hydrogen bonding associated with peptide links, loops in the main chain are also highly prevalent. As a result, the main chain's path in creating a loop is much less regular than the one found in a tight turn (Figure 1.23). Disulfide bridges can occasionally take the place of or add to peptide links in polypeptides. In the amino-acid sequence of a polypeptide, these bridges are created between the thiol-functional groups of two distinct cysteine (cys, C) residues that are separated from one another by at least two other amino acid residues. These can be compared to the polypeptide version of a "tie bar" or other similar reinforcing tool. There are known spiral forms that are both left- and right-handed, and each disulfide bridge's state is defined by a set of conformational angles (χ_1, χ_2, χ_3, $\chi_{2'}$, $\chi_{1'}$) (Figures 1.24 and 1.25).

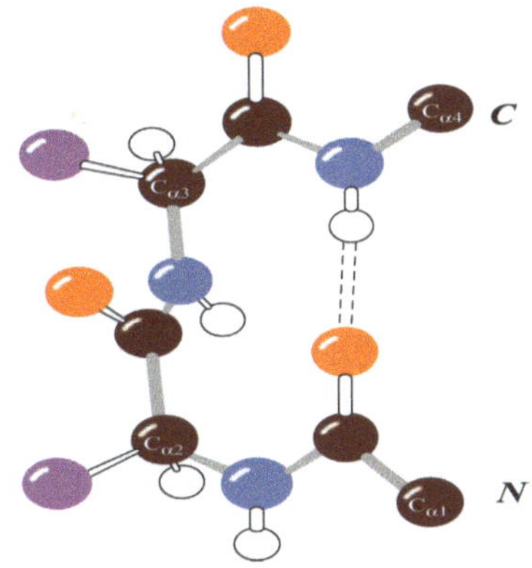

Figure 1.22. *Cartoon of β-turn (type II). Ball and stick representation of atoms and bonds with carbon (black), nitrogen (blue), oxygen (red), and side chain atom (purple), in order to illustrate how the C=O acceptor of the first n-th residue forms a hydrogen bond with the N–H bond donor of the (n + 3)-th residue defining an atom separation of 10 between acceptor O-atom and donor H-atom. This is the same of the type I turn. The main difference between Type I and Type II is the orientation of the peptide link joining the (n + 1)-th to the (n + 2)-th residue.*

Source: Andrew Miller and Julian Tanner. Essentials of Chemical Biology: Structure and Dynamics of Biological Macromolecules. Wiley. ISBN 978–0-470–84530–1. https://nattabiochem.wordpress.com/wp-content/uploads/2009/08/essentials_chemical_biology.pdf.

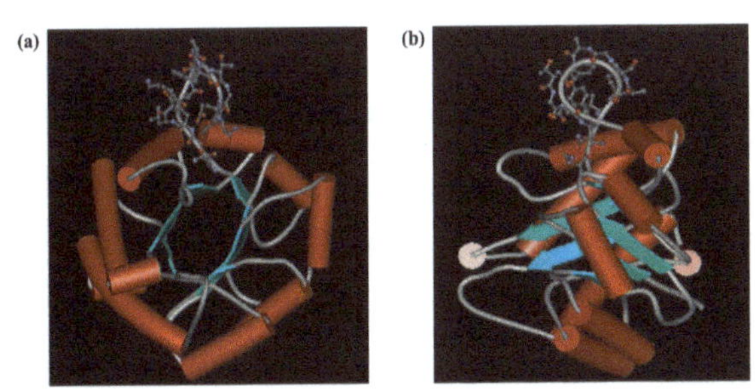

Figure 1.23. *Depiction of loop structure from triose phosphate isomerase (chicken muscle) (pdb: 1tim).*

Note: (a) Schematic display structure of triose phosphate isomerase (top view), atoms and bonds of amino acid side chains of key loop are rendered in ball and stick representation with carbon (gray), nitrogen (blue) and oxygen (red); (b) schematic display structure of triose phosphate isomerase (side view), atoms and bonds of amino acid side chains rendered as in (a).

Source: Andrew Miller and Julian Tanner. Essentials of Chemical Biology: Structure and Dynamics of Biological Macromolecules. Wiley. ISBN 978-0-470-84530-1. https://nattabiochem.wordpress.com/wp-content/uploads/2009/08/essentials_chemical_biology.pdf.

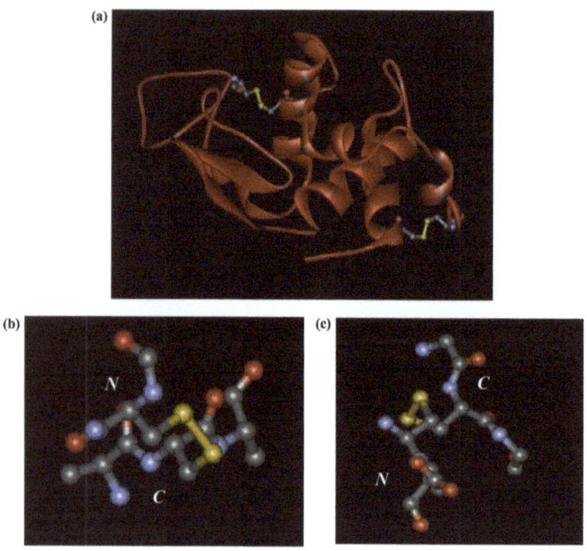

Figure 1.24. *Depiction of disulfide bridges from lysozyme (hen egg-white) (pdb:6lyz).*

Note: (a) Ribbon Display Structure of lysozyme (side view), atoms and bonds of amino acid side chains of cysteine residues involved in disulfide bridges are rendered in ball and stick representation with sulfur (yellow) carbon (gray), nitrogen (blue) and oxygen (red); (b) ball and stick representation of cysteine residues forming a right-handed disulfide bridge, atoms and bonds labeled as for (a); (c) ball and stick representation of cysteine residues from a left-handed disulfide bridge, atoms and bonds labeled as for (a).

Source: Andrew Miller and Julian Tanner. Essentials of Chemical Biology: Structure and Dynamics of Biological Macromolecules. Wiley. ISBN 978-0-470-84530-1. https://nattabiochem.wordpress.com/wp-content/uploads/2009/08/essentials_chemical_biology.pdf.

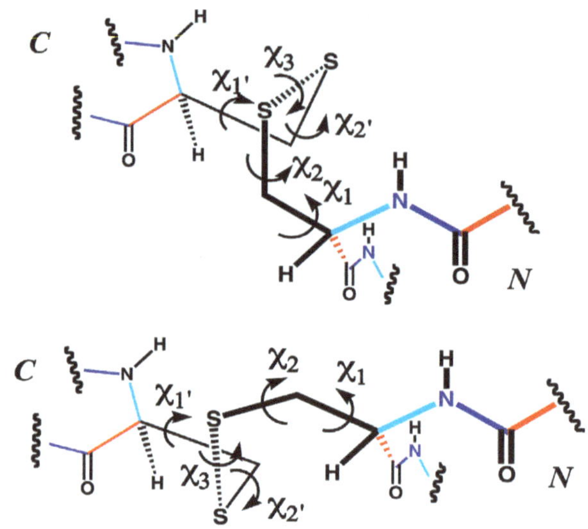

Figure 1.25. *Detailed structural description of disulfide bridges.*

Note: (a) Right-handed disulfide bridge showing all the main conformational angles; (b) Left-handed disulfide bridge similarly indicating all the main conformational angles. In both cases, N refers to a cysteine residue closest to the N-terminus of the polypeptide and C refers to a residue closest to the C-terminus.

Source: Andrew Miller and Julian Tanner. Essentials of Chemical Biology: Structure and Dynamics of Biological Macromolecules. Wiley. ISBN 978-0-470-84530-1. https://nattabiochem.wordpress.com/wp-content/uploads/2009/08/essentials_chemical_biology.pdf.

1.1.2.3. Alternative Secondary Structures

A polypeptide main chain may occasionally map out helical structures that are somewhat longer and more elongated than the α_R helix. For reasons that will become clear, these structures are referred to as the collagen helix (C) or the left-handed P_{II} helix. Given that most biological macromolecule structures or substructures are right-handed, this structure's left-handedness makes it unique. As seen in Figures 1.26 and 1.27, the polypeptide main chain is extended to resemble a β strand and an α_R helix in topography. Each helix turn contains roughly three amino acid residues, with an increase of approximately 3 Å between each residue, in contrast to 1.5 Å in the αR helix. Unusual φ and ψ angles are needed to form a PII helix. As a result, most polypeptides—aside from those containing a high percentage of glycine and proline residues—are unable to properly adopt an extended P_{II}-helical conformation.

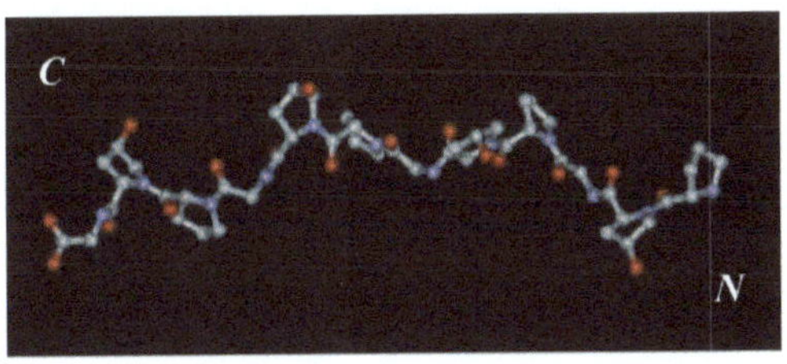

Figure 1.26. *Depiction of left-handed P_{II}-helix from collagen (pdb: 1bkv). Ball and stick representation of atoms and bonds with carbon (gray), nitrogen (blue) and oxygen (red).*

Source: Andrew Miller and Julian Tanner. Essentials of Chemical Biology: Structure and Dynamics of Biological Macromolecules. Wiley. ISBN 978-0-470-84530-1. https://nattabiochem.wordpress.com/wp-content/uploads/2009/08/essentials_chemical_biology.pdf.

Figure 1.27. *Cartoon depiction of the extended P_{II}-helix. Depiction emphasizes the left-handed character of this helix type.*

Source: Andrew Miller and Julian Tanner. Essentials of Chemical Biology: Structure and Dynamics of Biological Macromolecules. Wiley. ISBN 978-0-470-84530-1. https://nattabiochem.wordpress.com/wp-content/uploads/2009/08/essentials_chemical_biology.pdf.

1.1.3. Tertiary Structure

The specific three-dimensional arrangement of every amino acid in a single polypeptide chain is known as the tertiary structure, which is the next level above the secondary structure. Multiple non-covalent interactions hold this structure together, and it is typically native, active, and conformational. The tertiary structure of a polypeptide is the overall three-dimensional fold that is produced by the docking and mutual stabilization of three-dimensional secondary structural elements. Since many proteins are made up of a single polypeptide chain, this fold also serves as the protein's tertiary structure. Since this fold typically has a globular overall shape, proteins having this type of fold are referred to as globular proteins. The process through which proteins obtain their tertiary structure is referred to as protein folding. The debate over whether secondary structures form first and dock together to form tertiary structures (the framework model) or whether a crude tertiary structure forms first and is then stabilized through a process of "side-chain" negotiation (the molten-globule model) is one of the most contentious aspects of this highly controversial field of study of globular proteins and polypeptides. The latter model seems to be more widely accepted today, which is in line with the understanding that there are numerous "pathways" for protein folding, each of which is heavily reliant on the main structure of the specific protein or polypeptide of interest. Therefore, given a given primary structure, we are frequently unsure of what the globular tertiary structure will be. This is because we are still unsure of the precise mechanism by which the primary structure drives the formation of the tertiary structure.

The ribbon display structures (Figure 1.28) are the most easily comprehensible of the shorthand representations of globular protein/polypeptide three-dimensional structures that have been developed due to the general atomic complexity of protein architecture. In order to emphasize various facets of protein structure during analysis, there are additional representations as well, such as surface display, CPK, and schematic display structures (Figure 1.29). According to the Richardson classification, which is based on the topographical behavior of repetitive secondary structure elements in the tertiary structure, the illustrated structures depict all the major classes of globular proteins, namely small metal-rich proteins, small SS-rich proteins, anti-parallel α proteins, anti-parallel β proteins, and parallel α/β proteins (Figures 1.28 and 1.29). Hydrophobic amino acid residues are almost always present in the interior of globular proteins (e.g., phenylalanine (phe, F), valine (val, V), leucine (leu, L), etc.), while hydrophilic amino acid residues predominate on the exterior, which is composed of both hydrophobic and hydrophilic residues. For the time being, let's assume that globular proteins have a "soapy" exterior and a "waxy" interior. The significance of this will become apparent later. Its "waxy" interior is remarkably clear. Actually, hydrophobic amino acid residues' internal packing bears a striking resemblance to the crystalline state of organic solids. The distribution of amino acids within a polypeptide chain suggests that hydrophobic interactions and close-range non-covalent van der Waals forces are important for maintaining the chain's stability. On the other hand, non-covalent hydrogen bonds play a major role in the formation of both repetitive and non-repetitive secondary structures, with the exception of the disulfide bridge. It has recently come to light that additional forces—that is, weak polar forces—might also be crucial in maintaining the tertiary structure of proteins and polypeptides. It's possible that more forces are still to be found. Without a doubt, the structure of globular proteins and polypeptides is still a rich, intriguing, and sometimes contentious field of study.

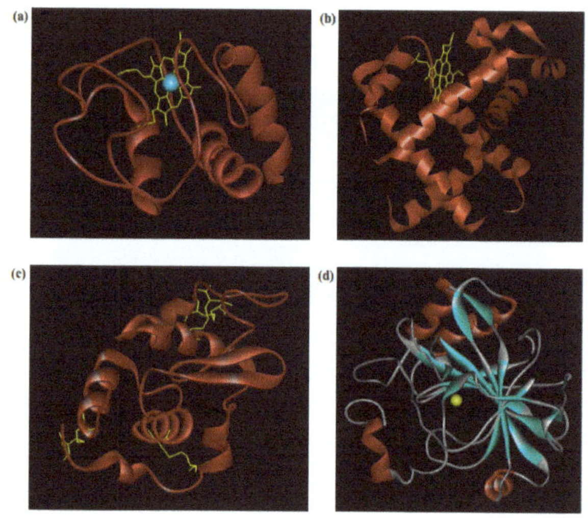

Figure 1.28. *Ribbon display structure cartoon depictions.*

Note: (a) Small metal rich protein cytochrome c (horse heart) (pdb: 1hrc). Polypeptide backbone is shown as a ribbon (red). Stick bonds constituting porphyrin macrocycle (prosthetic group) are highlighted (yellow). Van der Waal's sphere (blue) represents central iron ion; (b) Anti-parallel α-protein myoglobin (sperm whale) (pdb: 1mbi) with labeling system as for (a); (c) Small SS rich protein lysozyme (hen egg white) (pdb: 6lyz). Polypeptide backbone is shown as a ribbon (red), stick bonds of cysteine amino acid residues linked by disulfide bridges are also shown (yellow); (d) Anti-parallel β-protein carbonic anhydrase I (human erythrocyte) (pdb: 2cab). Polypeptide backbone is colored according to secondary structure with α-helix (red), anti-parallel β-sheet (light blue), loop structures (random coil) (light gray). Van der Waals sphere (yellow) represents central zinc ion.

Source: Andrew Miller and Julian Tanner. Essentials of Chemical Biology: Structure and Dynamics of Biological Macromolecules. Wiley. ISBN 978–0-470–84530–1. https://nattabiochem.wordpress.com/wp-content/uploads/2009/08/essentials_chemical_biology.pdf.

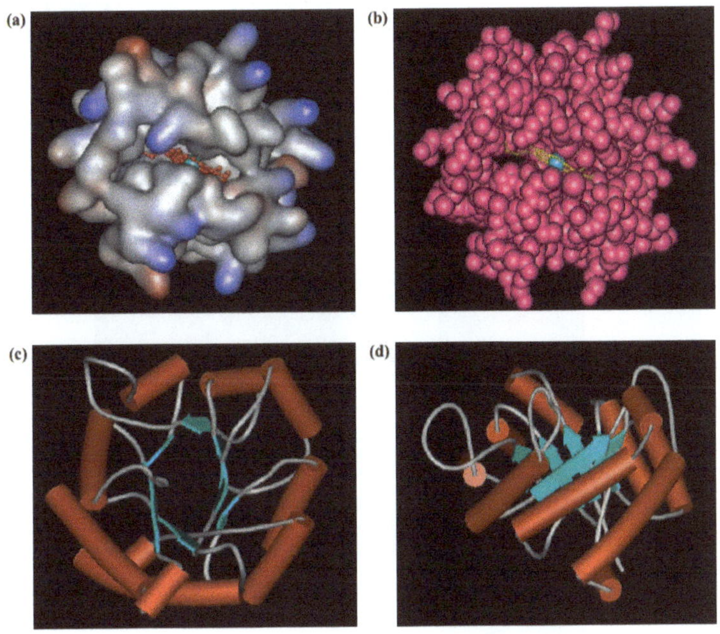

Figure 1.29. *Alternative cartoon depictions of proteins.*

Note: (a) surface display structure of small metal rich protein cytochrome c (horse heart) (pdb: 1hrc) showing Van der Waal's surface colored for positive charge (blue) and for negative charge (red). Ball and stick representations of iron-porphyrin macrocycle (prosthetic group) are shown (red) for each subunit with central iron ion rendered as Van der Waals sphere (light blue); (b) CPK structure of cytochrome c in which all polypeptide atoms are rendered as Van der Waals spheres (purple); (c) schematic display structure (top view) of parallel α/β-protein triose phosphate isomerase (chicken muscle) (pdb: 1tim) with α-helix shown as cylinders (red), β-strands as arrowed ribbons (light blue), loop structures (random coil) as rods (light gray); (d) schematic display structure (side view) of triose phosphate isomerase, otherwise as for (c).

Source: Andrew Miller and Julian Tanner. Essentials of Chemical Biology: Structure and Dynamics of Biological Macromolecules. Wiley. ISBN 978–0-470–84530–1. https://nattabiochem.wordpress.com/wp-content/uploads/2009/08/essentials_chemical_biology.pdf.

As functional proteins involved in respiration, metabolism, and communication, globular proteins and polypeptides are widely used in cells. Nonetheless, the great majority of proteins both inside and outside of cells serve as scaffolds or structural components. In these situations, protein polypeptides typically show an extended overall fold as opposed to a globular one, which is why proteins are referred to as fibrous proteins. Collagen, the primary structural element of bone, should be regarded as the most important of these. As shown in Figures 1.30 and 1.31, the collagen molecular fold is primarily formed by joining three extended left-handed P_{II} helices to form a right-handed PII triple helix, also known as the collagen triple helix. Each constituent P_{II} helix is connected to its neighbors by hydrogen bonds, providing both high lateral and longitudinal stability. Although they have not yet been thoroughly identified, other forces are also believed to be at play. Collagen molecules must have a disproportionately high amount of proline and glycine residues in comparison to other globular proteins/polypeptides because only polypeptides with glycine and proline residues can form extended P_{II} helices due to the unusual φ and ψ angles required. As previously mentioned, the total fold of a single polypeptide is the definition of the term "tertiary structure" in strict terms. Although a single polypeptide can form a collagen triple helix, the collagen triple helix that comprises the collagen fold is essentially made up of three distinct collagen polypeptides that are first connected by disulfide bridges.

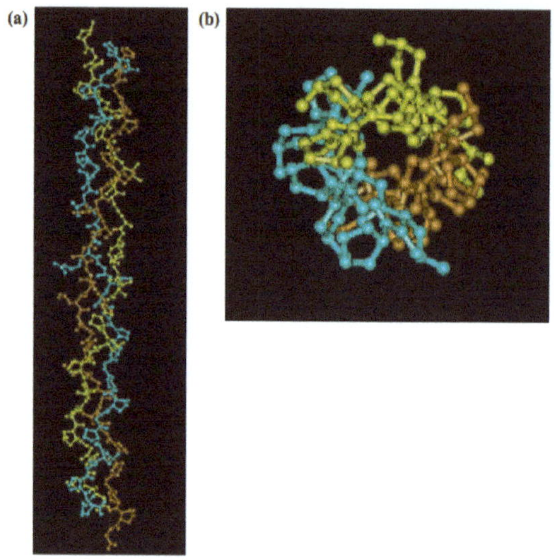

Figure 1.30. *Depiction of right-handed collagen triple helix. (pdb: 1bkv) (a) Ball and stick representation of atoms and bonds of triple helix (side view) with complete, individual polypeptide chain single colored (yellow, brown or blue) (side view); (b) Ball and stick representation of atoms and bonds of triple helix (top view), polypeptide chains rendered as in (a).*

Source: Andrew Miller and Julian Tanner. Essentials of Chemical Biology: Structure and Dynamics of Biological Macromolecules. Wiley. ISBN 978-0-470-84530-1. https://nattabiochem.wordpress.com/wp-content/uploads/2009/08/essentials_chemical_biology.pdf.

Consequently, it may be more accurate to refer to the collagen triple helix, which makes up the collagen molecular fold, as a type of quaternary structure rather than the tertiary structure of collagen. However, during the formation of bone, collagen molecules unite via their triple helices to form long bundles that solidify into the bone matrix. It could be argued that the collagen quaternary structure is better described as these bundles. Finally, we would like to draw attention to the fact that a small population of single polypeptide proteins, including dominant but not exclusive PII triple helical regions, have tertiary structures that are actually a combination of globular and fibrous protein folds. These are thought to be hybrid proteins that exhibit both fibrous and globular properties at the same time.

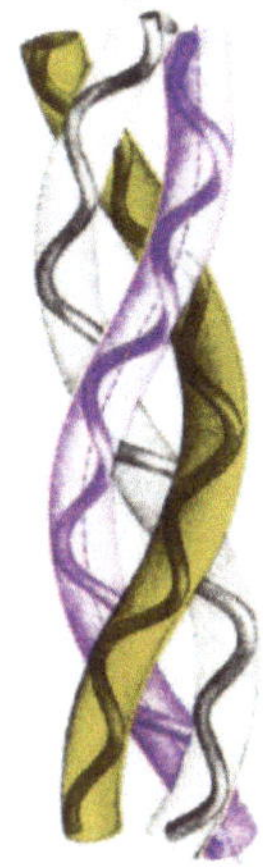

Figure 1.31. *Cartoon depiction of the right-handed collagen triple helix. Depiction emphasizes how three polypeptides in left-handed P_{II} helical conformations associate to form the triple helical structure wherein each polypeptide adopts a gentle right-handed rope-like twist in order to maximize stabilizing inter-chain hydrogen bond interactions. The structure is also stabilized by a sheath of ordered water molecules of solvation.*

Source: Andrew Miller and Julian Tanner. Essentials of Chemical Biology: Structure and Dynamics of Biological Macromolecules. Wiley. ISBN 978-0-470-84530-1. https://nattabiochem.wordpress.com/wp-content/uploads/2009/08/essentials_chemical_biology.pdf.

1.1.4. Quaternary Structure

The precise spatial arrangement and interactions of the quaternary structure represent the next "step up" between two or more polypeptide chains from the tertiary structure.

More complex multi-polypeptide proteins can be created by the interaction of different polypeptide chains; each polypeptide is referred to as a subunit. For the tetrameric protein hemoglobin, subunit interactions and interrelationships are shown (Figure 1.32). When it comes to globular proteins, polypeptide chain association makes it possible to supplement or even coordinate the activities of individual polypeptide elements, allowing the entire molecule to carry out a variety of biological tasks. The formation of quaternary structures in fibrous proteins increases their overall molecular strength.

Figure 1.32. *Quaternary structure of hemoglobin (human fetal) (pdb: 1a3n). This protein is comprised of 4 distinct polypeptides (subunits), α_1, α_2, β_1, and β_2 which are rendered two as ribbon display structures (red and yellow; right-side), one as a schematic display structure (left-side, rear), and one as CA stick display (left-side, front). Ball and stick representations of iron-porphyrin macrocycle (prosthetic group) are shown (green) for each subunit with central iron ion rendered as Van der Waals sphere (light gray). Subunits are all associated non-covalently with each other which is usual for polypeptide subunits that comprise a multi-polypeptide protein.*

Source: Andrew Miller and Julian Tanner. Essentials of Chemical Biology: Structure and Dynamics of Biological Macromolecules. Wiley. ISBN 978-0-470-84530-1. https://nattabiochem.wordpress.com/wp-content/uploads/2009/08/essentials_chemical_biology.pdf.

1.1.5. Prosthetic Groups

Numerous globular proteins and polypeptides, particularly those that catalyze chemical reactions, also contain non-peptidic structures that can bond or not with the polypeptide. We refer to these as prosthetic groups. All of these prosthetic groups interact with proteins to provide specific functions that might not be possible otherwise. An excellent illustration is the iron porphyrin ring found in the proteins cytochrome c, myoglobin, and hemoglobin, which permits the latter to function as a redox substrate and the former two to function as molecular oxygen carriers and scavengers during respiration.

1.2. CARBOHYDRATE STRUCTURE

One class of macromolecules that most people are aware of is carbohydrates, particularly in relation to our diets. Some people follow "low-carb" diets to shed pounds. On the other hand, before significant competitions, athletes frequently "carb-load" to make sure they have enough energy to compete at a high level. In actuality, carbohydrates are a necessary component of our diet and can be found naturally in grains, fruits, and vegetables. The body uses carbohydrates, especially glucose, a simple sugar that is a part of starch and a common ingredient in many staple foods, to produce energy. In addition, carbohydrates play other crucial roles in the lives of humans, animals, and plants. The stoichiometric formula for carbohydrates is $(CH_2O)_n$, where n is the number of carbons in the molecule. Put another way, the carbon, hydrogen, and oxygen ratio in carbohydrate molecules is 1:2:1. This formula also clarifies the meaning of the term "carbohydrate," as it consists of two components: carbon ("carbo") and water (hence, "hydrate"). Three subtypes of carbohydrates are distinguished: monosaccharides, disaccharides, and polysaccharides.

1.2.1. Monosaccharides

Monosaccharides are simple sugars, the most prevalent of which is glucose (*mono-* = "one"; *sacchar-* = "sweet"). The typical range of carbons in monosaccharides is three to seven. Most names of monosaccharides finish in *-ose*. A sugar is referred to as an aldose if it contains an aldehyde group (the functional group with the structure R-CHO) or a ketose if it contains a ketone group (the functional group with the structure RC(=O) R′). They can also be referred to as trioses (three carbons), pentoses (five carbons), or hexoses (six carbons), depending on the number of carbons in the sugar. To view an illustration of the monosaccharides, refer to Figure 1.33.

Glucose has the chemical formula $C_6H_{12}O_6$. Glucose is a vital energy source for humans. Energy from glucose is released during cellular respiration, and this energy is utilized to aid in the production of adenosine triphosphate (ATP). Using carbon dioxide and water, plants create glucose, which is then used by the plant to meet its energy needs. An overabundance of glucose is frequently stored as starch, which humans and other plant-eating animals catabolize—the breakdown of larger molecules by cells.

Other common monosaccharides are fructose and galactose, which are found in fruit sugars and milk sugars, respectively. Despite having the same chemical formula ($C_6H_{12}O_6$), fructose, galactose, and glucose are all known as isomers because of their distinct functional group arrangements surrounding asymmetric carbons, which cause them to differ structurally and chemically (Figure 1.34).

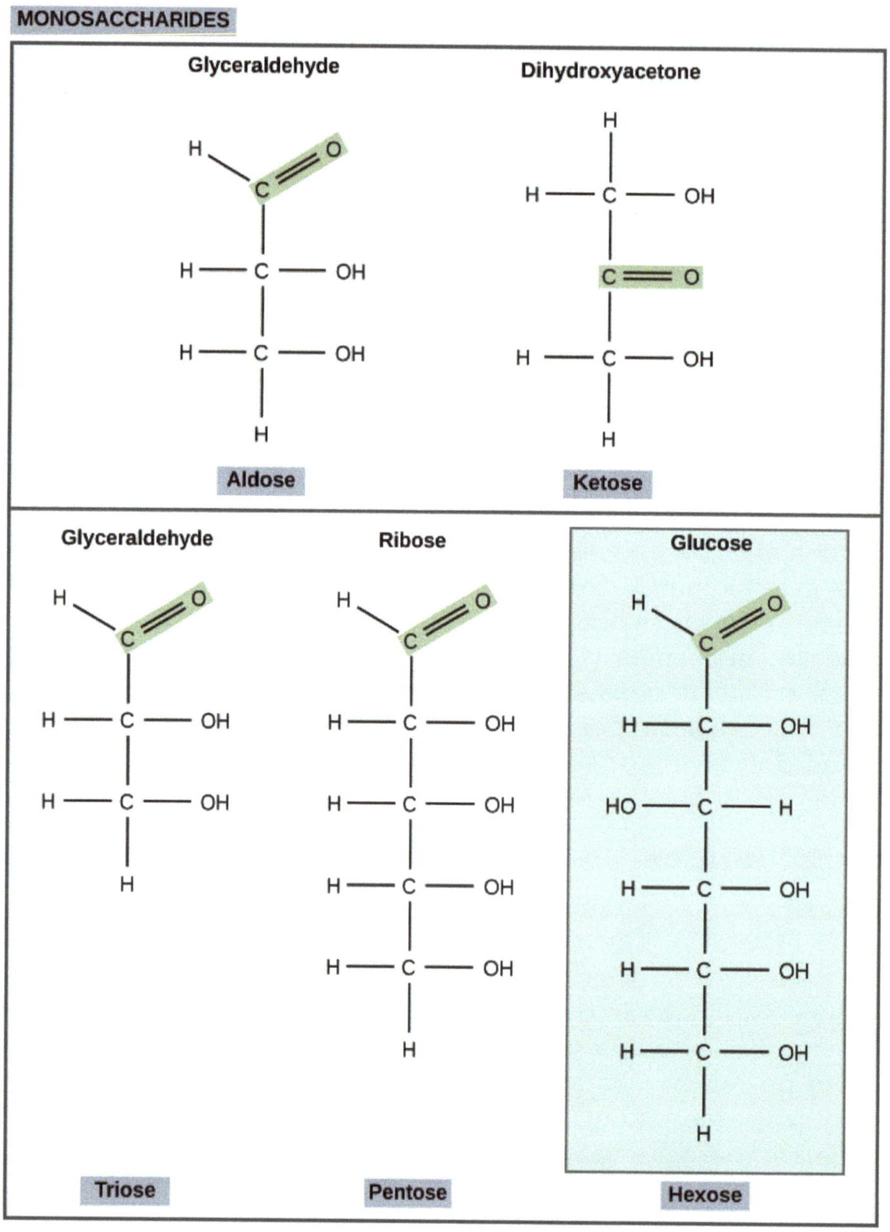

Figure 1.33. *Monosaccharides are classified based on the position of their carbonyl group and the number of carbons in the backbone. Aldoses have a carbonyl group (indicated in green) at the end of the carbon chain, and ketoses have a carbonyl group in the middle of the carbon chain. Trioses, pentoses, and hexoses have three, five, and six carbon backbones, respectively.*

Source: https://courses.lumenlearning.com/wm-biology1/chapter/reading-types-of-carbohydrates/.

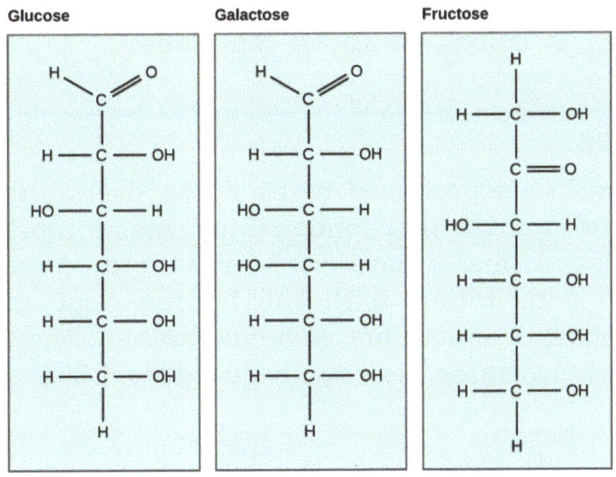

Figure 1.34. *Glucose, galactose, and fructose are all hexoses. They are structural isomers, meaning they have the same chemical formula ($C_6H_{12}O_6$) but a different arrangement of atoms.*

Source: https://courses.lumenlearning.com/wm-biology1/chapter/reading-types-of-carbohydrates/.

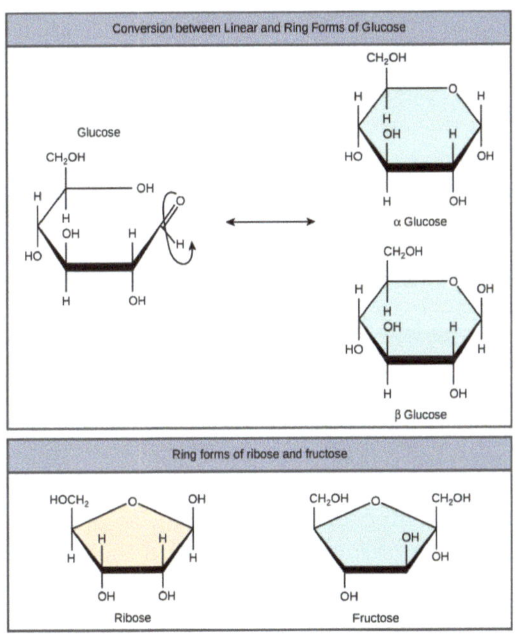

Figure 1.35. *Five and six carbon monosaccharides exist in equilibrium between linear and ring forms. When the ring forms, the side chain it closes on is locked into an α or β position. Fructose and ribose also form rings, although they form five-membered rings as opposed to the six-membered ring of glucose.*

Source: https://courses.lumenlearning.com/wm-biology1/chapter/reading-types-of-carbohydrates/.

Monosaccharides can exist as rings or as linear chains, but in aqueous solutions, rings are the most common form (Figure 1.35). When glucose forms a ring, the hydroxyl group (–OH) can be arranged in two different ways around the anomeric carbon (carbon

1 that becomes asymmetric during the ring formation process). The hydroxyl group in sugar is said to be in the alpha (α) position if it is located below carbon number 1, and in the beta (β) position if it is located above the plane.

1.2.2. Disaccharides

Disaccharides (*di–* = "two") are produced when a dehydration reaction occurs between two monosaccharides (also known as a condensation reaction or dehydration synthesis). A water molecule is released and a covalent bond is formed when the hydroxyl group of one monosaccharide joins forces with the hydrogen of another monosaccharide during this process. Glycosidic bonds are covalent connections that arise between two monosaccharides, or carbohydrates, as shown in Figure 1.36.

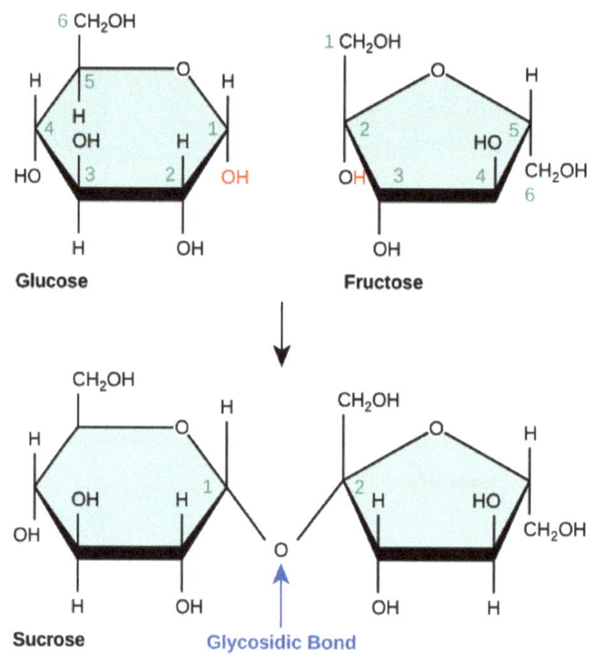

Figure 1.36. *Sucrose is formed when a monomer of glucose and a monomer of fructose are joined in a dehydration reaction to form a glycosidic bond. In the process, a water molecule is lost. By convention, the carbon atoms in a monosaccharide are numbered from the terminal carbon closest to the carbonyl group. In sucrose, a glycosidic linkage is formed between carbon 1 in glucose and carbon 2 in fructose.*

Source: https://courses.lumenlearning.com/wm-biology1/chapter/reading-types-of-carbohydrates/.

Alpha and beta glycosidic bonds, also known as glycosidic linkages, are the two types of glycosidic bonds. In the case of the first glucose, an alpha bond forms when the OH group is located below the ring plane and a beta bond forms when the OH group is located above the ring plane.

Lactose, maltose, and sucrose are examples of common disaccharides (Figure 1.37). Glucose and galactose are the monomers that make up the disaccharide lactose. It naturally exists in milk. Maltose, also known as malt sugar, is a disaccharide that results from the dehydration of two glucose molecules. Sucrose, or table sugar, is the most widely used disaccharide. It is made up of the monomers fructose and glucose.

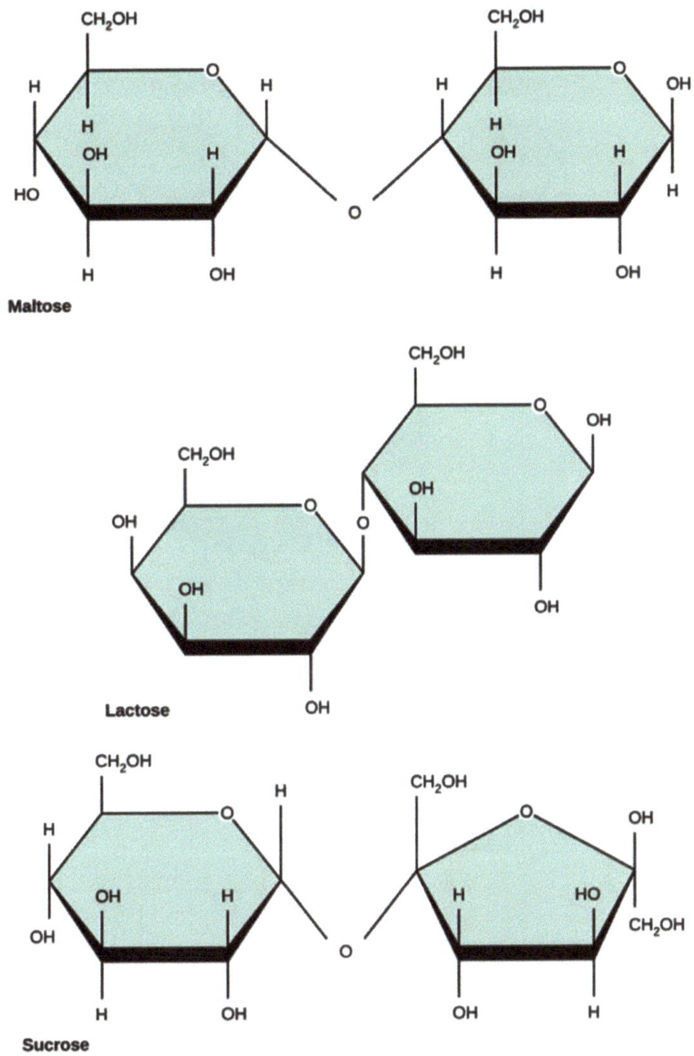

Figure 1.37. *Common disaccharides include maltose (grain sugar), lactose (milk sugar), and sucrose (table sugar).*

Source: https://courses.lumenlearning.com/wm-biology1/chapter/reading-types-of-carbohydrates/.

1.2.3. Polysaccharides

A polysaccharide is a lengthy chain of monosaccharides joined by glycosidic bonds (*poly*– = "many"). The chain could contain a variety of monosaccharide types and could be branched or unbranched. The number of monomers joined can determine whether the molecular weight is 100,000 Daltons or higher. The main types of polysaccharides are chitin, starch, glycogen, and cellulose.

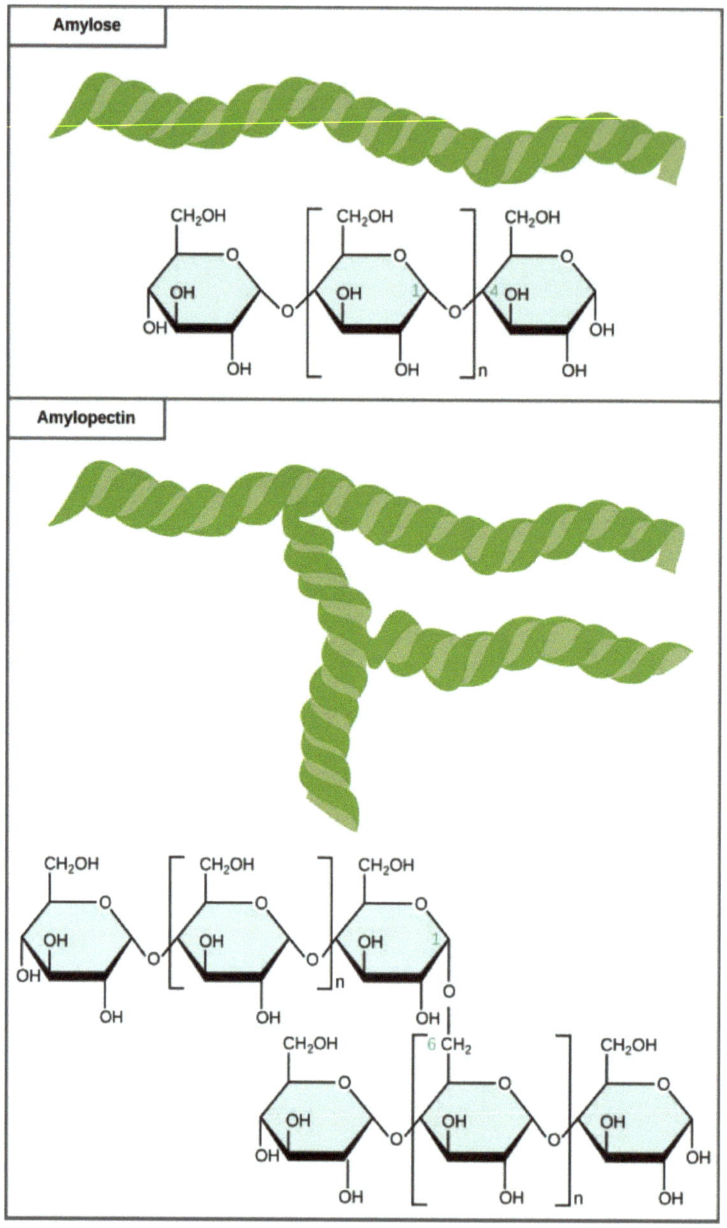

Figure 1.38. *Amylose and amylopectin are two different forms of starch. Amylose is composed of unbranched chains of glucose monomers connected by α 1,4 glycosidic linkages. Amylopectin is composed of branched chains of glucose monomers connected by α 1,4 and α 1,6 glycosidic linkages. Because of the way the subunits are joined, the glucose chains have a helical structure. Glycogen (not shown) is similar in structure to amylopectin but more highly branched.*

Source: https://courses.lumenlearning.com/wm-biology1/chapter/reading-types-of-carbohydrates/

Plants store carbohydrates as starch, which is composed of a combination of the glucose polymers amylose and amylopectin. Plants have the ability to synthesize glucose, and any extra glucose that is not immediately needed by the plant is stored as starch in a variety of plant parts, such as the roots and seeds. In addition to feeding the embryo during germination, the starch in the seeds can feed both people and other animals. Enzymes like salivary amylases break down the starch that humans eat into smaller molecules like glucose and maltose. The glucose can then be absorbed by the cells.

The building blocks of starch are glucose monomers connected by α 1–4 or α 1–6 glycosidic bonds. The carbon numbers of the two residues that have joined to form the bond are indicated by the numbers 1–4 and 1–6. Amylopectin is a branched polysaccharide (α 1–6 linkages at the branch points), whereas amylose is starch formed by unbranched chains of glucose monomers (only α 1–4 linkages), as shown in Figure 1.38.

In humans and other vertebrates, glucose is stored as glycogen, which is composed of glucose monomers. The highly branched molecule known as glycogen is the animal counterpart of starch and is typically found in muscle and liver cells. Glycogenolysis is the process by which glucose is released from glycogen when blood glucose levels fall. The most prevalent natural biopolymer is cellulose. Plants primarily use cellulose for their cell walls, which give the cell structural support. Paper and wood are primarily made of cellulosic materials. Glycose monomers joined by β1–4 glycosidic bonds form the building blocks of cellulose (Figure 1.39).

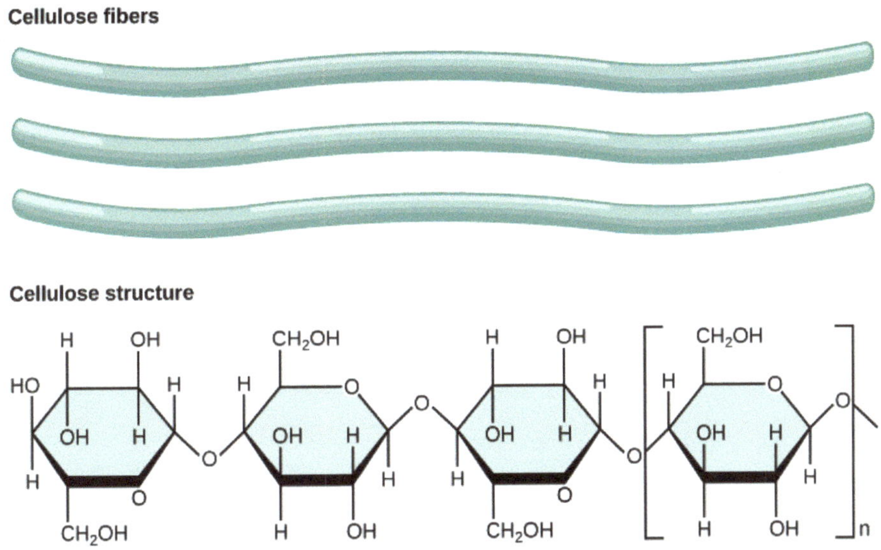

Figure 1.39. *In cellulose, glucose monomers are linked in unbranched chains by β 1–4 glycosidic linkages. Because of the way the glucose subunits are joined, every glucose monomer is flipped relative to the next one resulting in a linear, fibrous structure.*

Source: https://courses.lumenlearning.com/wm-biology1/chapter/reading-types-of-carbohydrates/.

Every other glucose monomer in cellulose is turned over, as illustrated in Figure 1.39, and the monomers are firmly packed as extended long chains. This is the source of cellulose's high tensile strength and stiffness, both of which are crucial for plant cells. Although human digestive enzymes cannot break down the β 1–4 linkage, plant-eating animals like cows, koalas, buffalos, and horses can digest and use cellulose-rich plant material as food thanks to the specialized flora in their stomachs. The rumen, a portion of a herbivore's digestive tract, is home to specific bacteria and protist species that secrete the enzyme cellulase in these animals. Additionally, bacteria that break down cellulose can be found in the appendix of grazing animals, which is why ruminants' appendices are so important to their digestive systems. Cellulases have the ability to degrade cellulose into glucose monomers, which the animal can utilize as fuel. Because termites' bodies contain other organisms that secrete cellulases, they are also capable of breaking down cellulose.

Animals use carbohydrates for a variety of purposes. The exoskeleton, or external skeleton, of arthropods—insects, crustaceans, and other animals—protects their internal body parts (see the bee in Figure 1.40).

Figure 1.40. *Insects have a hard outer exoskeleton made of chitin, a type of polysaccharide.*

Source: https://courses.lumenlearning.com/wm-biology1/chapter/reading-types-of-carbohydrates/.

The biological macromolecule chitin, a nitrogen-containing polysaccharide, is what makes up this exoskeleton. It is composed of modified sugar N-acetyl-β-d-glucosamine units repeated repeatedly. Fungal cell walls also contain a significant amount of chitin; as fungi belong to the domain Eukarya and are not related to either plants or animals.

1.3. STRUCTURE OF NUCLEIC ACIDS

The most crucial macromolecules for the continuation of life are nucleic acids. They contain instructions necessary for a cell to function as well as the genetic blueprint for the cell.

Deoxyribonucleic acid (DNA) and ribonucleic acid (RNA) are the two primary forms of nucleic acids. The genetic material present in all living things, including multicellular mammals and single-celled bacteria, is called DNA. It is present in the mitochondria, chloroplasts, organelles, and nuclei of eukaryotes. The DNA in prokaryotes is not protected by a membrane.

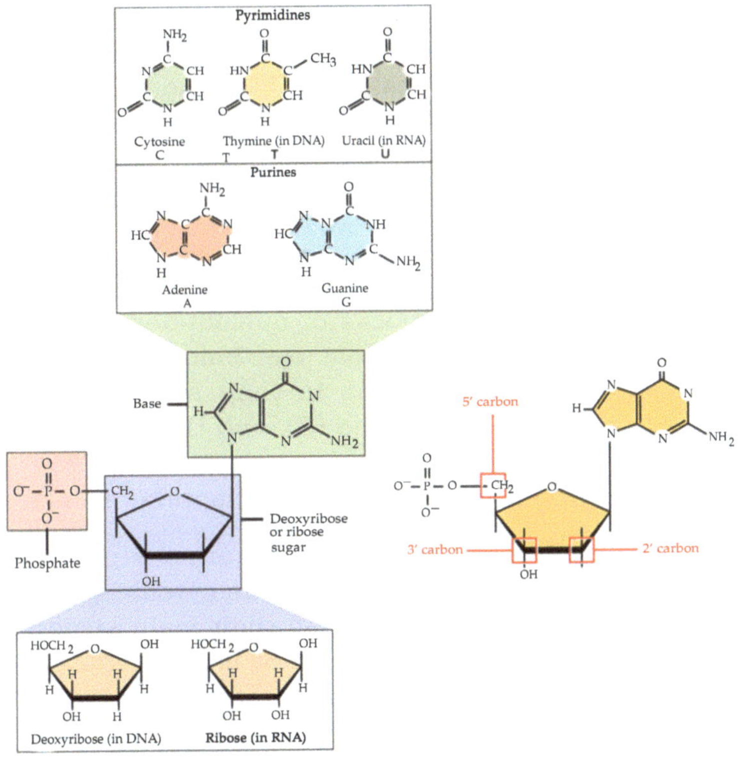

Figure 1.41. *A nucleotide is made up of three components: a nitrogenous base, a pentose sugar, and one or more phosphate groups. Carbon residues in the pentose are numbered 1′ through 5′ (the prime distinguishes these residues from those in the base, which are numbered without using a prime notation). The base is attached to the 1′ position of the ribose, and the phosphate is*

attached to the 5′ position. When a polynucleotide is formed, the 5′ phosphate of the incoming nucleotide attaches to the 3′ hydroxyl group at the end of the growing chain. Two types of pentose are found in nucleotides, deoxyribose (found in DNA) and ribose (found in RNA). Deoxyribose is similar in structure to ribose, but it has an H instead of an OH at the 2′ position. Bases can be divided into two categories: purines and pyrimidines. Purines have a double ring structure, and pyrimidines have a single ring.

Source: https://courses.lumenlearning.com/wm-biology1/chapter/reading-structure-of-nucleic-acids/.

A cell's whole genetic makeup is referred to as its genome, and the study of genomes is known as genomics. Chromatin, the material that makes up eukaryotic chromosomes, is formed when DNA combines with histone proteins in eukaryotic cells but not in prokaryotes. There could be tens of thousands of genes on one chromosome. While some genes code for RNA products, many genes contain the instructions needed to make protein products. All cellular functions are governed by DNA, which toggles genes "on" and "off."

RNA, the other kind of nucleic acid, is primarily used in the synthesis of proteins. The DNA molecules never leave the nucleus; instead, they communicate with the other parts of the cell through an intermediate. Messenger RNA (mRNA) is this middleman. Additional RNA subtypes that are involved in protein synthesis and regulation include rRNA, tRNA, and microRNA.

Nucleotides are the monomers that make up DNA and RNA. A polynucleotide, such as DNA or RNA, is created when the nucleotides combine. Three elements make up each nucleotide: a pentose (five-carbon) sugar, a nitrogenous base, and a phosphate group (Figure 1.41). A sugar molecule, which is attached to one or more phosphate groups, is connected to each nitrogenous base in a nucleotide.

Because they contain both carbon and nitrogen, the organic molecules known as nitrogenous bases—important building blocks of nucleotides—get their name. They are bases because they have an amino group that can bind an additional hydrogen atom, lowering the concentration of hydrogen ions in their surroundings and making them more basic. Adenine (A), guanine (G), cytosine (C), and thymine (T) are the four possible nitrogenous bases that can be found in each nucleotide of DNA. Adenine, guanine, cytosine, or uracil (U) can also be found in RNA nucleotides in place of thymine.

Purines are made up of guanine and adenine. A purine is primarily composed of two carbon-nitrogen rings. The pyrimidines, which include cytosine, thymine, and uracil, are characterized by a single carbon-nitrogen ring as their main structural element (Figure 1.41). There are various functional groups affixed to each of these fundamental carbon-nitrogen rings. The nitrogenous bases are simply referred to by their symbols A, T, G, C, and U in molecular biology shorthand. A, T, G, and C are found in DNA, but A, U, G, and C are found in RNA.

Deoxyribose is the pentose sugar found in DNA, whereas ribose is the sugar found in RNA (Figure 1.41). The presence of a hydroxyl group on the second carbon of ribose and hydrogen on the second carbon of deoxyribose distinguishes the two sugars from one another. The numbers 1′, 2′, 3′, 4′, and 5′ correspond to the carbon atoms in the sugar molecule (1′ is pronounced as "one prime"). A 5′–3′ phosphodiester linkage is created when the phosphate residue is joined to the hydroxyl groups of one sugar's 5′ carbon and the next nucleotide's 3′ carbon sugar's hydroxyl group. Unlike the other linkages that join monomers in macromolecules, the phosphodiester linkage is created by removing two phosphate groups rather than by a straightforward dehydration reaction. There could be thousands of these phosphodiester linkages in a polynucleotide.

1.3.1. DNA Double-Helix Structure

The structure of DNA is double helix (Figure 1.42). The backbone of DNA is made up of sugar and phosphate, which are located outside of the helix. In the interior, the nitrogenous bases are arranged in pairs and connected to one another by hydrogen bonds, much like stairs. In a double helix, 0.34 nm separates each base pair from the one after it.

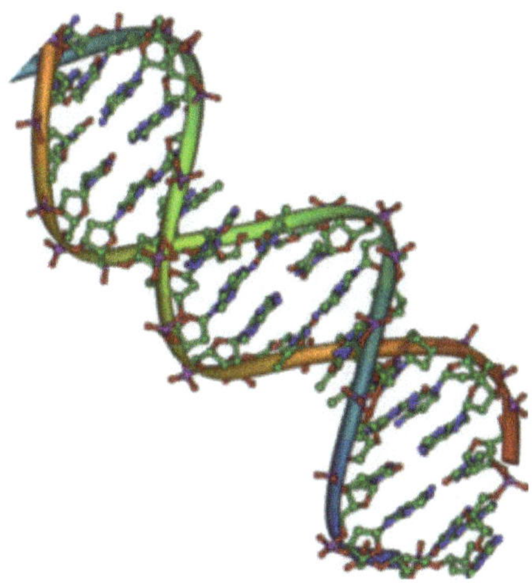

Figure 1.42. *DNA is an antiparallel double helix. The phosphate backbone (the curvy lines) is on the outside, and the bases are on the inside. Each base interacts with a base from the opposing strand.*

Source: https://courses.lumenlearning.com/wm-biology1/chapter/reading-structure-of-nucleic-acids/.

Because the helix's two strands run in opposing directions, the 5′ carbon end of one strand will come into contact with the 3′ carbon end of the strand that matches it. (This is known as antiparallel orientation and is crucial for many nucleic acid interactions, including DNA replication).

Base pairings of specific kinds are only permitted. For instance, a specific pyrimidine and purine combination is limited. As seen in Figure 1.43, this indicates that A can pair with T and G with C. The base complementary rule refers to this. Stated differently, the complementary nature of the DNA strands is apparent. The complementary strand would have the sequence TTAACCGG if one strand had the sequence AATTGGCC. Every strand in DNA is duplicated during replication, producing a daughter DNA double helix that has one copy of the parental DNA strand and one newly synthesized strand.

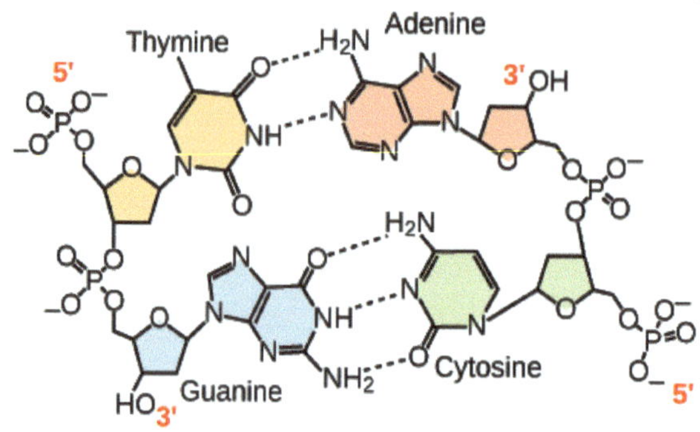

Figure 1.43. *In a double-stranded DNA molecule, the two strands run antiparallel to one another so that one strand runs 5′ to 3′ and the other 3′ to 5′. The phosphate backbone is located on the outside, and the bases are in the middle. Adenine forms hydrogen bonds (or base pairs) with thymine, and guanine base pairs with cytosine.*

Source: https://courses.lumenlearning.com/wm-biology1/chapter/reading-structure-of-nucleic-acids/.

1.3.2. RNA

Under DNA's guidance, ribonucleic acid, or RNA, plays a major role in the synthesis of proteins. Phosphodiester bonds bind the ribonucleotides that make up RNA, which is typically single-stranded. In the RNA chain, a ribonucleotide consists of a phosphate group, one of the four nitrogenous bases (A, U, G, and C), and ribose, or pentose sugar.

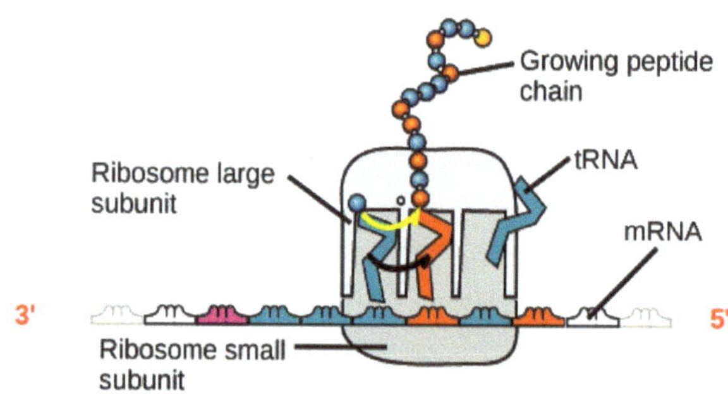

Figure 1.44. *A ribosome has two parts: a large subunit and a small subunit. The mRNA sits in between the two subunits. A tRNA molecule recognizes a codon on the mRNA, binds to it by complementary base pairing, and adds the correct amino acid to the growing peptide chain.*

Source: https://courses.lumenlearning.com/wm-biology1/chapter/reading-structure-of-nucleic-acids/.

Messenger RNA (mRNA), ribosomal RNA (rRNA), transfer RNA (tRNA), and microRNA (miRNA) are the four main forms of RNA. The first, mRNA, conveys the message from DNA, which manages every aspect of a cell's activity. A gene for a product is "turned on" and messenger RNA is created in the nucleus when a cell needs to synthesis a specific protein. The coding sequence of the DNA from which the RNA was copied is complementary to the RNA base sequence. In contrast, base U is present in RNA while base T is absent. The complementary RNA sequence is UUAACGCG if the DNA strand has the sequence AATTGCGC. The mRNA interacts with ribosomes and other cellular components in the cytoplasm (Figure 1.44).

Codons are groups of three bases that are read from the mRNA. One amino acid is coded for by each codon. The mRNA is read and the protein product is produced in this manner. The primary component of ribosomes on which mRNA binds is ribosomal RNA (rRNA). In addition to ensuring that the mRNA and ribosomes are properly aligned, the ribosome's rRNA catalyzes the formation of peptide bonds between two aligned amino acids through its enzymatic activity, peptidyl transferase. Among the four types of RNA, transfer RNA (tRNA) is the smallest, typically ranging from 70 to 90 nucleotides in length. It transports the proper amino acid to the location where proteins are synthesized. The proper amino acid can be inserted into the polypeptide chain thanks to base pairing between the tRNA and mRNA. The smallest RNA molecules, known as microRNAs, play a crucial role in controlling gene expression by obstructing the expression of specific mRNA messages.

1.4. MACROMOLECULAR LIPID ASSEMBLIES

Strangely, some people believe that lipids are only "dull," serving only as retentive membranes to delineate the borders of cells, the domains of different organelles, and the functional zones that exist within them. Lipids are essential for compartmentalization, but they also have a wide range of complex dynamic behaviors that support cellular metabolism and foster the systems of communication and cooperation among living cells in complex, multicellular organisms.

Lipids and the macromolecular lipid assemblies they form, such as membranes, are also important for compartmentalization. Since research into lipids and their macromolecular lipid assemblies has been comparatively neglected in comparison with other fields in chemical biology, there seems to be a lot still to learn and study. However, readers who are interested in chemical biology should understand that the impact of macromolecular lipid assemblies on biological function is just as significant and wide-ranging as that of biological macromolecules like proteins, carbohydrates, and nucleic acids. Thus, it is inevitable that further chemical biology studies on lipids and macromolecular lipid assemblies will be crucial to expanding our knowledge of how biology functions.

1.4.1. Monomeric Lipid Structures

In general, molecules with an intermediate molecular weight (MW 100–5,000 Da) that contain a significant amount of aromatic or aliphatic hydrocarbon can be referred to as monomeric lipids. The majority of lipids that are crucial to biology are found in a subclass of these large molecules called complex lipids. This subset's main constituents are sphingolipids, glycerophospholipids, and acylglycerols. There are other complex lipids that are known, but they either serve little structural purpose or have poorly understood biological functions. Long-chain fatty acids and triose glycerol combine to form acylglycerols. Figure 1.45 displays typical long-chain fatty acids found in nature. Although there are many different types of acylglycerols, the triacylglycerols—which are made up of fatty acids esterifying all three of the reactive hydroxyl groups of glycerol—are the most biologically significant (Figure 1.46). The primary lipid storage form in plants and higher animals is triacylglycerol. Triacylglycerols are referred to as triglycerides or simple triacylglycerols when they have identical fatty acid acyl chains. The term "mixed triacyl glycerols" is used when there are different fatty acid acyl chains involved (Figure 1.46).

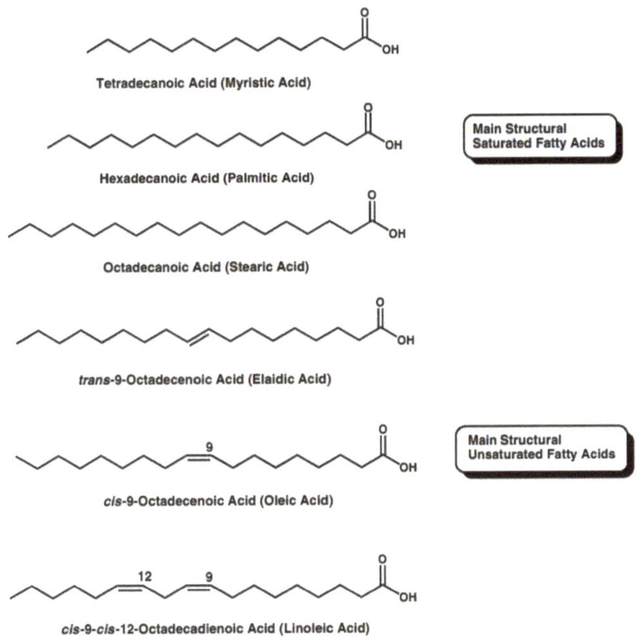

Figure 1.45. *Structures of main saturated and unsaturated fatty acids in structural lipids that form lipid assemblies.*

Source: Andrew Miller and Julian Tanner. Essentials of Chemical Biology: Structure and Dynamics of Biological Macromolecules. Wiley. ISBN 978-0-470-84530-1. https://nattabiochem.wordpress.com/wp-content/uploads/2009/08/essentials_chemical_biology.pdf.

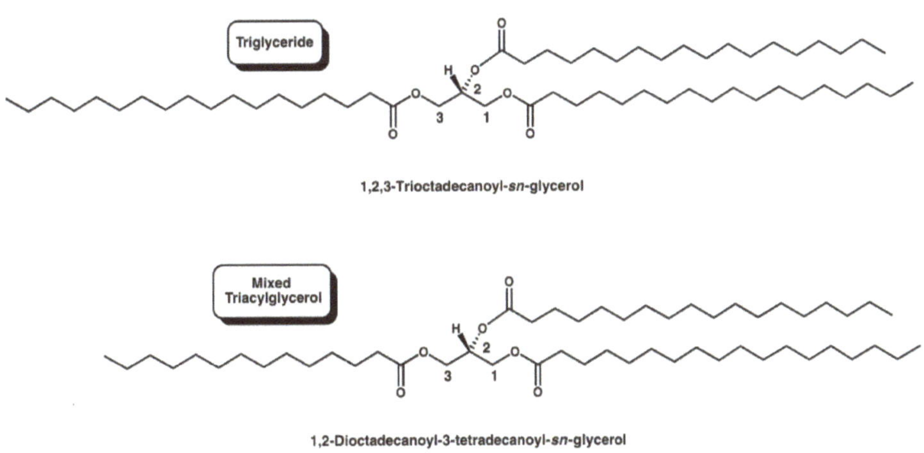

Figure 1.46. *Illustrative structures of main storage lipids, the triglycerides and the mixed triacylglycerols.*

Source: Andrew Miller and Julian Tanner. Essentials of Chemical Biology: Structure and Dynamics of Biological Macromolecules. Wiley. ISBN 978-0-470-84530-1. https://nattabiochem.wordpress.com/wp-content/uploads/2009/08/essentials_chemical_biology.pdf.

Glycerophospholipids can be conceptualized as triacylglycerol derivatives where a phosphate ester has taken the place of the carbon atom in the C-3 carboxylate ester. Due to the enormous number of potential glycerophospholipids, only the most well-researched and presumably most biologically significant are shown (Figure 1.47).

1.5 MACROMOLECULAR LIPID ASSEMBLIES

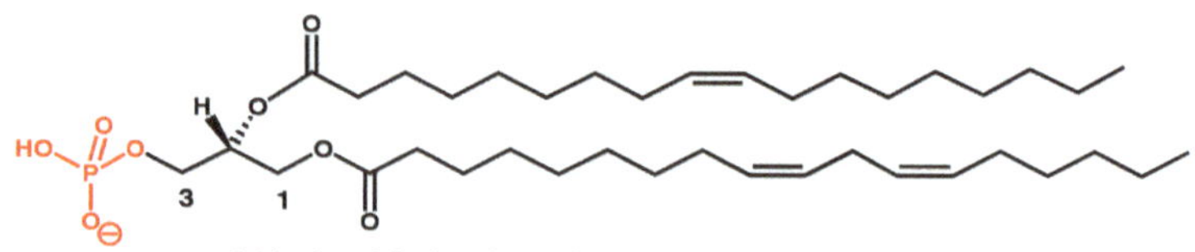

1-Linoleoyl-2-oleoyl-*sn*-glycero-3-phosphatidic Acid (LOPA)

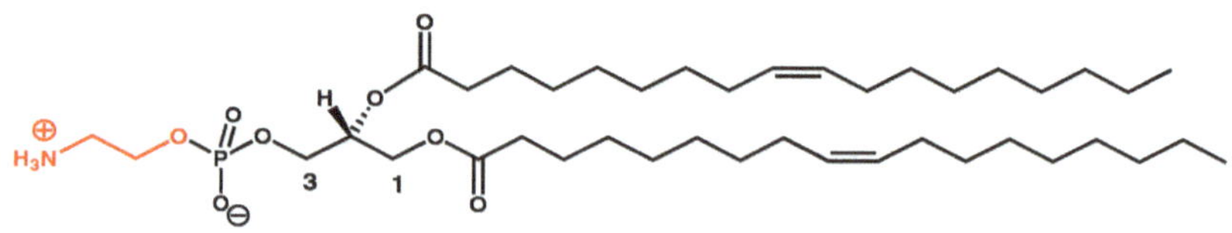

1,2-Dioleoyl-*sn*-glycero-3-phosphatidylethanolamine (DOPE)

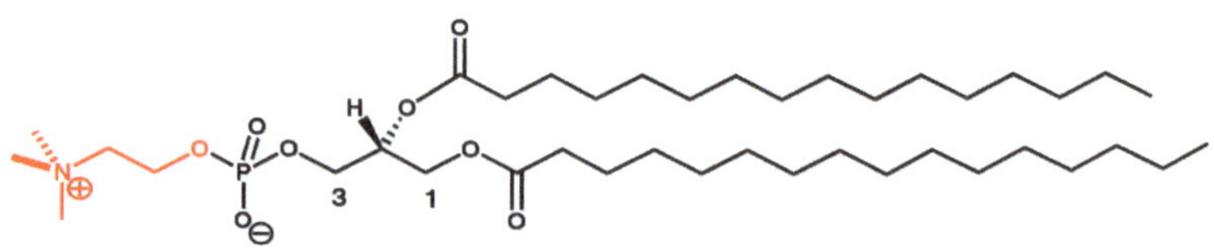

1,2-Dipalmitoyl-*sn*-glycero-3-phosphatidylcholine (DPPC)

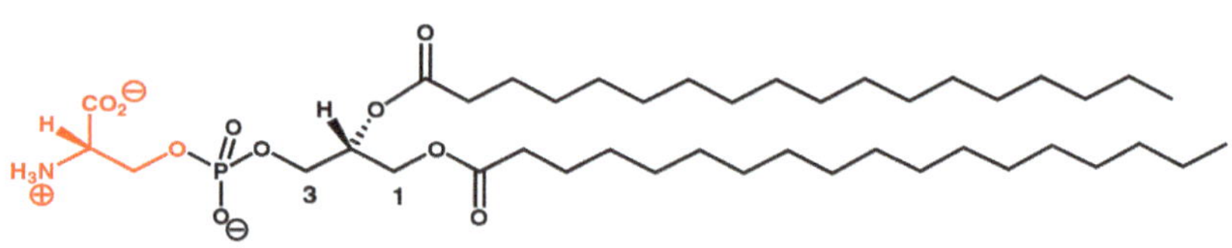

1,2-Distearoyl-*sn*-glycero-3-phosphatidylserine (DSPS)

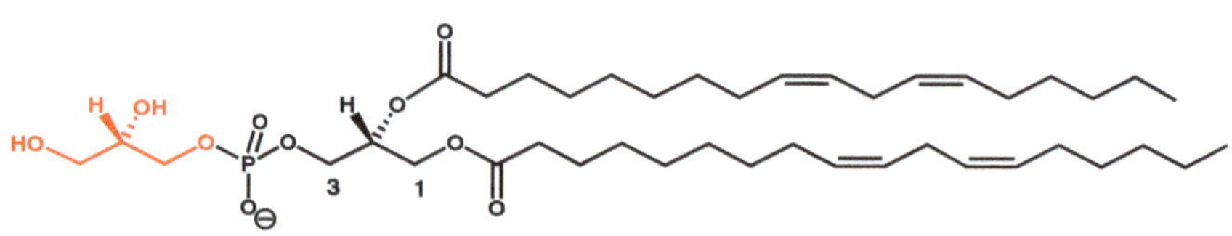

1,2-Dilinoleoyl-*sn*-glycero-3-phosphatidylglycerol (DLPG)

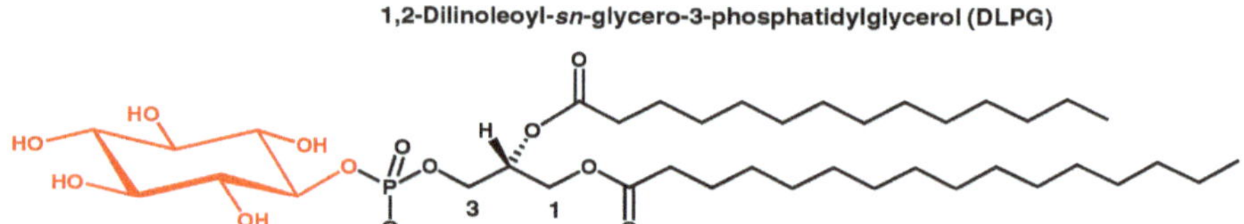

1-Palmitoyl-2-myristoyl-*sn*-glycero-3-phosphatidylinositol (PMPI)

Figure 1.47. *Representative structures of first rank structural lipids, the glycerophospholipids or phospholipids that are major and integral components of lipid assemblies.*

Source: Andrew Miller and Julian Tanner. Essentials of Chemical Biology: Structure and Dynamics of Biological Macromolecules. Wiley. ISBN 978-0-470-84530-1. https://nattabiochem.wordpress.com/wp-content/uploads/2009/08/essentials_chemical_biology.pdf.

The primary function of these glycerophospholipids is as components of biological membranes. Long-chain fatty acids and the basic sphingosine combine to form sphingolipids. Ceramides are produced when a fatty acid acylates the carbon atoms C-2 amine group. Phospholipid ester derivatization of the C-1 hydroxyl group yields phosphoceramides. Figure 1.48 illustrates members of both biologically significant types. Ceramides and phosphoceramides play a significant role in the lipids found in human skin and neural membranes. Phospholipids are the combined term for phosphoceramides and glycerophospholipids.

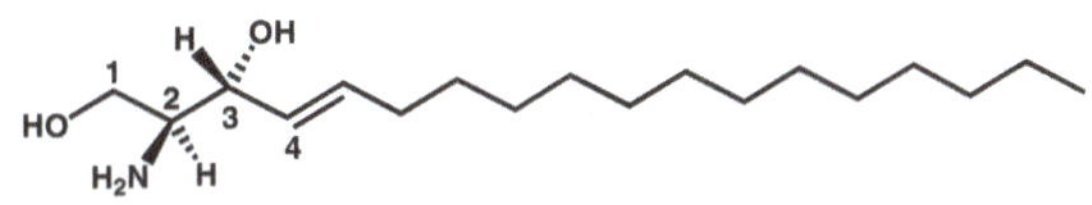

D-erythro-2-amino-4-octadecene-1,3-diol (Sphingosine)

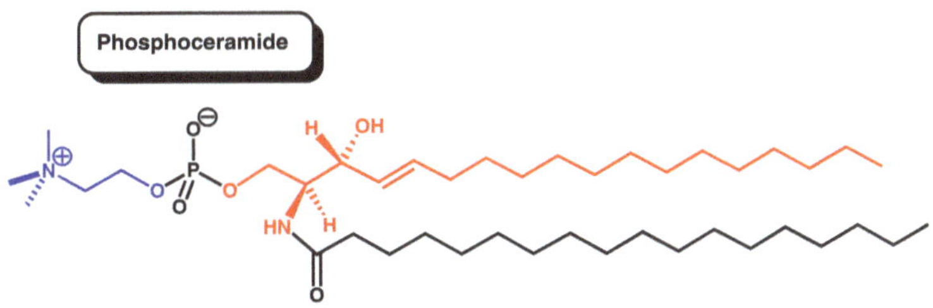

2-Stearoyl-1-phosphatidylcholine sphingosine (Stearoyl sphingomyelin)

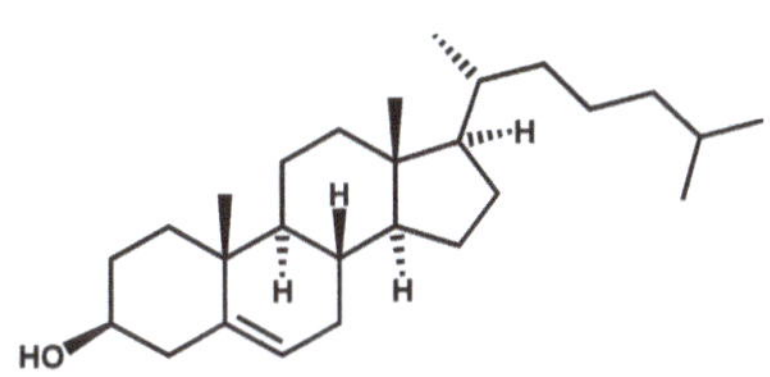

Cholest-5-en-3β-ol (Cholesterol)

Figure 1.48. *Representative structure of second rank structural lipids, the phosphoceramides that are formed from the amine diol sphingosine. These can be components of some lipid assemblies but in most cases are not dominant. The structure of cholesterol, another critical structural lipid is also shown.*

Source: Andrew Miller and Julian Tanner. Essentials of Chemical Biology: Structure and Dynamics of Biological Macromolecules. Wiley. ISBN 978-0-470-84530-1. https://nattabiochem.wordpress.com/wp-content/uploads/2009/08/essentials_chemical_biology.pdf.

1.4.2. Lyotropic Mesophases of Phospholipids

Phospholipids' amphiphilic nature gives rise to macromolecular lipid assemblies (Figure 1.49). In general, an extended chain region that is hydrophobic in nature and a compact polar region that is hydrophilic in nature are present in all phospholipids. The hydrophobic chain regions' propensity to self-associate and simultaneously exclude water in the presence of water results in macromolecular lipid assemblies, which can be thought of as large, extended non-covalent structures held together by van der Waals interactions and the hydrophobic effect. These assemblies adopt different phase states based on the type of phospholipid involved and the local environment. The term "lyotropic mesophases" also refers to these phospholipid structures in their different phase states. Using the Luzzati nomenclature, some of the most well-known lyotropic mesophases have been compiled (Table 1.2). We will talk about structures and mesophase transitions because they are important to understanding the biological behavior of lipids. All of the lyotropic mesophases may not be equally relevant in biology, and there can never be complete certainty about this given the current state of knowledge on lipids and their macromolecular assemblies.

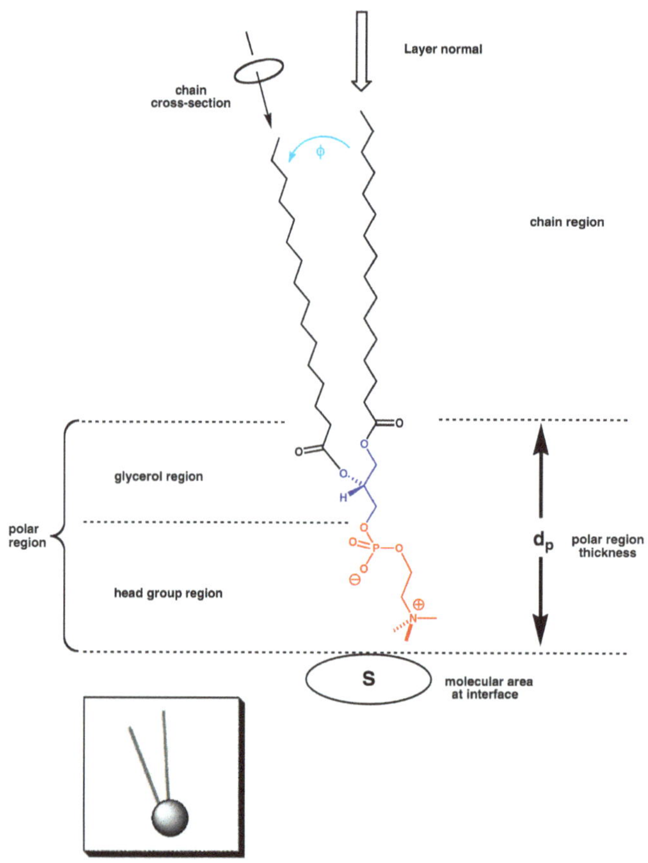

Figure 1.49. *Structure of dipalmitoyl L-α-phosphatidylcholine (DPPC) to illustrate the main molecular parameters and structural features that dictate the formation of crystalline lamellar phases of macromolecular lipid assemblies. Schematic representation of a phospholipid molecule is also shown (inset).*

Source: Andrew Miller and Julian Tanner. Essentials of Chemical Biology: Structure and Dynamics of Biological Macromolecules. Wiley. ISBN 978-0-470-84530-1. https://nattabiochem.wordpress.

com/wp-content/uploads/2009/08/essentials_chemical_biology.pdf.

Table 1.2. Summary of the Main Lyotropic Mesophases Giving Phase Type with Symmetry Name and the Nature of the Mesophase Structure in Each Case

Phase Type	Name	Phase Structure
Solid-like lamellar		
3D	L_c	3D crystal
2D	L_c^{2D}	2D crystal
	$P_{\beta'}$	rippled gel
1D	L_β	untilted gel
	$L_{\beta'}$	tilted gel
	$L_{\beta I}$	interdigitated gel
	$L_{\alpha\beta}$	partial gel
Fluid mesophases		
1D	L_α	fluid lammellar
2D	H	hexagonal
	H^C	complex hexagonal
	R	rectangular
	M	oblique
3D	Q	cubic
	T	tetragonal
	R	rhombohedral
	O	orthorhombic

Source: Andrew Miller and Julian Tanner. Essentials of Chemical Biology: Structure and Dynamics of Biological Macromolecules. Wiley. ISBN 978-0-470-84530-1. https://nattabiochem.wordpress.com/wp-content/uploads/2009/08/essentials_chemical_biology.pdf.

1.4.3. Solid-Like Mesophases

At low temperatures and/or low levels of hydration, all phospholipids have the potential to form one or more crystalline lamellar (L_c) phases. An actual crystal known as a three-dimensional lamellar crystal is produced when long and short-range order is identified in three dimensions. The close packing of two-dimensional phospholipid crystalline sheets produces the three-dimensional crystalline order (Figure 1.50). Phospholipid close packing and molecular configuration are defined in terms of several parameters in all crystalline and ordered states. These parameters are d_p, which is the thickness of the head group region, S, which is the surface area at the bilayer plane occupied by the individual phospholipid, φ, which is the tilt angle of the chain with respect to the bilayer plane, and ⊠, which is the mean cross-sectional area of a fatty acid alkyl chain perpendicular to the chain axis. Three-dimensional crystalline order disintegrates when the two-dimensional phospholipid crystalline sheets stop maintaining regular stacking arrangements with regard to one another. This results in a succession of two-dimensional crystalline sheets that are stacked erratically with regard to one another. Because they still retain a significant amount of crystalline order, these mesophases are referred to as 2D lamellar crystals.

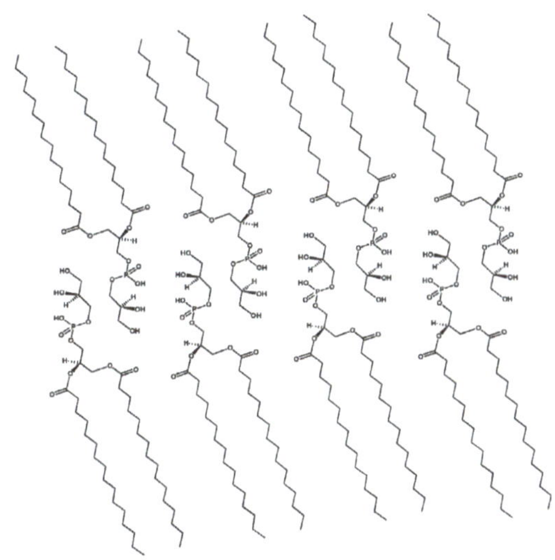

Figure 1.50. *Dipalmitoyl L-α-phosphatidylglycerol (DPPG) layer to illustrate crystal packing in 2Dlamellar layers. In lipid assembly terms this can be considered secondary/tertiary structure formation.*

Source: Andrew Miller and Julian Tanner. Essentials of Chemical Biology: Structure and Dynamics of Biological Macromolecules. Wiley. ISBN 978-0-470-84530-1. https://nattabiochem.wordpress.com/wp-content/uploads/2009/08/essentials_chemical_biology.pdf.

Gel states, or one-dimensional ordered lamellar phases, are created when phospholipids are stacked into bilayers, each of which is separated from the other by water. Similar to the 3D and 2D lamellar crystal phases, the fatty acid alkyl side-chains are stiff and extended, but they can rotate about their chain axes with difficulty. Depending on the tilt angle φ, there are several different types of 1D-ordered lamellar phases. These include L_{β} ($\varphi = 0$), L'_{β} ($\varphi > 0$), and an interdigitated phase $L_{\beta I}$ ($\varphi = 0$) where fatty acid alkyl side-chains from various monolayers overlap (Figure 1.51). There is another known 1D-ordered lamellar phase called $L_{\boxtimes}$, where the fatty acid alkyl side-chains change from linear to helical in shape.

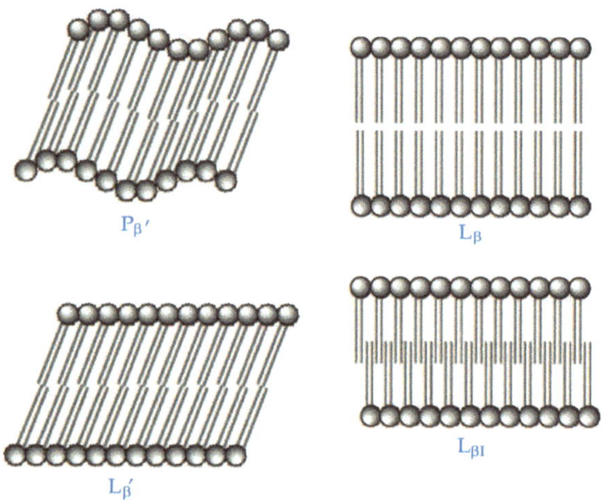

Figure 1.51. *Selection of 2D modulated ordered lamellar phases that are partially disordered but exhibit translational ordering in two dimensions. These gel phases exist halfway between crystalline L_c states and completely fluid phases such as the $L_{\alpha I}$ and H_{II} phases. In lipid assembly terms, these represent the equivalent of secondary/tertiary structure formation.*

Source: Andrew Miller and Julian Tanner. Essentials of Chemical Biology: Structure and Dynamics of Biological Macromolecules. Wiley. ISBN 978-0-470-84530-1. https://nattabiochem.wordpress.com/wp-content/uploads/2009/08/essentials_chemical_biology.pdf.

1.4.4. Fluid Mesophases

Transitions from solid-like mesophases to fluid-like mesophases occur under specific temperature and hydration conditions. Under biological terms, the physical conditions that give rise to such transitions are crucial. Most 3D fluid phases that have been found so far have cubic (Q) symmetry. Thus far, six cubic phases have been characterized; these seem to belong to two different families: one based on discrete lipid aggregates (micellar) and the other on periodic minimal surfaces (bicontinuous). Three-dimensional fluid phases can be classified as type II (inverse topology, water-in-oil) or type I (normal topology, oil-in-water) structures. Type I and type II cubic phases coexist frequently. For example, the primary bicontinuous phases (Q230, Q224, and Q229) all form type I and type II structures, which are made of water or fatty acid side-chains, respectively, and consist of two distinct interwoven but unconnected networks of channels or rods (Figure 1.52). Three-dimensional fluid phases of type II, rhombohedral (R) or tetragonal (T), are also identified. For instance, the RII phase is made up of regular stacks of planar, two-dimensional hexagonal arrays made of aqueous channels that eventually form a three-dimensional lattice. Similar to this, the planar two-dimensional square arrays that make up the T_{II} phase are stacked once more to create the three-dimensional lattice. Researchers studying chemical biology are currently debating whether or not cells could have cubic phases. In fact, there is a theory that suggests cellular organelle membranes could adopt the typical cubic phase, Q_I.

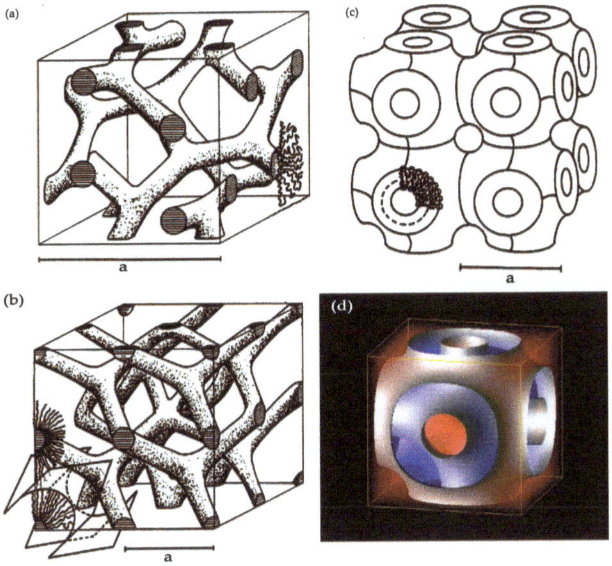

Figure 1.52. *Structural representations of the main cubic mesophases formed by lipid assemblies.*

Note: The mesophases shown are (a) Ia3d (Q^{230}) and (b) Pn3m (Q^{224}), both of which are Q_{II} fluid cubic phases; together with (c) Im3m (Q^{229}) which is a Q_I fluid cubic phase. Cartoon (d) provides an alternative representation of the Im3m (Q^{229}) mesophase. Subscript I refers to the normal phase (lipid "inside") and II to the reverse phase (lipid "outside"). In lipid assembly terms, these represent the equivalent of tertiary structure formation

Source: Andrew Miller and Julian Tanner. Essentials of Chemical Biology: Structure and Dynamics of Biological Macromolecules. Wiley. ISBN 978–0-470–84530–1. https://nattabiochem.wordpress.com/wp-content/uploads/2009/08/essentials_chemical_biology.pdf.

Having said that, membranes in cells are generally thought to adopt 1D fluid phases, specifically the fluid lamellar phases, L_α (Figure 1.53), which include the type I structure with normal topology, $L_{\alpha I}$. It is commonly accepted that, under typical physiological conditions, the $L_{\alpha I}$ phase is the default phase state of all biological membranes. It is also believed that, in certain situations, cellular membranes can take on specific 2D fluid phases, specifically hexagonal phases, H (Figure 1.54), such as the inverse topology type II structure, H_{II}. The hexagonal phases with normal and inverse topology, H_I and H_{II}, resemble stacked cylinders. Long alkyl acid side-chains are contained within the cylinders in the H_I phase, and polar head groups make up the cylinder surfaces; in the H_{II} phase, polar head groups surround the water cylinders, and long alkyl side-chains occupy the spaces between the cylinders (Figure 1.54). As of right now, the interconversion that occurs within biological membranes between the inverse hexagonal phase (H_{II}) and the $L_{\alpha I}$ fluid phase is thought to be essential to the dynamic behavior of all kinds of biological membranes, especially the ability for biological membranes to momentarily become more porous in the H_{II} phase before transitioning back to the much less porous $L_{\alpha I}$ phase. Furthermore, it seems that this interconversion plays a crucial role in promoting fusion events that involve biological membranes.

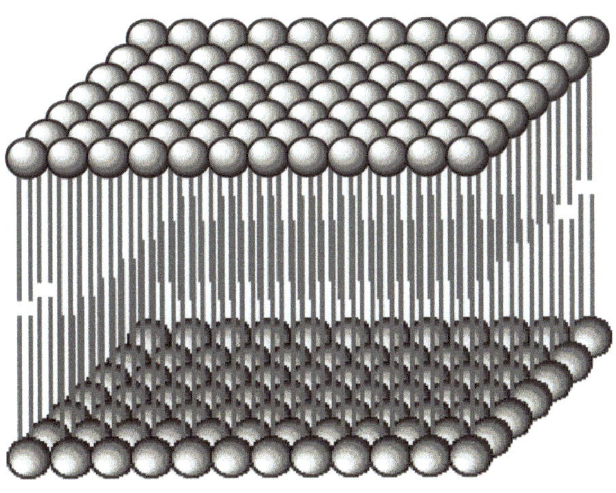

Figure 1.53. *Structural representations of the $L_{\alpha I}$ phase, the main fluid lamellar mesophase formed by lipid assemblies and the primary mesophase adopted by biological membranes. Subscript I refers to the normal phase (lipid "inside") and II to the reverse phase (lipid "outside"). In lipid*

assembly terms, this represents the equivalent of secondary/tertiary structure formation.

Source: Andrew Miller and Julian Tanner. Essentials of Chemical Biology: Structure and Dynamics of Biological Macromolecules. Wiley. ISBN 978–0-470–84530–1. https://nattabiochem.wordpress.com/wp-content/uploads/2009/08/essentials_chemical_biology.pdf.

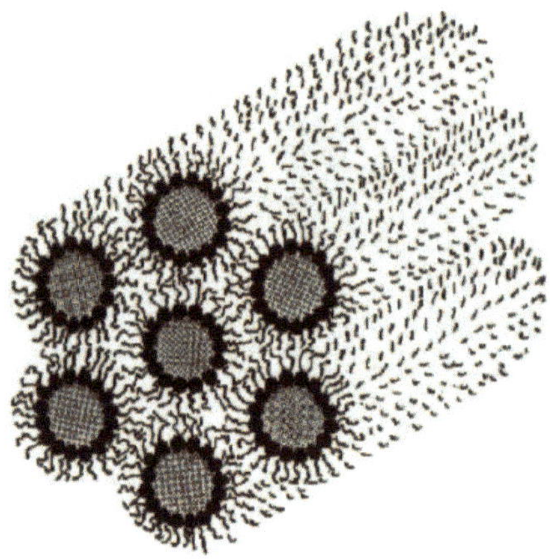

Figure 1.54. *Structural representations of the H_{II} phase, the main fluid hexagonal mesophase formed by lipid assemblies. Subscript I refers to normal phase (lipid "inside") and II to reverse phase (lipid "outside"). In lipid assembly terms, this represents the equivalent of secondary/tertiary structure formation.*

Source: Andrew Miller and Julian Tanner. Essentials of Chemical Biology: Structure and Dynamics of Biological Macromolecules. Wiley. ISBN 978–0-470–84530–1. https://nattabiochem.wordpress.com/wp-content/uploads/2009/08/essentials_chemical_biology.pdf.

ROLE MODEL

KAJ ULRIK LINDERSTRØM LANG

Kaj Ulrik Linderstrøm-Lang (November 29, 1896 – May 25, 1959) was a Danish protein scientist, who was the director of the Carlsberg Laboratory from 1939 until his death.

His most notable scientific contributions were the development of sundry physical techniques to study protein structure and function (especially hydrogen–deuterium exchange), and his definitions of protein primary, secondary, tertiary and quaternary structure.

Linderstrøm-Lang devoted himself unstintingly to protein science and trained a whole generation of eminent protein scientists, Linderstrøm-Lang maintained a fun atmosphere in his laboratory and a happy spirit that expressed itself in wonderful Christmas parties and frequent trips to the Tivoli amusement park in Copenhagen. Linderstrøm-Lang was also a writer, musician and story-teller, and was active in the resistance movement against the Nazi occupation of Denmark.

Early Work

Linderstrøm-Lang began as a physicist. Only a year after the publication of the Debye–Hückel theory, Linderstrøm-Lang applied it to proteins and contributed to define the term isoionic point. In particular, he formally considered the *ensemble* of protonation states. Linderstrøm-Lang began in the Carlsberg laboratory under its second director S. P. L. Sørensen (who invented the pH scale). 1949 volumetric studies showing that the interior of proteins has very few charges and, hence, is likely to be hydrophobic.

- Innovative Methods

Perhaps the most elegant method developed by Linderstrøm-Lang is the Cartesian diver for measuring density. A long tube containing oils of gradually increasing density was prepared. A droplet containing a protein mixture is introduced and falls until it reaches its density. Very small changes in the density of the droplet (e.g., those due to an ongoing enzymatic reaction) could be observed by movements of the droplet in the density gradient.

- Hydrogen-Deuterium Exchange

Kaj Ulrik Linderstrøm-Lang is considered the father of hydrogen–deuterium exchange.

- Contributions to Protein Structure and Stability

Linderstrøm-Lang is justly famous for his organization of protein structure into four levels: primary, secondary, tertiary and quaternary structure. He did so in his Lane

Medical Lectures, which were delivered at Stanford University in 1951 on October 2, 4, 8, 10 and 12, and later published by Stanford University Press.

- Contributions to Other Scientists

Linderstrøm-Lang contributed to the training of a whole generation of protein scientists, such as Frederic M. Richards, H. A. Scheraga, Christian B. Anfinsen, William F. Harrington, etc. He is also credited with the insight that the free energy of hydrophobic interactions does not depend only on energy (but also entropy), from his realization that mixing water and alcohol (which contains a hydrophobic methyl group) gives off heat.

- Awards and Honors

Linderstrøm-Lang was elected an International Member of the United States National Academy of Sciences in 1947, an International Honorary Member of the American Academy of Arts and Sciences in 1950, an International Member of the American Philosophical Society in 1951, and a Foreign Member of the Royal Society (ForMemRS) in 1956.

CLASS ACTIVITY

A mutation occurs, and cytosine is replaced with adenine. What impact do students think this will have on the DNA structure?

SUMMARY

- All living organisms are comprised of cells that may vary considerably in terms of size, shape and appearance; in complex multicellular organisms, many cells are organized into diverse, functional organs to perform a collective function. In spite of their wide morphological diversity, all cells of all living organisms, wherever they are located, are comprised of proteins, carbohydrates, nucleic acids and lipid assemblies. These together give a cell form and function.
- Protein structures are made by condensation of amino acids forming peptide bonds. The sequence of amino acids in a protein is called its primary structure. The secondary structure is determined by the dihedral angles of the peptide bonds, the tertiary structure by the folding of protein chains in space. The association of folded polypeptide molecules to complex functional proteins results in quaternary structure.
- The primary structure of a protein is the hierarchy's basic level and is the particular linear sequence of amino acids comprising one polypeptide chain.
- The secondary structure is the next level up from the primary structure and is the regular folding of regions into specific structural patterns within one polypeptide chain. Hydrogen bonds between the carbonyl oxygen and the peptide bond amide hydrogen are normally held together by secondary structures.
- Tertiary structure is the next level up from the secondary structure and is the particular three-dimensional arrangement of all the amino acids in a single polypeptide chain. This structure is usually conformational, native, and active, and is held together by multiple non-covalent interactions.
- Quaternary structure is the next 'step up' between two or more polypeptide chains from the tertiary structure and is the specific spatial arrangement and interactions.
- Carbohydrates are an essential part of our diet; grains, fruits, and vegetables are all natural sources of carbohydrates. Carbohydrates provide energy to the body, particularly through glucose, a simple sugar that is a component of starch and an ingredient in many staple foods. Carbohydrates also have other important functions in humans, animals, and plants.
- Nucleic acids are the most important macromolecules for the continuity of life. They carry the genetic blueprint of a cell and carry instructions for the functioning of the cell.
- The two main types of nucleic acids are deoxyribonucleic acid (DNA) and ribonucleic acid (RNA). DNA is the genetic material found in all living organisms, ranging from single-celled bacteria to multicellular mammals. It is found in the nucleus of eukaryotes and in the organelles, chloroplasts, and mitochondria. In prokaryotes, the DNA is not enclosed in a membranous envelope.

REVIEW QUESTIONS

1. What constitutes protein structure?
2. What are the four stages of protein structure?
3. What is the primary structure of a protein?
4. What are carbohydrates?
5. Name the two types of nucleic acids.

REFERENCES

1. Aspinall, G. O. (1982). *The Polysaccharides* (Vol. 1). New York: Academic.
2. Aspinall, G. O. (1983). *The Polysaccharides* (Vol. 2). New York: Academic.
3. Atkins, P. W. (1995). *Physical Chemistry* (5th ed.). Oxford: Oxford University Press.
4. Blackburn, G. M., & Gait, M. J. (1990). *Nucleic Acids in Chemistry and Biology*. Oxford: IRL Press at Oxford University Press.
5. Branden, C., & Tooze, J. (1999). *Introduction to Protein Structure* (2nd ed.). New York: Garland Publishing.
6. Cevc, G. (Ed.). (1993). *Phospholipids Handbook*. New York: Dekker.
7. Creighton, T. E. (1993). *Proteins* (2nd ed.). New York: Freeman.
8. Fersht, A. R. (1999). *Structure and Mechanism in Protein Science: A Guide to Enzyme Catalysis and Protein Folding*. New York: Freeman.
9. Kennedy, J. F., & White, C. A. (1983). *Bioactive Carbohydrates in Chemistry, Biochemistry and Biology*. Chichester: Ellis Horwood.
10. Kyte, J. (1995). *Structure in Protein Chemistry*. New York: Garland Publishing.
11. Lesk, A. M. (1991). *Protein Architecture: A Practical Approach*. Oxford: IRL Press at Oxford University Press.
12. Rees, D. A. (1977). *Outline Studies in Biology: Polysaccharide Shapes*. London: Chapman and Hall.
13. Sanger, F. (1952). The arrangement of amino acids in proteins. In M. L. Anson, K. Bailey, & J. T. Edsall (Eds.), *Advances in Protein Chemistry* (Vol. 7, pp. 1–67). Academic Press. doi:10.1016/S0065-3233(08)60017-0.
14. Schulz, G. E., & Schirmer, R. H. (1979). *Principles of Protein Structure*. New York: Springer-Verlag.
15. Sinden, R. R. (1994). *DNA Structure and Function*. San Diego: Academic.
16. Sternberg, M. J. E. (1996). *Protein Structure Prediction: A Practical Approach*. Oxford: IRL Press at Oxford University Press.
17. Sun, P. D., Foster, C. E., & Boyington, J. C. (2004). Overview of protein structural and functional folds. *Current Protocols in Protein Science, 17*(1), Unit 17.1. doi:10.1002/0471140864.ps1701s35.
18. Voet, D., Voet, J. G., & Pratt, C. W. (1999). *Fundamentals of Biochemistry*. New York: Wiley.

CHAPTER 2

Chemical and Biological Synthesis

LEARNING OBJECTIVES

After studying this chapter, you will be able to:

- Explain the chemical synthesis of peptides and proteins
- Discuss on chemical synthesis of nucleic acids
- Focus on the chemical synthesis of oligosaccharides
- Understand the chemical synthesis of lipids
- Describe the biological synthesis of biological macromolecules
- Identify the directed biological synthesis of proteins
- Define the biological syntheses of nucleic acids, oligosaccharides and lipids

KEY TERMS FROM THIS CHAPTER

Amino-acid residue
Biological synthesis
Chemical synthesis
High performance liquid chromatography (HPLC)
Immobilized metal affinity chromatography (IMAC)
Methyl-tert-butyl ether (MTBE)
Ninhydrin test
Polyacrylamide gel electrophoresis (PAGE)
Proline
Solid phase oligonucleotide synthesis (SPONS)
Tert-butyloxycarbonyl(t-BOC)
Benzhydryloxycarbonyl (BHOC)
Chemical biology
Gel filtration chromatography (GFC)
Hydrophobic interaction chromatography (HIC)
Ion exchange chromatography (IEC)
N-benzyloxycarbonyl
Peptide nucleic acids (PNAs)
Polyethylene glycol (PEG)
Sodium dodecyl sulfate (SDS)
Solid phase peptide synthesis (SPPS)
Trichloroacetic acid (TCA)

INTRODUCTION

For chemists interested in chemical biology, chemical synthesis and biological synthesis are two of the most potent tools available. The process of figuring out structure and dynamics starts in synthesis. There wouldn't be enough information to study structure without synthesis. Critical probes for studying dynamics are unavailable without synthesis. One distinctive aspect of chemical biology may be the need for both chemical and biological synthesis methods; this is where a fundamental principle of the field is based. Built up from easily accessible, small molecule starting materials, traditional chemical synthesis has an extraordinarily rich history and a proven track record of success in the directed synthesis of a multiplicity of molecules, from unusual inorganic compounds to complex and intricate organic molecules. The directed synthesis of biological macromolecules using "factory organisms" like the bacterium *Escherichia coli*, which use intricate anabolic intracellular pathways for the assembly of biological macromolecules, has a far shorter but no less notable track record in biological synthesis. Recombinant technologies and gene cloning methods play a major role in the directed biological syntheses of many biological macromolecules. While directed synthesis and the isolation of a target molecule are the general goals of both chemical and biological synthesis, there are clear differences between the two methods. As a result, while biological synthesis is not typically covered in general chemistry courses, some aspects of it may be covered.

2.1. CHEMICAL SYNTHESIS OF PEPTIDES AND PROTEINS

Consider creating a dipeptide from the individual amino acids as a starting point for comprehending the difficulties involved in peptide synthesis. Preventing the formation of side reaction products, avoiding α-carbon racemization, and maintaining the proper amino-acid residue sequence are all important. To guarantee the chemoselective formation of only the single, desired peptide link in coupling, N- and C-terminal protecting groups are necessary in order to satisfy all of these requirements. Therefore, the synthetic strategy can be thought of as a five-stage process overall (Figure 2.1):

i. protection of the α-amino group of the N-terminal residue
ii. protection of the carboxyl group of the C-terminal residue
iii. activation of the carboxy group of the α-N-protected amino acid
iv. coupling (peptide link formation) to give fully protected dipeptide
v. deprotection (as appropriate).

For the synthesis of most dipeptides, protection of amino-acid residue side-chain functional groups will also be appropriate in order to avoid additional undesirable side reactions.

For a dipeptide to be constructed into a tripeptide, Step (e) must only include the selective deprotection of the dipeptide's N-terminal α-amino group. Then, a third amino acid residue can be introduced by joining the dipeptide's free α-amino group with the free carboxyl group of a particular α-N-protected amino acid. Subsequent selective deprotection may result in the coupling of an additional specifically chosen α-N-protected amino acid, and so forth. The synthesis of di-, tri-, tetrapeptides, and other polymers is dependent on the specific deprotection of the α-N-protecting group. For there to be such selective deprotection, the deprotection conditions must be as orthogonal (different) from those necessary to remove other protecting groups from the synthesis as possible. Since the α-N-protecting group can be eliminated under circumstances that do not impact the other protecting groups, it can be concluded that this protecting group is orthogonal to the other protecting groups and susceptible to selective deprotection. To successfully complete solution phase peptide synthesis and prevent low yields and horrible product mixtures, orthogonal protecting groups and selective deprotection are essential. In addition, coupling—the formation of peptide links—must be carried out under extremely strict control; otherwise, α-carbon racemization—especially during coupling—becomes a major issue. This is due to the requirement for activation of each N-terminal protected amino acid chosen to extend the peptide chain in order to form peptide links (Figure 2.1). The α-carbon position becomes more acidic upon activation, increasing the likelihood of acid dissociation and subsequent racemization. Any peptide longer than a few amino acid residues becomes obviously extremely challenging for solution phase synthesis, which calls for such an approach (Lloyd-Williams et al., 1997).

2.1.1. Solid Phase Peptide Synthesis (SPPS)

The development of solid phase peptide synthesis (SPPS) has largely solved issues with solution phase peptide synthesis. The growing peptide chain in SPPS is anchored to a solid phase resin, which eases the iterative process of peptide bond formation, eliminates the need for crystallizations and purifications after each step of the synthesis, and in some ways simplifies protection/deprotection issues. However, the chemical principles of peptide link formation and peptide synthesis remain unchanged. In addition to winning Bruce Merrifield a Nobel Prize in 1986, SPPS has made peptide synthesis less difficult technically to the point where automation of the procedure is now possible (Merrifield, 1963).

Solid phase synthesis can be simplified to the following steps:

a. The first amino-acid residue (C-terminus) is coupled to reactive group X attached to an insoluble matrix (solid support) via its free carboxyl functional group. The amino-acid residue is otherwise α-N-protected (and at other functional groups as appropriate).

b. Theα-N-protecting group is selectively removed from theα-amino position then a second α-N-protected amino acid (also protected at other functional groups as appropriate) is introduced and activated for coupling with the first amino acid bound to the resin.

c. Once the peptide link is formed, the process of α-N-deprotection and amino-acid coupling continues for as long as required.

d. Finally, the desired length peptide is cleaved from the solid support by chemistry conditions suitable to cleave the linkage between solid support and peptide chain, and remove all the amino-acid side-chain protecting groups as well.

2.1.2. Solid Supports and Linkers for SPPS

To enable the two phases to react, the solid support needs to be thoroughly solvated. Although polystyrene served as the basis for the initial support, polyamide resins have largely replaced them due to their similar polarity to the peptide backbone. Due to their superior swelling capacity, which produces a larger solid/liquid interface, resins based entirely on cross-linked polyethylene glycol (PEG) or resins grafted onto low cross-linked polystyrene have been used more recently. A resin linker binds each expanding peptide chain to the solid support. The resin linker needs to possess adequate strength to withstand all of the α-N-deprotection and coupling processes that are a part of a peptide synthesis. At the conclusion of the synthesis, the resin linker needs to be quantitatively cleaved in order to release the full-length peptide. These days, the most widely used resins to replace the original polystyrene are Wang (via p-benzyloxy benzyl), Rink amide (an even larger benzyl based linker), or Super acid labile (2-chlorotrityl). This is because these resins have been modified to have fundamentally acid-sensitive resin linkers.

2.1.3. Coupling Protected Amino Acids in SPPS

Highly effective amino-acid coupling is necessary for SPPS. An amino acid that is protected by α-N is linked to the amino terminus of a resin-linked peptide chain that is growing in each coupling step.

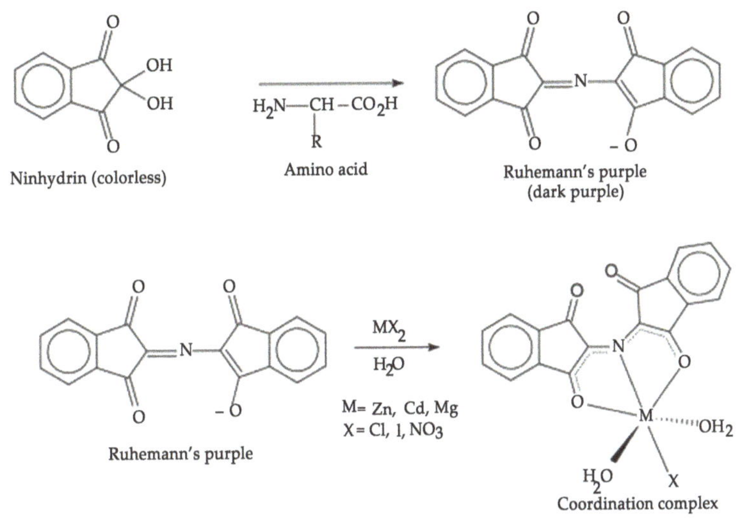

Figure 2.1. *Ninhydrin Test. Ninhydrin combines with α-N-primary amine functional groups to form a primary amine product that combines with a second molecule of ninhydrin to generate a chromogenic chromophore.*

Source: https://ars.els-cdn.com/content/image/3-s2.0-B0123693977002004-gr4.jpg.

To build up long peptide chains (>20 amino-acid residues) on resin before the resin linker is cleaved, each coupling step must occur at yields that are close to 100%. The ninhydrin test (Figure 2.1) can be used to track the degree of coupling. When ninhydrin and primary α-amines combine, a color that can be measured colorimetrically is produced that is either purple, blue, or blue-green. Proline is an example of a secondary amine that reacts less with ninhydrin, producing a reddish-brown color. Incomplete coupling can be diagnosed with the ninhydrin test, which indicates when a new coupling process needs to be carried out.

Coupling must occur without lowering the optical purity of each new α-N protected amino acid added during a chain extension cycle. Therefore, the coupling agents that are selected to activate free carboxyl groups during the formation of peptide links should have the ability to fully suppress racemization. First introduced as the dicyclohexylcarbodiimide (DCC), carbodiimide is a type of coupling agent. By forming highly reactive O-acylurea intermediates, this activates free carboxyl groups. These intermediates can be easily attacked by α-amino functional groups through nucleophilic attack, resulting in the formation of peptide links. However, hydroxylbenzotriazole (HOBt) was added to react with the O-acylurea in order to lower the racemization risk because DCC alone is insufficient to prevent racemization during coupling. The unitary use of 2-(1H-benzotriazol-1-yl)-1,1,3,3-tetramethyluronium hexafluorophosphate (HBTU), along with benzotriazole-1-yl-oxy-trispyrrolidino-phosphonium hexafluorophosphate (PyBOP), has largely replaced the combined use of DCC and HOBt in the process of activating amino-acid coupling in peptide synthesis.

2.1.4. Protection/Deprotection Strategies in SPPS

Merrifield's initial approach to protection involved giving amino-acid side-chain functional groups benzyl-protecting groups and using the tert-butyloxycarbonyl (t-BOC) group for α-N-protection on every amino acid added to growing peptide chains. When dichloromethane is mixed with trifluoroacetic acid (TFA), the t-BOC group is acid labile. As a result, after peptide synthesis was completed, the full-length peptide was separated from the resin and side-chain deprotection was carried out simultaneously using liquid hydrogen fluoride—hardly mild conditions for resin cleavage! Fortunately, a much gentler orthogonal protecting group strategy has since been developed for peptide synthesis with the introduction of 9-N-fluorenylmethyloxycarbonyl (FMOC) groups for α-N-protection. The FMOC group is stable to decompose in neutral or acidic conditions, but it is sensitive to mild base piperidine (20–50% v/v in DMF for 20 min). Consequently, separation of the full-length peptide from the resin can be tuned or adjusted to occur under similarly mild acid conditions. This allows for the protection of amino-acid side-chain functional groups during peptide synthesis using protecting groups sensitive to only mild acid (or other sets of orthogonal) conditions.

Currently, a variety of groups can be used to protect amino-acid side-chain functional groups. For example, the N-benzyloxycarbonyl (Z) amino protecting group can be broken down by strong acid or simple catalytic hydrogenolysis using a Pd/charcoal catalyst. On the other hand, t-BOC groups can shield ⊠-amino groups in lysine and the N1-position of tryptophan, while hydroxy and carboxyl functional groups can be shielded by tert-butyl masking. Protecting groups are labile in both situations when exposed to mild acids (HF in organic solvents or TFA). Furthermore, trityl group association frequently masks the primary amide side-chains of glutamine and asparagine, the thiol of cysteine, and the imidazole ring of histidine. In each of the situations mentioned above, tritiyl protection is also noticeably labile in response to mild acid treatment. Lastly, 2,2,5,7,8-pentamethylchroman6-sulfonyl chloride (PMC) or 2,2,4,6,7-pentamethyl-dihydrobenzofuran-5-sulfonyl chloride (PBF) is often used to protect the unusual guanidine functional group connected to the side-chain of arginine. Additionally, PMC and PBF are labile protecting groups that are slightly acidic. It is obvious that the release of a full-length peptide bound to resin can lead to the synthesis of extremely reactive species like butyl cations (tBu-cations and tBu-fluoroacetate, for example). To capture these species and avoid adverse reactions, scavengers like phenol, cresol, or thioanisole are also added to the cleavage cocktail. After the proper cleavage mixture has been evaporated, many byproducts are eliminated by precipitating the de-protected peptide product from solution with the addition of diethyl ether or methyl-tert-butyl ether (MTBE). The crude deprotected peptide must then be homogenized using high-performance liquid chromatography (HPLC) in reversed phase preparative mode.

2.1.5. Chemical Synthesis of Polypeptides

Although it can be used to synthesis polypeptides longer than 100 amino acid residues, SPPS has made relatively easy peptide synthesis possible for peptides with up to 40 amino acid residues. Clearly, biological synthesis is the best way to synthesize long polypeptides and proteins. Chemical synthesis does, however, sometimes have some notable advantages over biological synthesis, especially when the incorporation of odd amino acid residues—like d-amino acids, fluorescent-labeled amino acids, linker moieties, and other non-peptidic groups—is necessary. The two methods do not, however, have to be mutually exclusive. For example, it's possible that in the future, chemical biology applications will depend more and more on the synthesis of both chemicals and biological materials, or semi-synthesis.

2.1.6. Chemical Synthesis of Peptide Nucleic Acids

Purine and pyrimidine bases are joined by a polypeptide backbone to form synthetic molecules known as peptide nucleic acids (PNAs) (Figure 2.2). Typically, modified protocols from standard peptide synthesis are used to synthesize PNA monomers. While benzhydryloxycarbonyl (BHOC) is used to protect the exocyclic amino groups on adenine, cytosine, and guanine, the N-terminus is protected by FMOC. The backbone of PNA is made up of repeating N-(2-aminoethyl)-glycine units, which produces a base-to-base spacing akin to that of natural DNA or RNA with the typical phosphodiester backbone. Because bases line up perpendicular to the polypeptide backbone, PNA can form antiparallel double helices, similar to those found in DNA and RNA, in which hydrogen bonds between complementary strands are formed by Watson-Crick base pairs. PNA's uncharged backbone produces a variety of intriguing and helpful characteristics. PNA/DNA duplexes bind far more strongly than DNA double helices (DNA/DNA duplexes) due to the absence of charge-charge repulsion. It is also possible to form particularly stable triple helical complexes with DNA when the PNA is referred to as a homopyrimidine. PNA/DNA triplexes typically have a 2:1 PNA/DNA ratio.

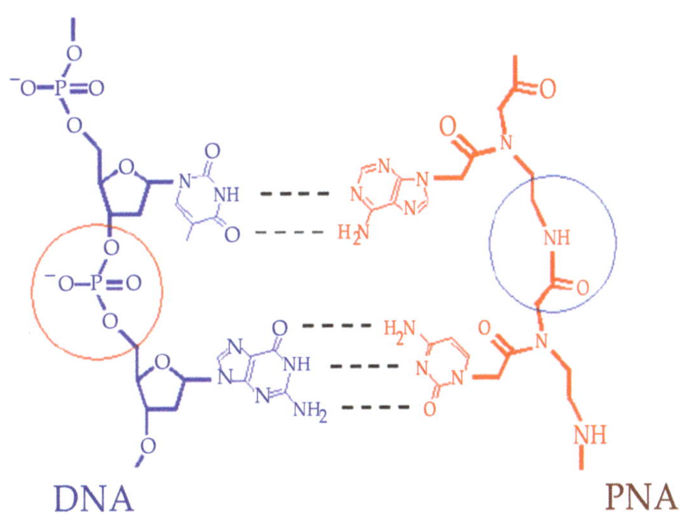

Figure 2.2. *Comparison of PNA and DNA.*

Source: https://www.researchgate.net/profile/Claire-Marie-Pradier/publication/221906936/figure/fig1/AS:305328735440897@1449807451274/Schematic-chemical-model-of-PNA-and-DNA-molecules-showing-the-different-backbone.png.

This structure requires a Watson–Crick PNA/DNA duplex annealed to a second PNA strand by Py.Pu.Py Hoogsteen base pairing. For many of the possible technological uses of PNA, the stability of the PNA/DNA triplex has proven essential. PNA can encroach on double helical DNA, joining forces with the complementary (antisense) strand to form a PNA/DNA triplex. PNAs can tolerate small amounts of purines in the sequence, even though homopyrimidines are the most stable form of the compound. PNAs offer a few useful advantages over conventional nucleic acids, including substantial lifetimes *in vitro* and *in vivo* and reasonable resistance to both nucleases and proteases. Additionally, PNAs made via chemical synthesis can be readily altered using a variety of methods described later with fluorescent groups and other labels. These molecules exhibit remarkable stability, strong binding to complementary nucleic acid sequences, and ease of derivatization, which could render them versatile tools for a range of uses.

2.2. CHEMICAL SYNTHESIS OF NUCLEIC ACIDS

Merrifield's solid phase peptide synthesis offered a fresh method for creating biological macromolecules chemically, and other disciplines swiftly adopted this technique. Another significant advancement in chemical synthesis that has had a significant influence on the biological sciences is solid phase oligonucleotide synthesis (SPONS). The primary obstacle in splice link synthesis (SPPS) was accurate peptide link formation; in oligonucleotide synthesis, the primary challenge has been accurate phosphodiester link formation. A significant portion of the advancements in SPONS stemmed from research conducted in Alexander Todd's lab in the 1950s on nucleotide and polyphosphate coenzyme synthesis. Khorana used DCC to activate phosphate-sugar coupling shortly after, in the early 1960s, and succeeded in synthesizing the first sequence-defined oligonucleotides. But Letsinger and associates created the solid phase approach in the late 1960s, which led to the development of the modern triester chemical approach.

2.2.1. Chemical Synthesis of Oligodeoxynucleotides

Today, oligodeoxynucleotide synthesis is almost always done using solid support. Initially, polystyrene cross-linked with divinylbenzene was used as the solid support, but now controlled-pore glass beads with specific porosities are preferred. The first protected deoxynucleotide's 3' -hydroxyl is pre-coupled to the solid support, and the oligodeoxynucleotide synthesis then proceeds in a solid phase, multistage process in the 3' →5' direction, similar to SPPS (Figure 2.3).

- The initial deoxynucleotide protected on the solid support has its 5'-hydroxyl group shielded with dimethoxytrityl (DMT). Removing the protection with trichloroacetic acid (TCA) reveals a free 5'-hydroxyl group and releases the orange DMT cation (which can be detected as an indicator of deprotection effectiveness).
- A 5' -DMT-protected deoxynucleotide with attached 3-phosphoramidite or H-phosphonate group is then activated for coupling to the 5' -hydroxyl group of the 3' -solid-support-bound deoxynucleotide by means of a coupling reagent known as tetrazole.

- Iodine is introduced to oxidize the diester link formed from coupling to a complete phosphodiester link (with P(V) oxidation state). Diester links formed with phosphoramidites or H phosphonate are unstable (due to the P(III) oxidation state).
- Acetic anhydride is then added to block or cap any unreacted 5' -hydroxyl groups so that sequence truncations are not incorporated in later steps.
- After forming a phosphodiester link and capping the 5'-hydroxyl group, the process involves deprotecting the 5'-DMT and coupling with 5'-DMT-deoxynucleotide-3'-phosphoramidites, followed by iodine oxidation and capping of any unreacted 5'-hydroxyl groups. This repetitive iterative process continues until the desired full-length oligo- or polydeoxynucleotide is achieved.

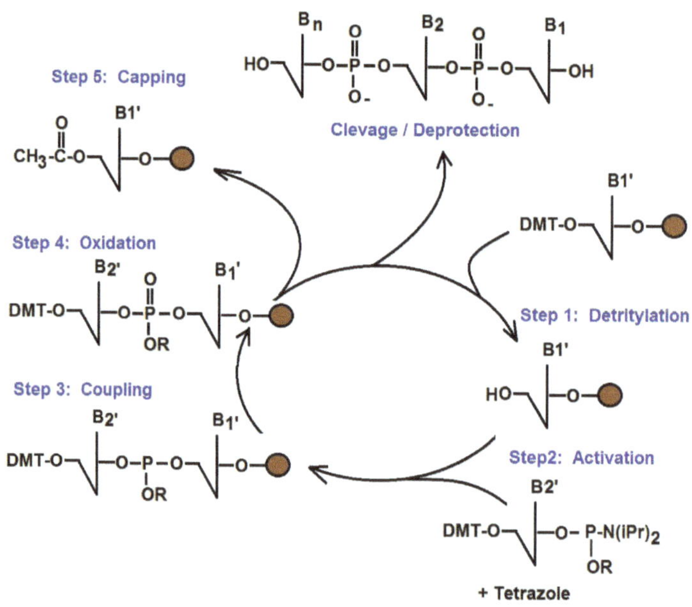

Figure 2.3. *Solid phase DNA synthesis.*

Source: https://www.biosyn.com/images/userfiles/image/Phosphoramidite%20Method.jpg.

After the oligo- or polydeoxynucleotide is fully synthesized, it is removed from the solid support and deprotected. Purification of the nucleotide then takes place, which can involve basic desalting or more commonly, polyacrylamide gel electrophoresis (PAGE) to ensure high purity. For higher yields and more complex modifications, HPLC is often used as the preferred purification method post-synthesis.

2.2.2. Chemical Synthesis of Oligonucleotides

For two main reasons, oligo- and polynucleotide synthesis presents a greater challenge than oligo- and polydeoxynucleotide synthesis. First of all, RNA molecules are extremely vulnerable to enzymatic and chemical degradation. During chain synthesis, it is important to safeguard the potentially reactive 2' -hydroxyl group present in each β-d-ribofuranose ring. The protecting group must, however, be both stable enough to permit differential protection of the 5'-hydroxyl group during oligo- or polynucleotide synthesis and labile enough to be eliminated after the synthesis. Stated differently, the 2'- and 5'-hydroxyl protecting groups need to be appropriately orthogonal. Now, 5'-DMT protection and 2'-tertbutyl dimethylsilyl (TBDMS) group protection can accomplish this. TCA is necessary every 5'-DMT deprotection step that occurs during oligo- or polynucleotide synthesis (just like it does for DNA). During oligo- or polynucleotide synthesis, TBDMS groups are comparatively strong; however, they are readily eliminated during final deprotection. A three-stage deprotection technique is applied after oligo- and polynucleotide synthesis is complete. First, triethylammonium thiophenoxide is used to remove methyl groups from the phosphodiester internucleotide links. Second, in order to remove the exocyclic amino-protecting groups and cleave the oligo- or polynucleotide from the solid support, a mixture of alcohol and ammonia is needed. Third, the TBDMS groups are highly selectively removed from 2' -hydroxyl positions using tetrabutylammonium fluoride (TBAF). The solid phase method allows for automated oligo- and polynucleotide synthesis, although the process is not as standard as that of DNA oligo- and polydeoxynucleotide synthesis.

2.2.3. Useful Deoxynucleotide/ Nucleotide Modifications

We will first discuss phosphorothioate links, which are produced when a single non-bridging oxygen atom in a particular phosphodiester link is swapped for a sulfur atom. These links begin with backbone modifications. Two diastereomers are possible at the chiral phosphate center formed by the new link. For a variety of reasons, phosphothioate links can be included in oligo-/polynucleotides or oligo-/polydeoxynucleotides. First, in contrast to the "hard" oxygen atom, the "soft" sulfur can shed light on the function of metal ions. Additionally, phosphorothioates are helpful in cell culture and even for functional studies in animals *in vivo* because they are far more stable to nucleases than regular phosphodiester links. Phosphorothioate linkages can be inserted into oligo- or polydeoxynucleotides using solid-state syntheses; however, a sulfurising agent must be used in place of iodine to introduce the non-bridging sulfur atom at each link. If both of the non-bridging oxygen atoms are replaced with sulfur atoms, achiral phosphorodithioate links will occur. Although phosphorodithioate links can cause significant deviations from standard oligo-/polynucleotide structures, they are generally very stable to hydrolysis and, as a result, are not used very often. On the other hand, it is indisputable that methylphosphonate linkages, which arise from the substitution of a methyl group for a non-bridging oxygen atom, are more intriguing. Since methylphosphonates are non-ionic in contrast to thioate and standard phosphodiester links, they can shed light on the significance of charge in oligo- and polynucleotide structures.

Solid phase synthesis makes it simple to incorporate sugar modifications into oligo- or polydeoxynucleotides or oligo-polynucleotides. These modifications have a variety of applications, most notably stabilizing oligo- or polynucleotide structures. The 2'-deoxy-β-d-ribofuranose rings that make up oligo- and polydeoxynucleotides can adopt both C2' and C3' endo conformations. By contrast, the C3' -endo conformation is the only one displayed by oligo-/polynucleotide β-d-ribofuranose rings. In light of this,

2'-substituents like fluoro, amino, methoxy, alloxy, etc., stabilize corresponding oligo- and polynucleotides against hydrolytic attack as well as conformational changes. This idea was most recently expanded to create locked nucleic acids (LNAs), which due to their conformational and metabolic stability may play a significant role in RNA therapeutics in the future. It should come as no surprise that non-natural base modifications can also be easily incorporated into an oligo- or polydeoxynucleotide or an oligo-/polynucleotide through solid phase synthesis. Probes with non-natural bases are helpful in distinguishing between structure and function. For example, 7-deazapurine nucleotides that lack the N-7 nitrogen are useful probes for those situations where this nitrogen acts as a hydrogen bond acceptor (e.g., Hoogsteen base pairing;). Instead of using base analogs, fluorescent or dye-labeled compounds can be used. For example, 2-aminopurine (AP) can replace adenine in dA/A nucleotides and pair with thymine in dT nucleotides without disrupting the DNA structure. Another option is 3-methyl-8-isoxanthopterin, which behaves like guanine in dG/G nucleotides and is created through chemical synthesis. These fluorescent probes have various applications. Additionally, chemical cross-linkers and photo-affinity cross-linkers can be added to oligo-/polydeoxynucleotides or oligo-/polynucleotides during or after synthesis. These can be used to attach molecules to surfaces or other compounds as needed. Some cross-linkers may alter nucleic acid functions by introducing covalent distortions, such as certain anticancer drugs like 'nitrogen mustards' and cis-platin.

2.3. CHEMICAL SYNTHESIS OF OLIGOSACCHARIDES

For the chemical synthesis of shorter nucleic acids and polypeptides, solid phase approaches are predominant; however, this is not the case for oligo- and polysaccharides. The intricacy of oligo- and polysaccharide primary structures is the issue. This is untrue in the case of homoglycans, but a lot of oligo- and polysaccharides linked to proteins or lipids to form glycoproteins or glycolipids can be extremely complex. The complexity arises from the fact that each monosaccharide can have at least two or more distinct "nucleophilic donor" hydroxyl groups, resulting in complex branching, whereas natural sugars only have one anomeric carbon that serves as the "electrophilic acceptor" for a glycosidic link. Therefore, two of the biggest obstacles to the chemical synthesis of oligo- and polysaccharides are primary structure branching and control of anomeric carbon stereochemistry. Therefore, a variety of chemo-selective protecting groups with the ability to perform orthogonal, selective deprotections are required for oligo- and polysaccharide chemical syntheses. To form glycosidic links with precise stereocontrol, additional reagents are needed, potentially with the help of nearby group protective groups. The fact that some coupling reagents and protecting groups may also be helpful for the protection and coupling of a limited variety of monosaccharide building blocks presents another difficult aspect of oligo- and polysaccharide chemical syntheses. It is obvious that the chemical synthesis of oligo- and polysaccharides is still very much a "work in progress," requiring the same simplifying principles that underpin the synthesis of nucleic acids and peptides. However, a few themes are beginning to emerge that are worthwhile going over with examples.

2.3.1. Protecting Groups

It is necessary to employ orthogonal protective groups that can be selectively removed in order to synthesize oligo- and polysaccharides chemically. Benzyl ether protecting groups are frequently used in the chemistry of oligo- and polysaccharides. As a "permanent" protecting group, benzyl ether is added to safeguard hydroxyl functional groups that aren't anticipated to be involved in the formation of glycosidic links. Therefore, during the synthesis of oligo- and polysaccharides, benzyl ether protected groups are strong enough to remain inert until they are finally eliminated by mild catalytic hydrogenolysis. In contrast, base-labile protecting groups—such as acetyl, benzoyl, and pivaloyl—which are frequently used as monosaccharide C-2 protecting groups—are used in the synthesis of oligo- and polysaccharides. As a result, these protective groups can aid in the control of stereochemistry by providing neighboring groups with participation during the formation of glycosidic links. Additionally, base-labile protecting groups are designed to offer momentary shielding during a synthesis in order to reveal a crucial hydroxyl group for a reaction at precisely the correct moment. Acid-labile protective groups, like the DMT group, are additional short-term protective groups that could be employed. Additionally, there is the TBDMS group, which TBAF may eliminate. TBDMS and DMT have both been used in the synthesis of oligonucleotides.

2.3.2. Creating Glycosidic Links

When two monosaccharides are coupled, the anomeric position in one (the glycosyl donor or the electronophilic acceptor) must be activated, and all functional hydroxyl group positions in the other (the nucleophilic donor or the glycosyl acceptor) must be protected, with the exception of the chosen hydroxyl group. Here is an illustration of this in its most basic form. Protective group manipulation must come first, and then coupling, which needs to be as stereocontrolled as feasible. In addition to helping with the current glycosidic link formation, protective groups are selected with future glycosidic link formations in mind. Selecting the appropriate coupling reagents and protecting groups is essentially an empirical process that requires a thorough understanding of oligo- and polysaccharide chemistry. Put another way, a select group of well-versed individuals still effectively control what will and won't work. In recent times, the following synthetic methodologies have emerged as leading ways to effect glycosidic link formation:

a. glycosyl trichloroacetamidate coupling
b. 1,2 anhydrosugar-thioglycoside coupling
c. n-pentenyl glycoside coupling.

The first (a) has been extremely useful in many solution phase oligo-/polysaccharide syntheses. Glycosyl donors (electrophilic acceptors) are activated under very mild conditions by catalytic amounts of trimethylsilyltriflate (TMSOTf) and other triflates including dibutyl boron triflate (DBBOTf), not to mention boron trifluoride in diethylether (BF_3/Et_2O). This method of glycosidic link formation is striking for control of the stereochemistry (α or β) of the anomeric center and hence of the glycosidic link. The second method (b) involves 1,2-anhydrosugars that are obtained easily from glycal precursors. Glycals are readily converted into corresponding anhydrosugars by epoxidation with dimethyl dioxirane (DMDO), only to react onwards with a thiolate to yield thioglycoside. These thioglycosides are robust glycosyl donors that are mobilized for coupling by treatment with thiophilic promoters such as di-tert-butyl peroxide/methyl triflate (DTBP/MeOTf). Once again, high α or β selectivity can be achieved in glycosylation reactions. Finally, the third method (c) involves n-pentenyl-glycoside- (NPG-) derivatized glycosyl donors that are activated by strong electrophilic reagents such as N-iodosuccinimide/triethylsilyltriflate (NIS/TESOTf). Coupling yields are typically excellent and so is control of stereoselectivity.

2.3.3. Solid Phase Oligosaccharide Chemistry

Since much of the solution phase chemistry readily converts to the solid phase, the field of solid phase synthesis is emerging surprisingly quickly. However, this approach to oligo-/polysaccharide synthesis is not routinely adopted due to incomplete α/β stereocontrol with glycosidic link formation and imperfect coupling yields. A polymeric version of the Merrifield resin is commonly employed as the solid support during the synthesis process. There are two common methods: either the glycosyl acceptor or the donor can be immobilized on the solid support at first. These two methods can diverge enough to necessitate the use of distinct coupling agents, protective groups, and even linkers to the solid support. Since the linker chemistry needs to be inert (orthogonal) to all other deprotection and coupling steps that occur during the specific synthesis, the linker's chemical makeup dictates all other protecting groups and coupling manipulations that may be performed during the entire synthesis. Therefore, it follows that other silyl protecting groups cannot be used elsewhere in the synthesis when a silyl ether is used as a linker. Thankfully, the process of silyl deprotection is clearly orthogonal to other oligosaccharide-associated chemistries, and these linkers exhibit remarkable resistance against the coupling conditions associated with the three primary coupling mechanisms previously discussed. If an active glycosyl donor is first immobilized on the solid support, silyl ether seems to be the most suitable type of linker. Otherwise, when an active glycosyl acceptor is first bound to solid support, thioglycoside and etherlinkers seem to be more suitable. Since thioglycoside linkers cleave easily when exposed to thiophilic reagents, N-bromosuccinimide/di-tert-butyl pyridine (NBS/DTBP) in MeOH is used to break them down into methylglycosides. If not, new release conditions for the ether linkers may apply, such as when ethane is catalyzed metathesis using Grubb's catalyst.

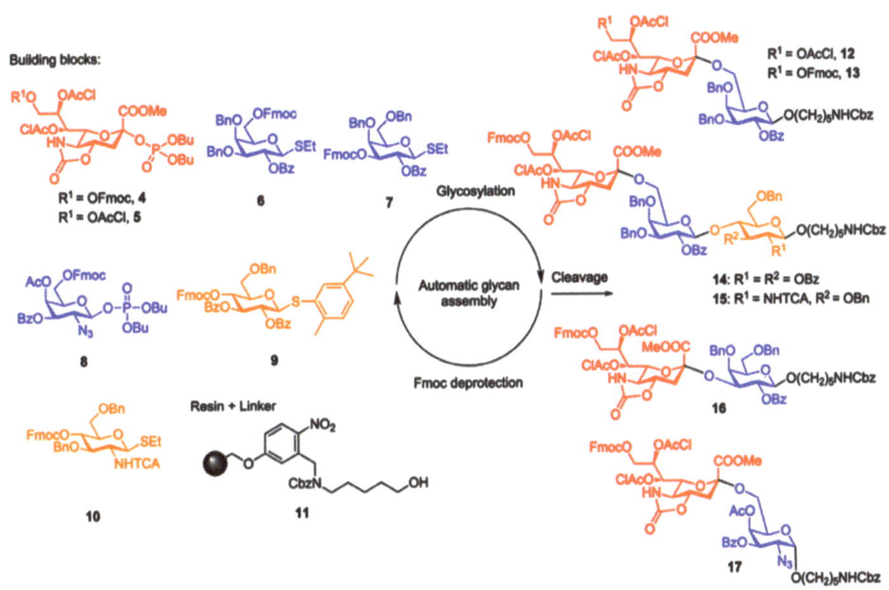

Figure 2.4. *Automated solid phase oligo-/polysaccharide synthesis.*

Source: https://www.researchgate.net/profile/Heung-Sik-Hahm/publication/276173361/figure/fig1/AS:616354080100363@1523961672934/Scheme-2-Automated-synthesis-of-oligosaccharides-with-a2–3-a2–6-sialic-acid.png.

A regular cycle of synthesis involves the cyclical introduction and coupling of a protected glycosyl acceptor or donor, respectively, depending on whether an active glycosyl donor is solid phase attached or an active glycosyl acceptor, as appropriate. Typically, the coupling is made possible by a specific step of selective deprotection. Despite advancements in solid phase oligosaccharide synthesis, the synthesis of oligosaccharides longer than four to five monosaccharide units continues to pose significant challenges, and, unlike oligopeptide or oligonucleotide synthesis, there is still no universal protocol in place. Solid phase oligosaccharide synthesis has an advantage over solution phase techniques in that it can synthesize libraries of di- and trisaccharides. These libraries can then be employed in screening investigations to look into various chemical biology issues. Furthermore, complete automation was used to carry out the final synthesis shown in Figure 2.4. Therefore, it shouldn't be too long before oligosaccharide synthesis becomes a routine process that is completely automated.

2.4. CHEMICAL SYNTHESIS OF LIPIDS

Generally speaking, many of the phospholipids that comprise macromolecular lipid assemblies in cells can be obtained naturally and are purified using high-performance liquid chromatography (HPLC), negating the need for their chemical synthesis. Chemical lipid syntheses are essentially subordinated to biological ones.

Though these synthetic approaches are important in lipid semi-synthesis should a fluorescent tag be required, etc., there is some value in illustrating the general method of synthesis of acylglycerols and phospholipids. Glycerol, an alkyltriol, contains hydroxyl functional groups that can be controlled to form acylglycerols. The fact that mono-and diacylglycerols readily migrate to nearby free hydroxyl groups, usually when catalyzed by heat, acid, or base, is a significant issue in the synthesis of pure acylglycerols. Therefore, for the chemical synthesis of defined acylglycerols, an appropriate protecting group strategy is crucial. For example, in an acidic environment, glycerol and acetone can react to form 1,2-isopropylidene (acetonide), which leaves one primary hydroxyl free for acylation later on. Alternately, benzaldehyde can react to produce 1,2- or 1,3-O-benzylidenesthat can be used for esterification after being separated by crystallization. The chosen fatty acids can then usually be activated by carbonyl diimidazole (CDI) to carry out acylation. Acetal deprotection can then, if necessary, be followed by additional esterification. This can be simplified to the controlled esterification of l-α-glycerophosphocholine in the case of phospholipid synthesis. The chosen fatty acids' CDI activation makes $CdCl_2$ (GPC) possible. Then, using the enzyme phospholipase D, headgroup transphosphatidylation can be used to create other phospholipids from this phosphatidylcholine. Although lipid synthesis might seem unimportant, the process of purifying lipids to homogeneity after synthesis is not insignificant. This makes reversed phase HPLC a potent technique, but it also requires the use of an evaporative light scattering (ELS) detector that has been specially designed to detect eluting lipid analytes that lack potent chromophores.

2.5. BIOLOGICAL SYNTHESIS OF BIOLOGICAL MACROMOLECULES

The term "biological synthesis" refers to the process of using living things' capacity for synthesis to produce targeted biological macromolecules of interest. Synthesis and purification must come first because research into the structure and function of biological macromolecules is heavily reliant on the availability of study material. Modern biological syntheses, especially those involving proteins and nucleic acids, may often necessitate the development of recombinant organisms and recombinant techniques (though this is typically not the case with carbohydrates and lipids). After that, the primary task is to purify the target lipid or biological macromolecule so that it is homogeneous in the presence of all other cellular constituents. As cells—recombinant or not—actually carry out the synthesis, purification ends up being our main concern. Therefore, the methods commonly employed in the extraction of lipids and biological macromolecules from organisms will be presented to the reader of chemical biology first.

2.5.1. Ion Exchange Chromatography

Molecules are separated using Ion Exchange Chromatography (IEC) based on variations in their accessible surface charge. The technique is very popular because it can be used to separate any molecule that is soluble in aqueous solution and charged, particularly in the purification of proteins. The eluting biological macromolecule (analyte) typically exhibits high recovery and high retention of biological activity due to the mild elution conditions. Ion exchange is classified into two classes. In anion exchange, molecules are separated based on the negative (anionic) charges they carry by means of a column that carries functional groups with a positive charge. Molecules are sorted based on the positive (cationic) charges they carry through ion exchange, where functional groups with a negative charge are carried by the column. The analyte molecules of interest and the column bind electrostatically. The most crucial elements are the charges that are accessible on the surface because binding only occurs over a brief distance. Anion and cation exchange resins are categorized as weak or strong; this is not a measure of binding strength, but rather of how pH affects the charge of the functional groups. The pH of the mobile phase affects the electrical charge that weak ion exchange resins retain or lose. No matter how much the mobile phase's pH changes, strong ion exchange resins keep their charge.

An ion exchange resin's functional groups need a counter-ion, which can be the sample analyte itself, buffer molecules, or salt molecules. When the analyte directly counter-ion interacts with the functional groups on the ion exchange resin, binding to an ion exchange column takes place. Ion exchange chromatography is typically performed by binding the analyte to the column in low salt conditions, then eluting the mixture by adding more salt to cause the analyte to be displaced back into the mobile phase and thus pick up a salt counter-ion from the mobile phase. In order to elute the analyte, a pH change can be utilized in conjunction with or instead of a salt gradient, especially when weak ion exchange is being used.

In ion exchange chromatography, for example, the pH of the mobile phase is crucial because the majority of charged amino acid residues on protein surfaces are poor ion exchangers. This generally indicates that at lower pH values, analyte molecules will be more positively charged, and at higher pH values, more negatively charged. When deciding which kind of ion exchange medium to use for purification, the isoelectric point (pI), or pH at which the molecule's net charge is zero, is a helpful place to start. Since a mobile phase at neutral pH would most likely result in surface positive charge, using an analyte with a high isoelectric point (basic) would cause it to be positively charged at neutral pH. Consequently, it would be ideal to apply the analyte to a cation exchange column and vice versa. Since the isoelectric point is the pH at which the net charge is zero—a value that does not always define net surface charge—it should only be used as a general guideline.

It is necessary to decide whether to use strong or weak ion exchange resins in addition to the type of exchange resin. Both can usually be used, but weak ion exchange resins lose their charge at very high pH levels (>10 for anion exchange and <4 for cation exchange). Because of the column's buffering ability, weak ion exchange media typically take longer to equilibrate. However, they are most helpful in the separation of analytes that bind firmly to strong ion exchange columns and do not elute even at high salt concentrations. This can occasionally be the case with phospholipids or big nucleic acids. In these situations, a different method of elution is to use weak ion exchange resin and then elution with base (for anion exchange) or acid (for cation exchange) to neutralize as much charge as possible on the resin. Chromatofocusing, which frequently uses tertiary or quaternary amines as the stationary phase, is basically a weak anion exchange variation. Roughly speaking, molecules will be divided based on their pI. As with anion exchange, the analyte is first bound at a pH higher than the target's pI. Next, a focusing buffer with a lower pH is added to the column; since the focusing buffer will titrate the buffering groups on the ion exchanger, no gradient is needed. This allows for the elution of very high protein concentrations in a small volume at peak widths as small as 0 point05 pH units. The method is both a significant preparatory technique and a potent analytical probe of surface charge. Establishing a pH that is appropriate for maximal binding is the ideal place to start when developing an ion exchange method. For anion exchange, use pH 8.5, and for cation exchange, pH 4.0. Try a 0–1 M NaCl gradient across 50–100 column volumes at first. The addition of 20% water-miscible solvent (methanol, glycerol, isopropanol, or acetonitrile) should help reduce hydrophobic interactions with the resin and solubilize the target analyte molecules if solubility of the target analyte of interest is a problem in the separation (poor recovery or peak tailing).

2.5.2. Hydrophobic Interaction Chromatography

Biological macromolecules are separated using hydrophobic interaction chromatography (HIC) based on the hydrophobic groups present on their surface. It exhibits many similarities with IEC; for example, HIC and IEC both have mild binding and elution conditions, are particularly helpful for protein purification, and are based on hydrophobicity rather than charge. With the exception of having a hydrophobic functional group, like a phenyl, on the surface, the HIC resin and IEC resin are similar. Consider HIC as a high-resolution technique similar to precipitation of ammonium sulfate. Because standard buffer conditions are insufficiently hydrophobic, the majority of proteins will not bind to an HIC column under these circumstances. Therefore, the standard procedure is to increase hydrophobic interactions while decreasing ionic interactions in the solution by adding a lyotropic salt, such as ammonium sulfate. A favorable increase in total entropy results from the release of water from hydrophobic surfaces when they bind to one another. Analyte precipitation is caused by lyotropic salts because they increase the ordered structure of water, which lowers entropy and makes up for it. Since HIC is an entropy-driven process, temperature has a significant impact on it; in general, a higher temperature causes a stronger binding to the column.

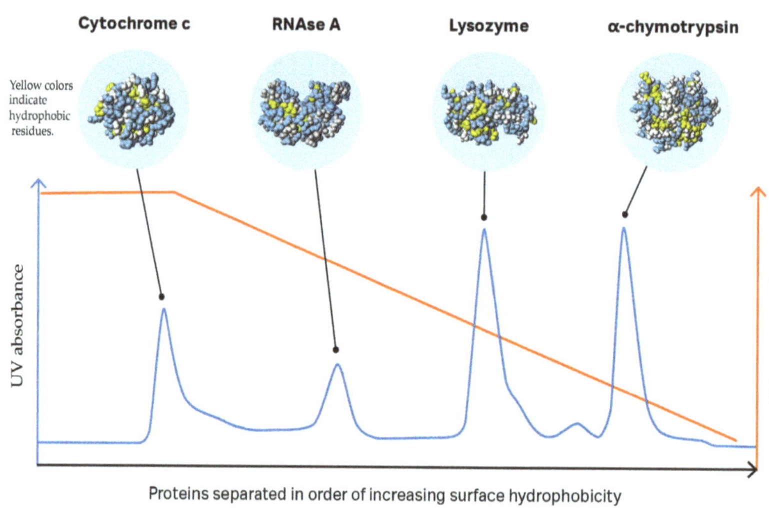

Figure 2.5. *Hydrophobic interaction chromatography (HIC).*

Source: https://cdn.cytivalifesciences.com/api/public/content/digi-3609-original-png.

The secret to HIC is to bind an analyte as strongly as possible under ammonium sulfate precipitation, which can be easily ascertained in certain pilot experiments, but not so highly as to precipitate the analyte. HIC should function well at a comparatively low ammonium sulfate concentration if the analyte is relatively hydrophobic in comparison to the other contaminants. In order to achieve a salt concentration high enough to bind the target, it is best to use ammonium sulfate "cuts" in the beginning if the target is hydrophilic in order to remove the majority of the hydrophobic contaminants. The analyte is then eluted using a gradient of decreasing ammonium sulfate salt concentration. Negative chromatography should be explored as a purification technique if the analyte is highly hydrophilic. This will ensure that other molecules bind to the target and prevent the target from binding. Even in the absence of ammonium sulfate, some very hydrophobic proteins, like membrane proteins, will attach to the column. In this instance, the analyte may be eluted using an increasing concentration of a chaotropic salt (such as guanidinium chloride, urea, or isothiocyanate salts). For HIC, proteins make the best analytes.

2.5.3. Reversed Phase Chromatography

In the reverse phase chromatography, a polar mobile phase (often an aqueous solution) and an extremely hydrophobic stationary phase—both of which are modifiable—are used for separation. The most popular technique for analyzing biological macromolecules and lipids, as well as for preparative separation of small molecules such as lipids, oligo- and polypeptides, oligo- and polynucleotides, or oligo- and polydeoxynucleotides, is high-performance liquid chromatography (HPLC). Because the highly hydrophobic stationary phase and the organic elution solvents tend to cause irreversible denaturation of all but the strongest proteins, HPLC is not frequently used for protein analysis. HIC, on the other hand, does not cause denaturation because the protein typically maintains its proper conformation due to the high concentration of salt and the surface's weak hydrophobicity. Analytes are typically bound to the stationary phase in high polarity solutions (e.g., water), following which a gradient with a water-miscible solvent (typically acetonitrile, methanol, or isopropanol) is used to lessen the polarity of the mobile phase. The mobile phase's pH is crucial because certain analyte functional groups behave as hydrophilic when charged but as hydrophobic when uncharged. As a result, pH variations can have a big impact on retention on a reversed phase column. In order to improve the interaction between charged groups and the surface, ion-pairing agents are frequently employed. Ion-pairing agents consist of a hydrophobic region that binds to the stationary phase and a hydrophilic region that interacts ionically with any charges on the target. Usually, these agents are trifluoroacetic acid, TFA, or formic acid. The approximate retention of the target on the column should be determined using a relatively steep, broad gradient in the first stages of trying reversed phase chromatography for separation. The resolution can then be increased by reducing the start and end points to the range of interest. It is frequently noticed that molecules elute at a lower solvent strength as the gradient slope decreases. This is because elution is not strictly binary in nature; the target can begin to slowly move down the column at a solvent strength that is slightly less than the full elution from the column. This effect is more noticeable with shallower gradients, which results in earlier elution.

2.5.4. Gel Filtration Chromatography

Molecules are separated based on their size and shape using gel filtration chromatography (GFC). Other names for it include gel permeation chromatography and size exclusion chromatography. Gel filtration chromatography operates on a different principle than other column chromatography modes in that the interactions between the analyte, eluant, and support should ideally all be equal, i.e., Every attempt should be made to avoid any potential interactions between the resin and the analyte. The stationary phase's particles, which each have a distribution of pore sizes, allow for separation. Bigger molecules are unable to pass through the pores and move swiftly through the column, while smaller molecules must fit through increasingly tiny openings, causing a separation based on molecular weight. The pore size determines the separation range; for small biomolecules (>10 kDa), large analytes (<200 kDa), and even the mass of small plasmids, organelles, or viruses, certain pore sizes work best for separation. Since analytes never actually alter their physical surroundings or bind to the resin during chromatography, this method is the "softest" separation technique available.

GFC is frequently applied as the last "polishing step" in a purification process. "Desalting" (also known as "buffer exchange") columns are a particular kind of special gel filtration chromatography column. Large analytes are entirely excluded by the resin, but salt, solvent, and buffer molecules are not excluded. To ensure that the analyte elutes well ahead of the sample salts and buffers that could otherwise get trapped in the column's pores, the column is first allowed to acclimate to the appropriate buffer before the analyte is applied. This method is frequently helpful for unstable molecules where dialysis would take too long, and it is typically faster for small volumes than dialysis. Because the columns can be calibrated using known proteins and an approximate mass can be determined directly from the time of elution, GFC even allows researchers to gain insight into the quaternary structure of proteins.

2.5.5. Hydroxyapatite Chromatography

A crystalline form of calcium phosphate ($(Ca_5(PO_4)_3OH)_2$), hydroxyapatite is occasionally employed in the purification of analytes, especially proteins and DNA. The stationary phase of this resin is crystalline, in contrast to all the others, and the analyte binds to it in a way that is partially ionic and partially surface calcium ion specific. The method is frequently employed as a last polishing step as well as to separate proteins that co-purify using a variety of other techniques. Other benefits of this include the stationary phase's ability to be autoclaved, which makes it useful in strictly sterile work. On the other hand, flow rates are slow, and the columns don't last very long (one issue is that carbon dioxide frequently binds to HA, causing a crust on the column's surface). Proteins are typically adsorbed onto the column using a low ionic strength phosphate buffer, and then they are eluted using an increasing phosphate gradient; typically, 0.5 M phosphate is enough to elute every protein that has been adsorbed. Because of significantly higher flow rates and operational column pressures, ceramic hydroxyapatite may now be used in place of the original crystalline hydroxyapatite.

2.6. DIRECTED BIOLOGICAL SYNTHESIS OF PROTEINS

A large portion of current chemical biology research revolves around the biological synthesis of proteins. All proteins must be purified before being biologically synthesized in the modern world from one organism to another. Growing recombinant "factory organisms" and using recombinant techniques are often necessary when very large quantities of proteins (mg–g levels) are needed. The reader of chemical biology will be introduced to the most recent methods for biological protein synthesis and purification in this section. Importantly, proteins differ greatly in their physical and structural characteristics; as a result, even though biological protein syntheses are generally similar, there can be significant differences in the finer points.

2.6.1. Wild-Type or Recombinant Sources

Proteins can be obtained from two different sources before they are purified: recombinant sources or natural sources where the protein is present at its native level. Before the development of molecular biology and recombinant technologies, natural sources were the only available option. Nevertheless, they are still frequently utilized when the main structure of the protein of interest is unknown. In cases where factors like proper post-translational modification are crucial or where protein expression is naturally high, natural sources are also utilized. A genetically modified "factory organism" must be created in order to overproduce the desired protein in order to employ the recombinant approach to protein synthesis as an alternative. Although there are frequently issues with protein solubility and proper protein processing in *Escherichia coli* (*E. coli*), strains of the bacterium are generally used as factory organisms. Other factory organism systems consist of specific mammalian cell lines, yeast cells, and insect cell lines. There are benefits and drawbacks to every system.

2.6.2. Expression in E. Coli; Early Purification

The target protein can be separated from the other cell components in a multistep process once a wild-type or recombinant organism reaches full growth. In order to effectively release and, if feasible, start the process of fractionating out the protein of interest, cell walls must first be damaged. To break down the cells in microbial extracts, sonication, homogenization (either with a bead mill or under high pressure), or the addition of lysozyme—a protein bio-catalyst or enzyme—are frequently employed. The polysaccharide component of the bacterial cell wall coat is catalyzed to weaken by lysozyme. Then, soluble and insoluble fractions from the microbial extracts may be separated by centrifugation. At this point, non-ionic detergents like triton X-100 in buffer are widely used to maximize the release of the target soluble protein. Protease inhibitors are necessary even at this point to stop the target protein from being broken down by natural biocatalysts, or enzymes that are appropriate for proteolysis. These inhibitors can be purchased as broad-specificity cocktail forms or as individual inhibitors that block metalloproteases, acid and serine proteases, and other enzymes. To preserve protein stability and activity, an appropriate buffer must be selected; typically, this means keeping the pH between 7.0 and 8.0, adding a relatively high concentration of salt (0.1 to 0.5 M), and possibly adding sucrose or glycerol (10%) to help the protein dissolve. Cysteine bridge formation is prevented by maintaining a reducing environment, which can be achieved through the use of dithiothreitol (DTT) or β-mercaptoethanol (βME). The preparation of a cell-free extract from a eukaryotic system generally requires more consideration because techniques like differential centrifugation may be used to partially fractionate intracellular organelles before the release of the protein or interest.

Precipitation is now frequently used as a practical first step in the purification of proteins to separate and concentrate the target protein. Polyethylene glycol (PEG), ammonium sulfate, and organic sulfate are the three primary precipitation techniques employed. Sulfur will attach to water molecules at high concentrations, lowering the quantity of water available to protect hydrophobic regions on a protein surface. The degree of saturation is represented by the percentage, which is used to define sulfate concentrations in terms of percentage "cuts" (0–20 % w/v, 20–40 % w/v, etc.). Typically, the protein of interest will be adjusted to aggregate wholesale in a single percentage "cut" (i.e., at a single, roughly saturated sulfate level). The pellet is then prepared for additional purification after the aggregate is collected by centrifugation. Precipitating proteins can also be done with organic solvents, most commonly acetone or ethanol. Precipitation can be caused by gradually increasing the solvent's hydrophobicity through concentration increases in the solvent, which will encourage intermolecular electrostatic interactions. PEG precipitation operates according to a comparable theory. Selective heat-induced denaturation is also occasionally employed, in which the protein of interest is left intact in solution while contaminants are denatured by gradually rising temperature.

The two main categories of protein purification techniques are affinity chromatography and classical chromatography. Affinity chromatography depends on the selective and reversible binding of the target to an affinity ligand immobilized on an insoluble matrix, whereas classical chromatography depends on variations in the chemical and physical properties of molecules. Classical chromatography can still be used alone or in conjunction with affinity chromatography purification steps, even though recombinant techniques have greatly increased the relative use of affinity chromatography. Polyacrylamide gel electrophoresis (PAGE) using sodium dodecyl sulfate (SDS) is commonly used to monitor or follow the purification of proteins.

2.6.3. Affinity Chromatography

Specific biomolecular recognition between

the target protein and a molecule attached to the matrix is used in affinity chromatography. Many varieties of affinity chromatography are helpful in the purification process. Affinity chromatography has been especially helpful thanks to recombinant technologies, as affinity purification tags can be incorporated into recombinant proteins to facilitate purification.

2.6.3.1. Immobilized Metal Affinity Chromatography (IMAC)

Proteins can be specifically coordinated with immobilized metal ions to enable their isolation. The most popular application of IMAC involves engineering the protein to have a sequence cluster consisting of four to eight histidine residues. This allows for straightforward purification, typically through interaction with immobilized Ni^{2+} that is located on the matrix surface. Although Ni^{2+} binds to polyhistidine sequences less firmly than Cu^{2+}, doing so allows for more purification selectivity. Metals (typically Cu^{2+}, Zn^{2+}, Ni^{2+}, Co^{2+}, or Fe^{2+}) are typically immobilized by chelating to nitrilotriacetic acid on the solid matrix support. Imidazole, which competes with the protein of interest's histidine clusters for coordination with the matrix-associated metals, is then typically used to enable elution.

Recombinant techniques are used to engineer proteins to carry polyhistidine tails (N-terminal or C-terminal His tag) in frame with their primary structure amino acid sequence. This confers several benefits. The protein can frequently be purified by IMAC using a very straightforward method thanks to the His tag. Additionally, since the affinity sequence is so short, antibodies can be produced against the target protein without removing the tag. Even insoluble proteins can be purified by first pre-solubilizing them in urea or guanidinium chloride, and then using metal affinity column purification to unfold the unfolded protein under the same guidelines. Refolding can then be tried after the purification of the affinity column. However, employing His-tag purification has certain disadvantages. For example, the presence of the His tag may deform the target protein's structure and possibly even its function in a noticeable way. There are situations when the His-tagged protein of interest is less soluble than the protein of wild type. Moreover, the protein of interest and the IMAC columns may leak heavy metal ions, which may cause oxidative side-chain damage to the amino acids. The general prohibition against using reducing agents like DTT or high concentrations of βME during IMAC purification may exacerbate this issue. Occasionally, four chelating bonds have been used to immobilize Co^{2+} instead of Ni^{2+} in order to improve the specificity of His-tag binding to metal ions and lessen the amount of heavy metal ions that leak into the eluant. Even natural histidine clusters can be used by this system, as in the purification of chicken lactate dehydrogenase.

2.6.3.2. Glutathione-S-Transferase Tags

One of the first recombinant affinity tags was glutathione-S-transferase (GST). Typically, GST is involved in the catalysis of the transfer of glutathione tripeptide onto substrates, either endogenous or exogenous, that have groups or side chains that are electrophilic. The protein of interest can be fused to GST, which is typically originally from Schistosoma japonicum, to facilitate purification using immobilized reduced glutathione. GST can give solubility to certain proteins that are insoluble when expressed alone, despite the tag's size. An additional advantage of utilizing GST is the fusion protein's easy-to-use activity assay. Active GST will transfer reduced glutathione onto 1-chloro-2,4-dinitrobenzene (CDNB), a reaction that can be easily tracked using UV–visible spectroscopy. GST is naturally in dimer form, so the fusion protein is likely to also be dimeric. Therefore, attaching a GST tag may disrupt the innate quaternary structure of certain proteins unless it is removed. This can be achieved by linking the GST tag to the protein of interest using a brief cleavable amino-acid residue spacer sequence.

2.6.3.3. Maltose Binding Protein Tags

This tag utilizes the natural maltose binding protein (MBP) from *E. coli*. The protein being studied can be attached to the maltose-binding protein, allowing for purification through affinity to immobilized amylose. The protein can then be eluted using maltose. MBP, unlike GST, is monomeric and one of the largest tags at 42 kDa. MBP is highly effective in promoting the solubility of insoluble or poorly soluble proteins of interest.

2.6.3.4. Biotinylation of Proteins

The biotin-avidin/streptavidin interaction is known to be extremely strong and can be utilized for protein purification. There are two methods to achieve this: either the protein can be biotinylated in a lab setting using biotin-ester reagents, or a naturally biotinylated sequence can be used during protein synthesis. *In vitro* biotinylation can be challenging as it may result in the modification of multiple lysine residues, making it less effective for protein purification. Site-specific biotinylation is achievable by creating a fusion protein with a biotinylated peptide tag, such as the C-terminal residues of BCCP. This allows for specific biotinylation of the target sequence at lysine 89 in *E. coli* with the help of biotin holoenzyme ligase enzyme. The fusion protein can then be isolated using an avidin-affinity column. The binding strength of biotin to natural tetrameric avidin/streptavidin is nearly irreversible, with a K_d value of approximately 10^{-13} M. However, new avidin resins have been developed that utilize a resin-bound or modified monomeric form of streptavidin, allowing for elution of the purified protein using biotin.

2.6.3.5. Intein Tags

Inteins are protein segments that can naturally self-splice, separating from a parent protein but leaving a complete daughter protein chain behind by forming a new peptide bond. This process has been utilized to create proteins that can self-splice, allowing a protein of interest to be separated from an attached affinity tag. For example, a protein may be engineered with an intein affinity tag, which can later be induced to detach after purification, resulting in a highly purified protein of interest. An example of an intein affinity tag is a chitin-binding domain combined with a self-splicing intein derived from yeast. Self-cleavage of the intein can occur in the presence of thiols such as DTT, βME, or cysteine, without the need for additional enzymes. The beauty of this system lies in its ability to perform the self-cleavage reaction on an affinity column immediately following purification on cellulose beads. This allows for the elution of the protein of interest without any attached protein affinity tags. Additionally, fusion proteins can be engineered with an intein protein affinity tag attached at either the N- or C-terminus of the protein of interest as needed.

2.6.3.6. Other Affinity Tags and Radio-labeling of Proteins

In addition to the ones listed above, there are a plethora of additional affinity tags that could be used. This is a brief overview of the potential diversity. The first is Protein A, which has a strong binding affinity for IgG immunoglobulins. On the other hand, fusion proteins can employ IgG Z domain mimics to attach to protein A columns. Next is the 15-amino-acid peptide known as the S-peptide tag, which forms a strong bond with a protein that is derived from pancreatic ribonuclease A. A peptide with 10 amino acid residues called the Strep-tag binds tightly to streptavidin. The next generation, known as strept-tag II, is a nine-amino acid sequence that binds to streptavidin derivatives (streptactin). Desthiobiotin, an analog of biotin, is used in both situations to elute fusion proteins from affinity columns. Moving on, the flag tag is a small hydrophilic peptide (N-AspTyrLysAspAspAspAspLys-C) that may be fused with a protein of interest such that the fusion protein can be purified using an anti-flag affinity resin. Next, the calmodulin binding peptide is a relatively small peptide (4 kDa) that binds to immobilized calmodulin protein and elutes from the affinity column under very mild conditions. Indee, binding only requires the presence of Ca^{2+}, while elution is made possible in the same buffer in the presence of EGTA (a particularly effective Ca^{2+} chelator). By contrast, the T7 tag consists of an 11-amino-acid residue peptide from the T7 10-protein and purification is effected using an anti-T7 monoclonal antibody affinity column. However, elution from this column requires very acidic conditions – pH 2.2! Finally, the cellulose binding domain tag (CBD tag) (~12 kDa) binds strongly to homoglycan cellulose or chitin columns. The fusion protein may then be eluted with ethylene glycol, a very soft method of elution, after which the ethylene glycol may be easily removed by dialysis. A considerable advantage of this tag is that it is particularly cost effective since the resin is very inexpensive, robust and readily reusable.

The CBD tag method is suitable for labeling proteins with radioactivity, but another option is to create fusion proteins with tag sequences recognized by kinases, such as N-ArgArgAlaSerVal-C. This sequence triggers cAMP-dependent-protein-kinase-mediated phosphorylation of the tag serine residue with [32P]. This method requires the protein of interest to be phosphorylated *in vitro* with the help of the appropriate kinase after purification. While effective, there is a shift towards using recombinant technology-based approaches for radio-labeling proteins in cells rather than *in vitro*. In situ labeling with [35S] is preferred over [32P] as it is gentler and longer-lasting due to its longer radioactive half-life.

2.7. BIOLOGICAL SYNTHESES OF NUCLEIC ACIDS, OLIGOSACCHARIDES, AND LIPIDS

If a comprehensive understanding of biological macromolecule and lipid assembly structures and functions is to be achieved, these are also equally significant to the biological syntheses of proteins.

2.7.1. Biological Synthesis of Nucleic Acids

Many of the chromatographic techniques covered in this chapter can be used to separate and fractionate nucleic acids from natural sources if the goal is only the simple isolation of nucleic acids from cells. In cells, cationic polyamines and proteins are commonly linked to DNA, and both can be eliminated using a phenol solution. As an alternative, proteins can be denuded using proteases, which are proteins that break down peptide bonds, or they can be detached using chaotropes or detergents. After that, ethanol may be used to precipitate nucleic acids free of proteins. Pancreatic DNAses, which are proteins that break down phosphodiester links in DNA, can be used to separate cellular RNA from DNA, or RNAses, which are also proteins that break down phosphodiester links in RNA, can be used to separate cellular DNA from RNA. While the most common method for separating nucleic acids is electrophoresis, HA chromatography and HPLC are also very helpful. HA chromatography is a useful method for separating double helical DNA from cellular RNA and proteins because HA columns have a unique tendency to bind double-stranded DNA molecules very strongly. Affinity chromatography modifications can also be helpful. For instance, the 3-end of eukaryotic mRNA normally contains a poly(A) sequence. As a result, immobilized-poly(U) on a solid support can be an effective method for separating mRNA from extracts of cells. Of course, there are instances in which obtaining and purifying whole cellular DNA or RNA is necessary. But occasionally, just one DNA or mRNA molecule is needed. In these situations, the molecule needs to be located, amplified, and then purified to remove any remaining cell debris. In this second case, cloning and other recombinant techniques are heavily utilized in the biological synthesis of such DNA or RNA molecules before purification protocols.

2.7.2. Biological Synthesis of Oligosaccharides

Although oligo- and polysaccharides are frequently linked to membrane proteins, intracellular proteins (not found in bacteria), and possibly some extracellular proteins as well, they can also be purified using many of the same techniques. Lipo-polysaccharides (LPSs), which are incredibly common in some bacterial cell walls and have the potential to be highly immunogenic, are polysaccharides that are also closely linked to lipids. When necessary, oligo- and polysaccharides are coprecipitated with proteins and nucleic acids using alcohol, and protein linkages are broken down chemically. Ion exchange chromatography can then be used to separate charged proteins and nucleic acids from the uncharged or weakly charged oligo- or polysaccharides. Oligo-/polysaccharides with neutral or weak charges will pass through columns, while nucleic acids and proteins will bind and be removed from the sample. The oligo-/polysaccharides can then be analyzed and purified using specialized anion exchange chromatography and HPLC techniques. Additionally, affinity chromatography can be performed using immobilized lectin, such as the jack bean protein concanavilin A, which specifically binds to glucopyranose and mannopyranose monosaccharide residues.

2.7.3. Biological Synthesis of Lipids

Usually, lipids are synthesized biologically in cells and extracted from large lipid formations using organic solvents. Neutral lipids are extracted using non-polar solvents like diethyl ether or chloroform, while membrane-associated lipids require more polar solvents such as ethanol or methanol to break intermolecular hydrogen bonding. Commercially used phospholipids are typically purified from egg or bovine sources, while cholesterol is usually purified from egg or wool grease (derived from sheep). Sphingolipids are mostly purified from egg or mammalian tissue sources, and phosphatidylinositols are extracted from soybean or bovine sources. Synthetic lipids are often made from a glycerol-3-phosphocholine precursor that is purified from soybean lecithin. After extraction from a natural source, individual lipids can be purified to purity by using HPLC.

CASE STUDY

FORMAL STEREOSELECTIVE C-H AMINATION

Stereoselective C-H amination is a very attractive strategy for the conversion of simple low-cost chemical starting materials to high-value chiral amine building blocks and has therefore attracted much interest in organic chemistry. α-Methylbenzylamine is a substructure of many complex organic molecules and can therefore be a valuable intermediate in their synthesis.

The most successful chemical strategies reported so far have involved intramolecular and transition-metal catalyzed reactions with several methods using both chemical and enzymatic catalysts. A cascade of two enzymes was reported recently for the terminal amination of fatty acid methyl esters employing a recombinant biosynthetic cascade in a whole cell system. Previous work had already shown that whole-cell biotransformation is able to produce chiral amine from secondary alcohol, albeit in low conversion of 3%.

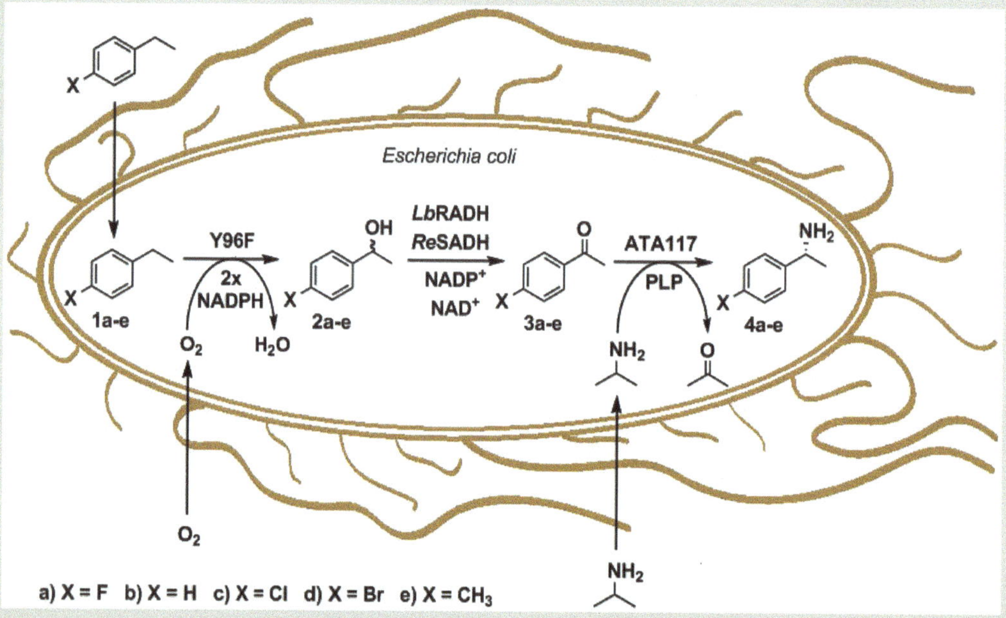

Figure 2.6. *E. coli BL21(DE3) cells harboring a self-sufficient P450 monooxygenase (Y96F) which produces scalemic mixtures of alcohols with a general preference for the (R)-enantiomer, (R)- and (S)-selective alcohol dehydrogenases (LbRADH using NADP+ and ReSADH using NAD+ as co-factor), and an w-transaminase (ATA117 requiring an amine donor and pyridoxal 5'-phosphate (PLP)) capable to convert 4-substituted ethylbenzenes 1a-e into alcohols 2a-e, ketones 3a-e and finally amines 4a-e (Flitsch et al., 2016).*

Source: https://learning.acsgcipr.org/wp-content/uploads/2022/07/31d1d2d70ec01e276c9d1fd5f3683c02.png.

Given that natural biosynthetic cascades often suffer from limited substrate ranges, CHEM21 researchers were interested in generating a de novo biosynthetic multi-enzyme cascade that is guided by retrosynthetic considerations in its design and constructions. The target enzymes for the cascade were carefully chosen to be complementary in substrate recognition. By introducing the pathway into *Escherichia coli*, CHEM21 researchers generated a biocatalyst which converts ethylbenzenes 1a-e to predominantly (R)-1-phenylethanamines 4a-e respectively (Figure 2.6) with conversions of up to 26% and ee values of 97.5%. Under the present reaction conditions, no additional co-factor except for the amine donor IPA and molecular oxygen was required. The CHEM21 approach is the first fully enzymatic approach to generate chiral amines at the benzylic position from alkyls.

The flexible generic nature of the design of the cascade will allow for the easy substitution of individual enzymes with homologs or mutants and thereby provide a cassette based modular approach for the design and construction of alternative cascades for the enantioselective C-H amination of other substrates.

CLASS ACTIVITY

SYNTHESIS OF ASPIRIN

- **Objective:** To synthesize aspirin (acetylsalicylic acid) from salicylic acid and acetic anhydride and to understand the principles of esterification and purification.
- **Materials:** Salicylic acid, Acetic anhydride, Sulfuric acid (catalyst), Ice bath, Distilled water, Ethanol, Filter paper, Beakers, flasks, and other laboratory glassware
- Procedure:
    - Reaction Setup:
    - Heating:
    - Cooling and Crystallization:
    - Filtration:
    - Purification:
    - Characterization:
- Discussion: Discuss the chemical reaction mechanism, the role of the catalyst, and the importance of purification techniques in chemical synthesis.

SUMMARY

- Chemical synthesis and biological synthesis are amongst the most powerful tools available to chemists interested in chemical biology. In synthesis begins the process of determining structure and dynamics.
- SPPS relies on very efficient amino-acid coupling. Each coupling step involves the linking of an α-N-protected amino acid with the α-N-deprotected amino terminus of a growing resin linked peptide chain.
- SPPS has allowed relatively facile peptide synthesis for peptides of up to 40 amino-acid residues, though it can be used to synthesize polypeptides in excess of 100 amino-acid residues in length.
- Whilst solid phase approaches dominate the chemical syntheses of the shorter nucleic acids and polypeptides, the same is not true for oligo-/polysaccharides. The problem is the complexity of oligo-/polysaccharide primary structures.
- Chemical synthesis of oligo-/polysaccharides demands the use of orthogonal protecting groups capable of selective removal. Benzyl ether protecting groups have seen widespread use in oligo-/polysaccharide chemistry.
- Coupling of monosaccharides requires activation of the anomeric position in one monosaccharide (electrophilic acceptor or glycosyl donor) and protection of all functional hydroxyl group positions except for the selected hydroxyl group in the other monosaccharide (nucleophilic donor or glycosyl acceptor).
- Biological synthesis is used here to describe harnessing the synthetic potential of living organisms for directed synthesis of biological macromolecules of interest.
- Ion exchange chromatography (IEC) separates molecules according to differences in their accessible surface charge. The method is very widely used, especially in protein purification, as separation may be applied to any molecule that is charged and soluble in an aqueous system.
- Hydroxyapatite is a crystalline form of calcium phosphate $(Ca_5(PO_4)_3OH)_2$ that is sometimes used in analyte purification, particularly of DNA and proteins.

REVIEW QUESTIONS

1. Describe the ideal properties of a solid support used in SPPS.
2. Explain the role of protecting groups in amino acid coupling.
3. Why are protecting groups necessary in peptide synthesis?
4. Compare solid-phase synthesis and solution-phase synthesis of polypeptides.
5. What are the common protecting groups used in oligodeoxynucleotide synthesis?
6. Provide examples of common modifications to nucleotides and their applications.
7. Explain the concept of solid-phase oligosaccharide synthesis.
8. What is ion exchange chromatography, and how does it work?
9. Explain the principle of hydrophobic interaction chromatography (HIC).

10. What distinguishes reversed-phase chromatography (RPC) from other chromatographic techniques?

REFERENCES

1. Anderson, G. W., & McGregor, A. C. (1957). T-butyloxycarbonylamino acids and their use in peptide synthesis. *Journal of the American Chemical Society, 79*, 6180–6183. doi:10.1021/ja01578a043.
2. Both, P., Busch, H., Kelly, P. P., Mutti, F. G., Turner, N. J., & Flitsch, S. L. (2016). Whole-cell biocatalysts for stereoselective C-H amination reactions. *Angewandte Chemie International Edition*, *55*, 1511–1513. doi:10.1002/anie.201509386.
3. Carpino, L. A. (1957). Oxidative reactions of hydrazines. IV. Elimination of nitrogen from 1,1-disubstituted-2-arenesulfonhydrazides. *Journal of the American Chemical Society, 79*, 4427–4431. doi:10.1021/ja01575a020.
4. Carpino, L. A., & Han, G. Y. (1972). 9-fluorenylmethoxycarbonyl amino-protecting group. *The Journal of Organic Chemistry, 37*, 3404–3409. doi:10.1021/jo00985a020.
5. Dirksen, A., Meijer, E., Adriaens, W., & Hackeng, T. M. (2006). Strategy for the synthesis of multivalent peptide-based nonsymmetric dendrimers by native chemical ligation. *Chemical Communications*, *15*, 1667–1669. doi:10.1039/B600673H.
6. du Vigneaud, V., Ressler, C., Swan, J. M., Roberts, C. W., & Katsoyannis, P. G. (1954). The synthesis of oxytocin. *Journal of the American Chemical Society, 76*, 3115–3121. doi:10.1021/ja01642a088
7. Kelly, P. P., Eichler, A., Herter, S., Kranz, D. C., Turner, N. J., & Flitsch, S. L. (2015). Active site diversification of P450cam with indole generates catalysts for benzylic oxidation reactions. *Beilstein Journal of Organic Chemistry*, *11*, 1713–1720. doi:10.3762/bjoc.11.185.
8. Lloyd-Williams, P., et al. (1997). *Chemical Approaches to the Synthesis of Peptides and Proteins*. Boca Raton: CRC Press.
9. McKay, F. C., & Albertson, N. F. (1957). New amine-masking groups for peptide synthesis. *Journal of the American Chemical Society, 79*, 4686–4690. doi:10.1021/ja01573a049.
10. Merrifield, R. B. (1963). Solid phase peptide synthesis. I. The synthesis of a tetrapeptide. *Journal of the American Chemical Society, 85*, 2149–2154. doi:10.1021/ja00897a025.
11. Nishiuchi, Y., Inui, T., Nishio, H., Bódi, J., Kimura, T., Tsuji, F. I., & Sakakibara, S. (1998). Chemical synthesis of the precursor molecule of the Aequorea green fluorescent protein, subsequent folding, and development of fluorescence. *Proceedings of the National Academy of Sciences of the United States of America, 95*, 13549–13554. doi:10.1073/pnas.95.23.13549.
12. O'Reilly, E., Kohler, V., Flitsch, S. L., & Turner, N. J. (2011). Cytochromes P450 as useful biocatalysts: Addressing the limitations. *Chemical Communications*, *47*, 2490–2501. doi: 10.1039/C0CC04489J.

13. Robin, A., Kohler, V., Jones, A., Ali, A., Kelly, P. P., O'Reilly, E., Turner, N. J., & Flitsch, S. L. (2011). Chimeric self-sufficient P450cam-RhFRed biocatalysts with broad substrate scope. *Beilstein Journal of Organic Chemistry*, 7, 1494–1498. doi:10.3762/bjoc.7.171.

14. Vigneaud, V., Ressler, C., Swan, C. J. M., Roberts, C. W., & Katsoyannis, P. G. (1953). The synthesis of an octapeptide amide with the hormonal activity of oxytocin. *Journal of the American Chemical Society, 75*, 4879–4880. doi:10.1021/ja01114a537.

CHAPTER 3

Molecular Biology as a Toolset for Chemical Biology

LEARNING OBJECTIVES

After studying this chapter, you will be able to:

- Understand the key concepts in molecular biology
- Discuss the tools and techniques in molecular biology
- Describe cloning and identification of genes in DNA
- Integrate cloning and expression
- Define site-directed mutagenesis

KEY TERMS FROM THIS CHAPTER

Agarose gel electrophoresis (AGE)
Biological systems
Chemical synthesis
DNA segments
Molecular biology
Plasmid manipulations
Prokaryotic DNA
Sodium dodecyl sulfate (SDS)
Aminopurine
Cassette mutagenesis
DNA ligases
Exogenous pDNA
Nicotinamide adenine dinucleotide (NAD)
Plasmids
Restriction enzymes

INTRODUCTION

Molecular biology is the study of living organisms' molecules. The biological functions of an organism are attributed to the chemical reactions occurring within its body. The "central dogma of biology" controls transcription, translation, and replication, among other bodily functions. The building blocks of these processes (DNA, RNA, and proteins) are related to molecular biology. The study of biomolecules' three-dimensional structures is the primary goal of molecular biology. X-ray diffraction and electron microscopy are two methods used to obtain a detailed model of their structures.

Our understanding of biological systems has been greatly enhanced by the interdisciplinary fields of molecular biology and chemical biology. Chemical biology studies and manipulates biological processes using chemical tools and techniques, while molecular biology concentrates on the molecular mechanisms of biological processes. A potent toolkit for examining and modifying biological systems at the molecular level is made possible by the integration of molecular biology techniques into chemical biology.

3.1. KEY CONCEPTS IN MOLECULAR BIOLOGY

Molecular biology is an extremely diverse and forward-thinking field in and of itself, and it is a potent tool for the directed synthesis of biological macromolecules, particularly nucleic acids and proteins. But far too frequently, chemists have drawn the boundary at the intersection of molecular biology and protein science. This is partly because of the terminology used when the field moved toward biology.

On the other hand, scientists have also greatly underestimated the ways in which molecular biology can advance chemistry. A serious reader of chemical biology cannot afford to make the same errors. In the quest to comprehend how biology functions at the molecular level, molecular biology—with its constantly expanding and improving array of recombinant techniques—is just as vital to the practice of chemical biology research as chemical synthesis.

3.1.1. The Central Dogma of Molecular Biology

The central dogma of molecular biology, first proposed by Francis Crick, states that DNA is transcribed into RNA, which is then translated into protein. While the first step can be reversible in certain cases, such as with the reverse transcription of RNA to DNA in some viruses, there is no known instance of genetic code flowing back from polypeptide sequences to DNA. This highlights the importance of DNA in biological processes, as it serves as the starting point for the synthesis of RNA and proteins. Therefore, the first step in any biological study should be to identify, isolate, and clone specific DNA segments from the genome of interest, typically genes. These genes contain open reading frames that encode genetic information for proteins, as well as other regulatory elements within the DNA structure. These initial steps are crucial for various activities, such as expressing and purifying proteins coded for by specific genes or conducting DNA sequencing (McCarthy & Holland, 1965).

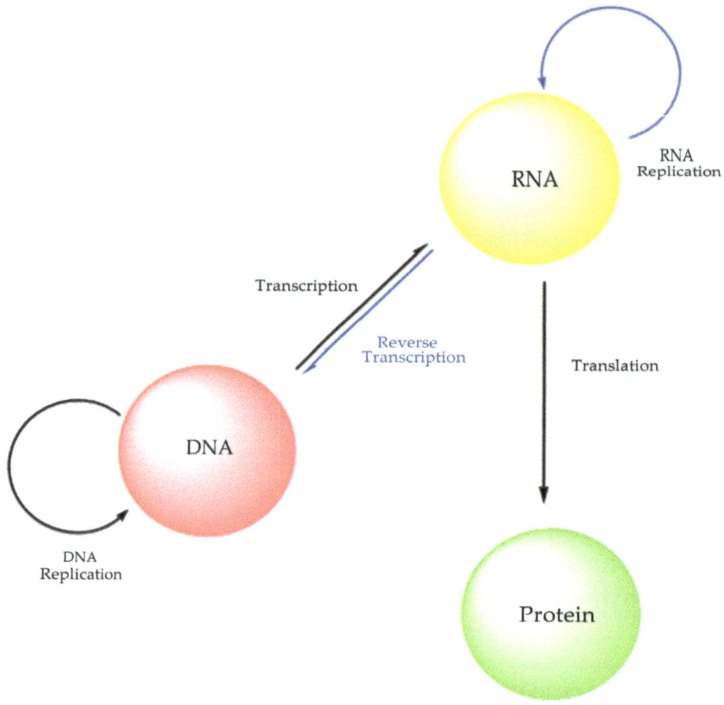

Figure 3.1. *The central dogma of molecular biology. Black lines represent information transfers that occur frequently in cells. Blue indicates the exception of reverse transcription that occurs in some viruses, and RNA-directed RNA polymerase catalyzed RNA replication that also occurs in some plants and viruses only. Nucleic acids can be both stores and recipients of genetic information, whereas proteins are always recipients only of genetic information.*

Source: Andrew Miller and Julian Tanner, Essentials of Chemical Biology Structure and Dynamics of Biological Macromolecules, ISBN 978–0-470–84530–1.

3.1.2. The Difference Between Prokaryotic and Eukaryotic Genes

Prior to delving into the topic of DNA sequence isolation techniques, it is imperative to acknowledge several essential distinctions between prokaryotic and eukaryotic DNA. Organisms without a distinct nucleus are referred to as prokaryotes, which includes bacteria and cyanobacteria. They have a single circular DNA chromosome, and smaller circular DNA fragments known as plasmids that are dispersed throughout the cell and can also code for proteins are also possible. Prokaryotes are almost always unicellular organisms that grow into colonies on agar plates. Eukaryotic cells, on the other hand, have a true nucleus that houses several DNA chromosomes (DNA bonded to and condensed around nuclear proteins) in addition to a large number of other intracellular compartments. In addition, these cells are noticeably bigger than prokaryotic ones. Eukaryotes comprise multicellular organisms like all animals and plants as well as some unicellular organisms like yeast. Even though prokaryotes and eukaryotes have the same genetic code, there are important molecular differences between the two groups because eukaryotes use a far more complex method of processing mRNA after transcription (a process known as post-transcriptional processing), and for very good reasons.

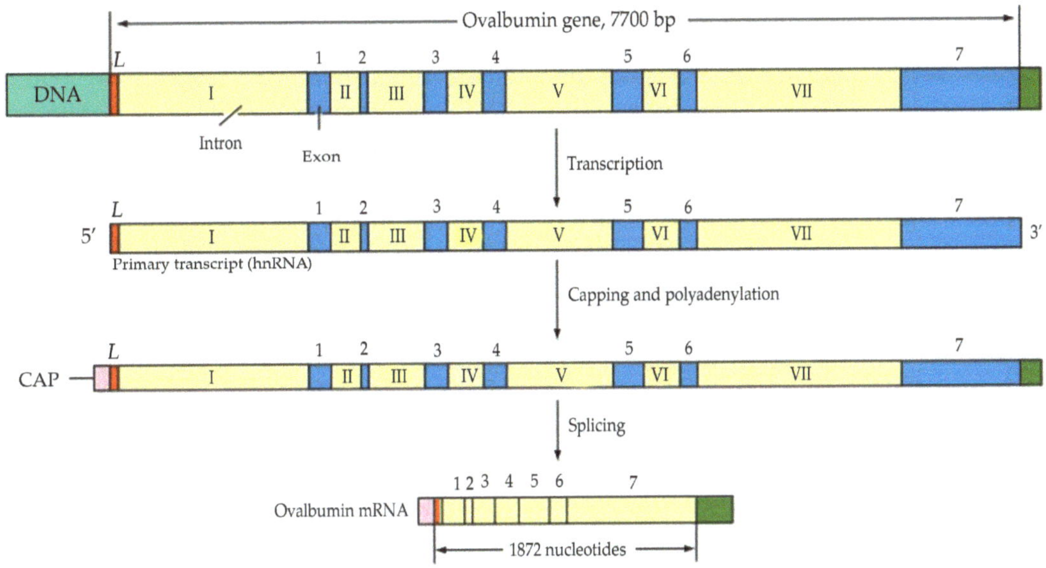

Figure 3.2. *Processing of eukaryotic pre-mRNA to mature mRNA as demonstrated for the chicken ovalbumin gene. Most higher eukaryotic genes have both introns (which do not code for protein) and exons (which usually do code for the protein). There is significant posttranscriptional processing to form mature mRNA prior to translation.*

Source: Andrew Miller and Julian Tanner, Essentials of Chemical Biology Structure and Dynamics of Biological Macromolecules, ISBN 978–0-470–84530–1.

The immediate products of mRNA transcription, or primary mRNA transcripts, are typically translated into protein in prokaryotes without any changes. Translation may even start at the fifth end of the mRNA molecule before transcription is finished at the third end. When an mRNA transcript is transcribed from a DNA operon—a contiguous group of genes whose protein products typically have related functions, like metabolism—it is possible for one transcript to contain the contiguous genetic code for several different proteins. Therefore, prokaryotic genomic DNA sequences can be translated directly into protein sequences. In contrast, prokaryotes cannot be said to be in the same category. One of the main causes of this is that exons, which are segments of the DNA sequence that code for the protein of interest and are therefore typically not intended for direct translation with exons, make up higher eukaryote genes. Exons are represented in the mature RNA. It is impossible for transcription and translation to occur simultaneously in eukaryotes, as this is only possible in prokaryotes, where transcription primarily occurs in the nucleus and translation occurs in the cytoplasm.

Initially, there are two essential posttranscriptional processing events:

1. The primary eukaryotic mRNA transcript is capped with a 5–5 triphosphate bridge linked to a 7-methylguanosine nucleoside, a process known as 5 capping.
2. Post-capping, the 3 end is modified by the addition of a 3 -poly(A) tail, a process known as 3 polyadenylation.

The next step is the intron splicing procedure. To produce a mature mRNA transcript with a base sequence corresponding to the contiguous base sequences (5 →3) of the original DNA exons in the eukaryotic gene, introns must be removed and exons "knitted back together." Few genes in simpler eukaryotes, like yeast, have introns. In higher eukaryotes, however, the situation is very different. Genes typically contain eight introns, with a total sequence length that is four to 10 times longer than the coding sequence found in the exons. Why higher eukaryotic genomes have such remarkable redundancy is still unknown. Higher eukaryotes show splicing variants, which increase complexity by varying splicing events based on factors such as cell type, expression time, and environmental cues.

Therefore, given the complexity of eukaryotic gene structure, how can we hope to obtain functionally useful genes for additional molecular biology manipulation? Mature mRNA transcripts provide the answer to this question. One could consider these mature transcripts from eukaryotes to be the functional equivalent of a prokaryotic primary mRNA transcript. The entire genetic code for a protein of interest is contained in processed, intron-free nucleotide sequences found in mature mRNA transcripts.

3.1.3. The Creation of cDNA Libraries

With the help of a reverse transcriptase enzyme (derived from the Moloney murine leukemia virus, MMLV), mature mRNA transcripts (sense strand) from eukaryotic cells can be purified and then reverse transcribed into complementary DNAs (cDNAs). These cDNAs will anneal with the mRNA transcripts through Watson-Crick base pairing to produce anti-parallel DNA/RNA duplexes or double helices. Every mature mRNA transcript has a poly(A) tail, which serves as a helpful handle for every reverse transcriptase reaction. After that, RNAse enzymes (which are specific to the hydrolysis of RNA phosphodiester links) must be used to break down DNA/RNA duplexes. A sense strand of DNA must then be constructed on each cDNA single strand in order to create equivalent, more stable antiparallel DNA/DNA duplexes, which can then be produced with the aid of a DNA polymerase enzyme. In this case, the DNA polymerase reaction depends on the poly(T) tail present in each cDNA molecule. Overall, the collection of DNA/DNA duplexes that results from reverse transcription and DNA polymerization of a sufficiently diverse population of distinct mature mRNA transcripts isolated from an organism can represent relatively complete eukaryotic genomic information for that organism, all in a functional form perfect for additional molecular biology manipulation. A cDNA library is this collection of DNA molecules. Using a DNA ligase enzyme, adapter DNA sequences are ligated onto both ends of each DNA/DNA duplex as the last step of preparation before usage. These adapters include the necessary sequence elements to transform each DNA/DNA duplex in the cDNA library into a genomic component. These genomic components can then be combined with recombinant technique paraphernalia to form the full toolkit of genomic components related to molecular biology.

3.2. TOOLS AND TECHNIQUES IN MOLECULAR BIOLOGY

Isolating, cloning, and identifying genomic information in an appropriately useful DNA form is the key to manipulation in molecular biology.

3.2.1. Plasmid DNA Vectors

Almost every modification of genomic or cDNA in molecular biology involves the plasmid. DNA that has the ability to replicate without the help of chromosomal DNA is referred to as plasmid. Prokaryotes are the original home of plasmid DNA (pDNA) systems, which were soon discovered to be excellent candidates for vectors. A vector is an independent DNA construct that can be rearranged to create recombinant pDNA constructs for cloning and/or heterologous gene expression. It can be created from genomic DNA or cDNA library sources. A pDNA-based vector ought to have a replication origin in addition to one or more selectable markers. The origin of replication is a non-coding DNA sequence that, when the time is right, indicates where pDNA replication should start. Genes that code for a selectable trait, like antibiotic resistance, are called selectable markers. Cloning vectors and expression vectors are the two main categories of vectors. The latter can be used to control the *in vivo* expression of a desired protein or non-coding RNA. Only 10 kb heterologous DNA inserts are stable with normal plasmids. Phagemids are another type of plasmid that has a single-stranded phage origin for single-stranded DNA preparation in addition to a plasmid origin of replication. Cosmids are a kind of plasmid that combines features of λ-based vectors and plasmids; they are useful for creating genomic DNA libraries because they can typically hold much larger DNA sequences (35–45 kb).

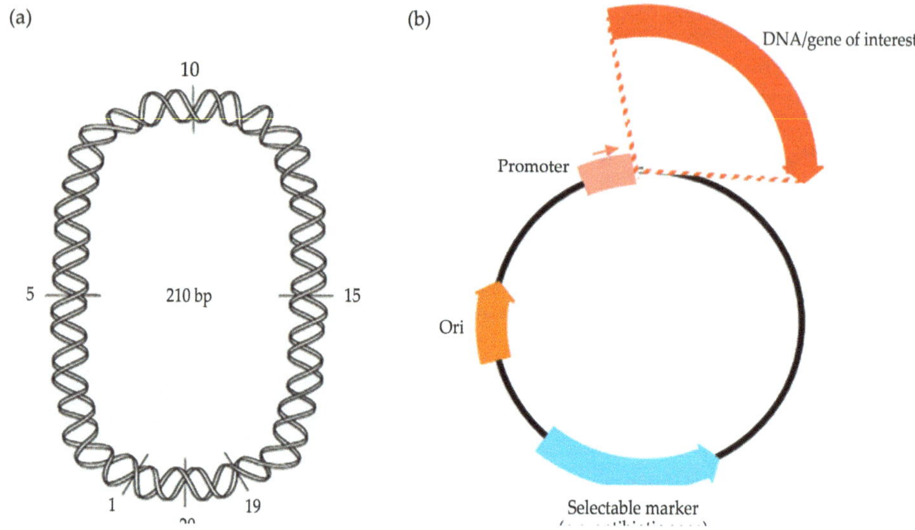

Figure 3.3. *Schematic to show a cloning vector, the simplest form of pDNA.*

Note: (a) diagram of closed circular DNA (cccDNA) (210 bp) to show how double-stranded DNA (dsDNA) links up to generate circular plasmid DNA (pDNA) molecules; (b) simplified diagram of pDNA cloning vector with selectable marker, non-coding origin of replication (Ori) and non-coding promoter element required for the control of expression of heterologous (foreign) DNA (DNA/gene of interest) eventually introduced downstream of the promoter (sense strand 3 -direction).

Source: Andrew Miller and Julian Tanner, Essentials of Chemical Biology Structure and Dynamics of Biological Macromolecules, ISBN 978–0-470–84530–1.

3.2.2. Restriction Enzymes

Enzymes known as restriction enzymes hydrolyze two phosphodiester links—one for each strand—to cut double-stranded DNA (dsDNA) without changing the deoxynucleotides that are attached. The word restriction comes from the discovery of these enzymes as a defense mechanism employed by bacteria to limit infection by specific bacteriophages, viruses that are specific to a particular bacterium. There are currently over 900 restriction enzymes known to exist from different organisms, of which 30–40 are frequently employed in a standard molecular biology lab. Often called restriction endonucleases, they differ from exonucleases in that they cut DNA ends rather than the phosphodiester links that they hydrolyze within a DNA sequence. Usually ranging in length from 4 to 12 bp, restriction enzymes function as "molecular DNA scissors" to cut dsDNA at sequence-specific restriction sites. Certain restriction endonucleases cut both DNA strands at the same location throughout the duplex, producing blunt-ended DNA fragments. Other restriction endonucleases cleave the DNA strands offset, leaving 5' or 3' deoxynucleotide overhangs at the ends of the cut that are defined in length, producing cohesive or sticky-ended DNA fragments. Palindromic sequences are typically recognized by restriction endonucleases, though this isn't always the case. The well-known Eco RI restriction enzyme, for instance, can recognize the sequence 5' -d(pGAATTC)-3,' which in duplex DNA is rotationally symmetric between dA and dT. Both sense and complementary DNA strands have cleavage sites offset between dG and dA, which results in sticky-ended DNA fragments with a 5' overhang measuring four deoxynucleotides in length. Xho I is the name of another restriction enzyme that is comparable.

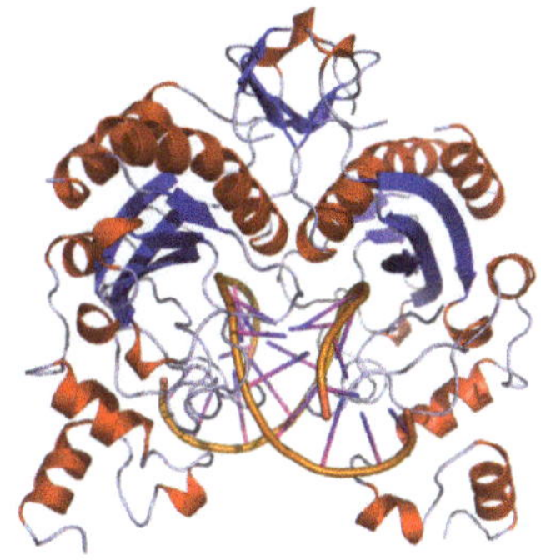

Figure 3.4. *EcoRV crystal structure complexed with double-stranded DNA.*

Source: https://upload.wikimedia.org/wikipedia/commons/thumb/f/f0/EcoRV_Crystal_Structure.rsh.png/800px-EcoRV_Crystal_Structure.rsh.png.

3.2.3. DNA Ligasses

Covalent ligation of heterologous duplex DNA fragments requires DNA ligases. Ligase enzymes help to rejoin DNA fragments that have been restricted (by restriction endonucleases). T4 DNA ligase is an ATP-dependent ligase enzyme that can facilitate both blunt-ended and sticky-ended ligation processes. The *E. coli* ligase, which doesn't need ATP but rather nicotinamide adenine dinucleotide (NAD) cofactor.

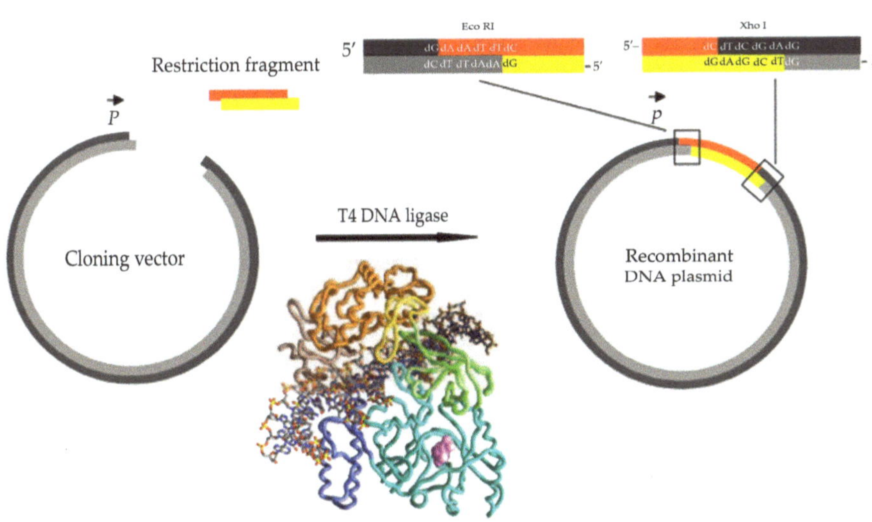

Figure 3.5. *Heterologous DNA restriction digest fragments created from two restriction endonucleases (Eco RI; Xho I) are allowed to interact with restriction digest fragments from a cloning vector.*

Source: Andrew Miller and Julian Tanner, Essentials of Chemical Biology Structure and Dynamics of Biological Macromolecules, ISBN 978-0-470-84530-1.

The enzymes restore phosphate and regenerate the phosphodiester bonds that restriction had severed in both situations. Sticky-end ligation should be selected whenever possible because it is generally much simpler and easier than other methods of ligation. In the event that sequence complementarity prevents this from happening for any reason, the following should be considered. If two non-complementary 5 sticky ends need to be joined, the overhangs should be filled in with Klenow polymerase, an enzyme that is derived from DNA polymerase I but lacks the 5 →3 exonuclease activity. This will result in duplex DNA. T4 DNA polymerase should be used to fill in overhangs where there are two or three sticky ends to join in order to create duplex DNA. Both scenarios result in blunt-ended duplex DNA fragments that can be joined together as needed through blunt-ended ligation mediated by DNA ligase.

3.2.4. Hosts

An organism carrying recombinant pDNA carrying heterologous DNA is called a host. When approaching molecular biology, the reader of chemical biology ought to have already realized that E. Even though the vector may eventually be used in other organisms like yeast or cell lines, *E. coli* is typically the workhorse organism used in the majority of molecular biology experiments. The reasons for this are partially historical: *Escherichia coli* was one of the first organisms to have its entire genome sequenced at the DNA level, has undergone extensive study, and has a comparatively clear understanding of the mechanism governing gene expression control. Additionally, *Escherichia coli* was the source of many of the original plasmids that were used as vectors, and a wide variety of bacterial strains are easily accessible and useful in a range of situations. *Escherichia coli* has also been weakened to a very low-risk microbe for molecular biology purposes. *Escherichia coli* does, however, have certain drawbacks. As an example, *E. coli* lacks the ability to influence gene splicing, and it can lead to a variety of additional issues that will be covered later on when using it for protein biological synthesis. Other cells that serve as hosts include yeast, insect, mammalian, and other types of cells. In *Escherichia coli*, the majority of genetic modifications are still carried out purely arbitrarily.

3.2.5. Cellular Transformation

The process of transformation introduces exogenous pDNA into host cells. Cells are prepared to accept additional exogenous pDNA by pre-treating them with an ice-cooled calcium chloride solution. The host cells have undergone cellular transformation and are now recombinant cells following a brief heat shock of 42°C for two minutes. Electroporation is an additional method for transforming cells. In electroporation, pDNA is combined with ice-cold cells, which are then put in a cuvette with a strong potential difference applied across it. This encourages a transient rise in the permeability of cell membranes, providing pDNA with a chance to enter the cytoplasm in a portion of host cells. After transformation, the bacteria are first cultured in growth medium (broth) for a long enough period of time to establish the expression of the selectable marker's genes and pDNA gene expression. Next, bacteria that are compatible with the selectable marker gene can be sown on an agar selection plate, allowing only transformed host cells with independently expressing pDNA to endure.

3.2.6. Selection

Genes that code for resistance to antibiotics like ampicillin, tetracycline, and kanamycin are commonly used as selectable markers. To enhance the selection process, other selectable markers, like those associated with the inactivation of β-galactosidase (β-gal) activity (lacZ selection), may also be employed. In this instance, the intact lacZ gene, which codes for the first 146 amino acids of β-gal, must be present on pDNA for the selection process to work. The chromogenic substrate 5-bromo-4-chloro-3-indolylβ-galactopyranoside (X-gal), which is colorless until it is hydrolyzed by the lacZ-expressing β-gal enzyme, is used to measure the expression of the lacZ gene. The lacZ gene is the site of successful ligation of heterologous DNA (from genomic or cDNA), thus ligation must prevent X-gal hydrolysis from occurring. In light of this, host cell colonies that were identified as white after transformation and plating on agar, X-gal-containing selection plates should have recombinant pDNA, whereas host cell colonies that turned blue should only have intact lacZ-expressing pDNA without any heterologous DNA included. The selection of LacZ is not entirely dependable due to two factors. Firstly, the β-gal activity is not always completely inactivated by insertions into the lacZ gene, and sometimes blue colonies do contain recombinant pDNA. Second, occasionally white colonies are white because of β-gal inactivation brought on by removing a small portion of the lacZ gene during clonal growth, even though they do not contain recombinant pDNA.

3.2.7. pDNA Purification

All forms of DNA are generally much more easily worked with and thermally stable than globular proteins. The traditional method of purifying pDNA from cells involved causing initial lysis, usually with high pH and sodium dodecyl sulfate (SDS) present. While most of the cellular proteins precipitate out of the cell wall material, genomic and pDNA remain in solution. Potassium acetate is used to neutralize the lysate after centrifugation. After using phenol, chloroform, and isoamyl alcohol to extract the cleared lysate containing the plasmid DNA, RNAse is usually added to eliminate any contaminating RNA. Ultracentrifugation caesium chloride gradient purification is an option when extremely pure pDNA is needed. The traditional method uses phenol, which is poisonous and unpleasant to handle, and is very laborious and time-consuming. As a result, the widespread use of alkaline lysis in conjunction with the miniprep system has greatly simplified the preparation and purification of pDNA. The procedure is carried out as follows: Clonal cells are first grown, usually overnight, at a small volume (several milliliters). The bacterial pellet is then again suspended in RNAse-containing buffer, which breaks down RNA. After that, cells are lysed with sodium hydroxide and SDS, which denatures proteins and opencircle genomic DNA but preserves supercoiled pDNA.

After neutralizing the sodium hydroxide with potassium acetate buffer, potassium dodecyl sulfate precipitates, which is then used to precipitate genomic DNA. Following clearing of the supernatant, the glass fiber filter frit made of silica is passed through, allowing pDNA to attach to the silica surface in the presence of the chaotropic reagent guanidinium isothiocyanate (GuNCS). In the absence of GuNCS, the frit is then cleaned and pDNA is eluted using a straightforward aqueous solution. There is enough DNA available for additional recombinant manipulations with this miniprep system of pDNA isolation and purification. If a specific pDNA is needed in large quantities, the procedure can be easily scaled up. The products of a PCR amplification reaction can also be purified using comparable guanidinium–silica frit preparation techniques, which eliminate excess deoxynucleotides and enzymes from the reaction.

3.2.8. Nucleic Acid Gel Electrophoresis

The process of moving charged molecules through a solid-phase medium while being affected by an electrical potential difference is known as electrophoresis. Agarose or polyacrylamide gels are the most common solid-phase media used in molecular biology. Agarose gel electrophoresis (AGE) uses agarose, a naturally occurring heteroglycan that has been previously described. The fundamental solid-phase medium for polyacrylamide gel electrophoresis (PAGE) is polyacrylamide, a synthetic polyamide polymer. PAGE is commonly used for the separation of smaller polydeoxynucleotides/ polynucleotides or oligodeoxynucleotides/ oligonucleotides (<100 bases), while AGE is typically used for the separation of larger polydeoxynucleotides/polynucleotides (>100 bases). Therefore, agarose is typically the preferred solid-phase medium for the separation of large nucleic acids. Compared to polyacrylamide gels, agarose gels are incredibly easy to make. A gel-slab mounted with a plastic comb is used to create sample wells. First, agarose is mixed with either tris acetic acid/ ethylenediaminetetraacetic acid (TAE) or tris boric acid/ethylenediaminetetraacetic acid (TBE) buffer. The heterogeneous mixture is then heated to dissolve the agarose and cooled to gelation. Compared to TBE, TAE produces larger agarose "pore sizes." As a result, TBE gels are typically used to resolve smaller DNA fragments and TAE gels to resolve DNA larger than 1,000 bases. But because DNA recovery from TBE–agarose gels is typically worse, TAE is also frequently utilized for the resolution of smaller DNA fragments.

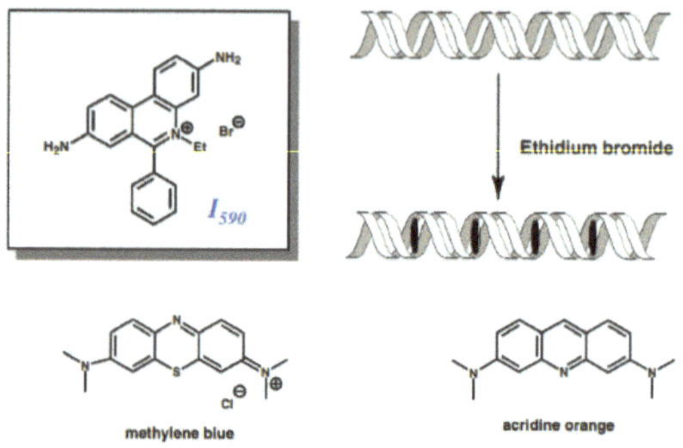

Figure 3.6. *The structure of ethidium bromide (EtBr) (left) is planar aromatic.*

Source: Andrew Miller and Julian Tanner, Essentials of Chemical Biology Structure and Dynamics of Biological Macromolecules, ISBN 978-0-470-84530-1.

Samples are combined with a loading buffer, which includes mobility dyes to facilitate sample loading and electrophoretic process monitoring, as well as glycerol or Ficoll to raise sample density so that samples sink easily into the wells, in order to run an agarose gel. Ethidium bromide staining can be used to visualize DNA following electrophoresis. Base pair intercalators, such as ethidium bromide, insert or intercalate between base pairs (maximum 1 ethidium bromide/6.5 base pairs) to significantly increase φ_F and, consequently, the intensity of fluorescence at I_{max} (595 nm). Ethidium bromide has strong mutagen properties. Thus, various staining techniques have been developed. These include stainings such as acridine orange, methylene blue, and silver. While ethidium bromide staining and silver staining are equally sensitive, the former can provide a permanent record of the gel while the latter cannot. Compared to ethidium bromide, methylene blue staining is 40 times less sensitive, but it also damages DNA less and is not carcinogenic. Since double-stranded DNA (dsDNA) fluoresces red under UV light and single-stranded DNA (ssDNA) fluoresces green, acryline orange can be used to distinguish between the two types of DNA. It is not too difficult to retrieve DNA fragments and samples from electrophoresis gels. Originally, there were many different ways to recover DNA from agarose gels; however, the GuNCS miniprep method has oftentimes replaced these methods. Here the strategy is to literally cut out the desired DNA band(s) and suspend in guanidinium thiocyanate buffer to dissolve the agarose gel before completing the rest of the purification procedure.

3.2.8.1. Factors Affecting Migration of Nucleic Acid

Nucleic acid migration can be influenced by a variety of factors, including the size of the gel pores, the voltage applied, the buffer's ionic strength, and the concentration of any intercalating dye used during electrophoresis, such as ethidium bromide.

3.2.8.1.1. Size of DNA

DNA is filtered through the gel based on molecule size, with smaller molecules passing through more quickly. The rate at which double-stranded DNA travels is roughly inversely proportional to the logarithm of base pair count. However, this relationship breaks down with very large DNA fragments, and standard agarose gel electrophoresis cannot separate them. The mobility of larger circular DNA may be more strongly impacted by the gel's pore size than that of linear DNA, and the resolution limit is dependent on the composition of the gel and the field strength. Pulse field gel electrophoresis (PFGE) is required for the separation of very large DNA fragments. Band inversion, in which large molecules move more quickly than small molecules, can occur in field inversion gel electrophoresis (FIGE), a type of pulsed field gel electrophoresis (PFGE).

3.2.8.1.2. Conformation of DNA

The way that DNA moves can be greatly influenced by its conformation; for example, supercoiled DNA moves more quickly than relaxed DNA because it is more compact and tightly coiled. Multiple forms of DNA may be present in a typical plasmid DNA preparation; the gel from the plasmids' electrophoresis would typically show a main band, which would be the negatively supercoiled form, with additional forms of DNA possibly appearing as minor, fainter bands. These minor bands can be either of the following: single-stranded DNA, which can occasionally appear depending on preparation methods, or nicked DNA (open circular form), which normally runs slower than supercoiled DNA. The relaxed closed circular form can also move ahead of the supercoiled DNA. However, the rate at which the different forms move can vary depending on the electrophoresis conditions used. For instance, linear DNA can move more quickly or more slowly than supercoiled DNA, and the gel's pore size can have a greater impact on the mobility of larger circular DNA than it does on linear DNA. The size of a circular DNA, such as a plasmid, can therefore be measured more precisely after it has been linearized by restriction digest, unless supercoiled DNA markers are employed.

In a dose-dependent manner, increased cross-linking-induced DNA damage will also lessen electrophoretic DNA migration.

3.2.8.1.3. Concentration of Ethidium Bromide

If ethidium bromide is present in the gel during electrophoresis, its concentration has a greater effect on circular DNA than linear DNA. All naturally occurring DNA circles are underwound, but the presence of ethidium bromide during electrophoresis can alter the movement of the DNA molecule in the gel because it intercalates into circular DNA and changes its charge, length, and superhelicity. DNA can transition from a negatively supercoiled molecule into a fully relaxed form and, at maximum intercalation, into a positively coiled superhelix by increasing the amount of ethidium bromide intercalated into the molecule. It is possible to separate circular DNA with various supercoiling topologies using agarose gel electrophoresis.

3.2.8.1.4. Gel Concentration

The gel's concentration controls its pore size, which has an impact on DNA migration. The percentage concentration of the gel affects the DNA's resolution. Larger DNA molecules can be separated by lowering the gel concentration, but smaller DNA molecules can be separated more effectively when the agarose concentration is increased. Large 5–10 kb DNA fragments can be well separated or resolved using a 0.7% gel concentration in a standard agarose gel electrophoresis, while small 0–2 kb DNA fragments can be well resolved using a 2% gel concentration. Very small fragments can be separated using a gel concentration of up to 3%, but a vertical polyacrylamide gel would be better suited for resolving small fragments. However, high percentage gels are frequently brittle and may not set evenly. High concentration gels require longer run times—sometimes even days. FIGE or PFGE should be used when running high percentage agarose gels. Gels with a low percentage (0.1% to 0.2%) are brittle and prone to breaking. Gels with a percentage of 1 are typical for numerous uses.

3.2.8.1.5. Applied Field

The rate of DNA migration at low voltages is proportional to the applied voltage, i.e., The DNA moves more quickly at higher voltages. Nevertheless, the mobility of high-molecular-weight DNA fragments increases differentially with increasing electric field strength, resulting in a decrease in the effective range of separation and a lower resolution at high voltage. It is advised to use 5 to 8 V/cm for standard gel electrophoresis when resolving DNA larger than 2 kb in size. Another limitation on voltage is that it heats the gel, which can result in melting if high voltage is applied to it for an extended length of time, especially for agarose gel with a low melting point.

DNA mobility, however, can alter in an unstable field. DNA of a given size may become much less mobile at a given cycling frequency in a field that is periodically reversed. In PFGE, this phenomenon may lead to band inversion, in which bigger DNA fragments travel more quickly than smaller ones.

3.2.8.2. Mechanism of Migration and Separation

During electrophoresis, the DNA is moved toward the positively charged anode by the negative charge of its phosphate backbone. Nevertheless, there is no separation by size when there is no gel matrix present because the migration of DNA molecules in solution during electrophoresis is independent of molecular weight. As streaming counterions traveling in the opposite direction cut off the hydrodynamic interaction between different parts of the DNA, there is no mechanism to produce a velocity-length dependence on a scale larger than the screening length of roughly 10 nm. This sets it apart from other processes where long-range hydrodynamic interaction plays a significant role, like sedimentation or diffusion.

Therefore, during electrophoresis, the gel matrix is in charge of separating DNA based on size; however, the exact mechanism behind this separation is not totally understood. There are several models explaining how biomolecules separate in gel matrixes. One popular model is the Ogston model, which views the polymer matrix as a sieve made up of a randomly distributed network of interconnected pores. Because larger molecules are more likely to be impeded and slowed down by collisions with the gel matrix, globular proteins or random coil DNA can pass through connected pores large enough to accommodate their passage. As a result, during the sieving process, molecules of different sizes can be separated.

However, the Ogston model fails to work for large molecules because the pores are much smaller than the molecule's size. The most widely used model for DNA molecules larger than 1 kb is the reptation model, or one of its variations. According to this model, DNA can pass through pores as an elongated molecule and crawl through them in a manner akin to a snake, a process known as "reputation." This changed into a biased reptation model at higher electric field strengths, where the molecule's leading end became strongly biased forward and pulled the rest of the molecule along with it. DNA that is larger than a specific size cannot be separated in the fully biased mode because the mobility has reached saturation. In practice, however, perfect parallel alignment of the chain with the field is not seen since that would imply equal mobility for both long and short molecules. The internal fluctuations of the chain are taken into consideration in the biased reptation model's further refinement.

The mobility of DNA in PFGE has also been explained by the biased reptation model. After a field is applied, the orientation of the DNA is gradually increased by reptation; the amount of time it takes for the orientation to reach its steady-state velocity depends on the size of the molecule. Larger molecules take longer to reorient when the field is altered, making it possible to distinguish between long chains that are unable to reach their steady-state velocity and short chains that travel mostly at a steady speed. There are, however, additional models.

During electrophoresis, more subtle dynamics were revealed by real-time fluorescence microscopy of stained molecules. The DNA exhibited significant elasticity, alternating between stretching in the direction of the applied field and contracting into a ball, or becoming hooked into a U-shape when it became entangled in the polymer fibers. We can refer to this observation as the caterpillar model. A different model suggests that the DNA entangles itself in the polymer matrix, and that the larger the molecule, the greater the likelihood of entanglements and impeded movement.

3.2.8.3. Visualization

Ethidium bromide, or EtBr for short, is the most widely used dye for agarose gel electrophoresis, which is used to visualize DNA or RNA bands. When intercalated into the major groove of DNA (or RNA), it fluoresces under UV light. When DNA is run through an EtBr-treated gel and exposed to UV light, bands containing more than 20 ng of DNA can be clearly seen. Because EtBr is known to be a mutagen, there are safer substitutes like GelRed, made by Biotium, which attaches itself to the minor groove.

Another dsDNA stain made by Invitrogen is called SYBR Green I. It is more costly than EtBr, but it is also 25 times more sensitive and might be safer because there is no information on its toxicity or mutagenicity in humans.

SYBR Safe is a SYBR Green variation that has been demonstrated to have toxicity and mutagenicity levels low enough to be classified as nonhazardous waste under U.S. Federal laws and regulations. It is substantially more expensive than SYBR Green, but it has sensitivity levels that are comparable to those of EtBr. Nonetheless, the expenses associated with disposing of EtBr can quickly surpass the initial cost differential in nations where the safe disposal of hazardous waste is required.

Before adding the mixture to the gel, scientists combine DNA with negatively charged loading buffers because EtBr-stained DNA is not visible in natural light. Loading buffers are helpful because they co-sedate with DNA, which means they travel at the same speed as DNA of a specific length, and they are visible in ambient light as opposed to UV light for EtBr-stained DNA. Common loading buffer dyes include xylene cyanol and bromophenol blue. They react with DNA fragments of 500 bp and 300 bp in length, respectively, at roughly the same rate, but the exact location changes with the gel's percentage. Other less common progress markers are Orange G and Cresol Red, which have respective runs of 50 and 125 bp.

Another method for achieving visualization is to move DNA from a SDS-PAGE to a nitrocellulose membrane and then expose it to a hybridization probe. We call this procedure "Southern blotting."

When using fluorescent dyes, the gel is exposed to ultraviolet light after electrophoresis; this is typically done by setting the gel on a light box and wearing protective gear to reduce UV radiation exposure. The imaging equipment that captures an image of the gel after it has been exposed to UV light is typically included in the illuminator apparatus. Because ethidium bromide has intercalated with DNA, it fluoresces reddish-orange when DNA is present. It is also possible to cut the DNA band off of the gel and dissolve it to extract the pure DNA. After that, the gel can typically be captured on camera, either digital or Polaroid. Despite the reddish-orange fluorescence of the stained nucleic acid, images are typically displayed in black and white. When DNA is damaged by UV light, it can become less effective to manipulate it later on, during ligation and cloning, for example. Greater harm is caused by shorter wavelength UV radiation (302 or 312 nm); for instance, exposure for as little as 45 seconds can drastically lower transformation efficiency. Therefore, exposure to shorter wavelength UV radiations should be limited if DNA is to be used for downstream procedures; instead, higher wavelength UV radiation (365 nm), which causes less damage, should be used. Shorter UV light wavelengths can be used for brief periods of time if capturing the gel image is essential, as higher wavelength radiations result in weaker fluorescence. The DNA may be shielded from harm by adding guanosine or cytidine to the electrophoresis buffer at a concentration of one milligram. An alternative is to use a blue light excitation source in conjunction with a blue-excitable stain like GelGreen or SYBR Green.

3.2.9. DNA Sequencing

The chain-termination sequencing method is the industry standard for sequencing DNA. Using the complementary DNA strand as a template, DNA polymerase is employed in this technique to rebuild duplex DNA for brief periods of time. The selection of the sequencing primer, which anneals by Watson-Crick base pairing to its complementary sequence in the complementary DNA strand and subsequently serves as an initiation point for DNA sense strand polymerization and duplex DNA rebuilding, determines the region of interest for sequencing.

Setting up four simultaneous reactions with equal, defined concentrations of primer, complementary DNA strand, and four deoxynucleotide triphosphate (dNTP) substrates was the standard procedure for many years. Only the low mole percent of a single radioactively labeled dideoxynucleotide (ddNTP) substrate—either ddATP, ddGTP, ddCTP, or ddTTP—distinguished each reaction from the others. Since these ddNTPs don't have a 2 or 3 hydroxyl group, adding them to a growing chain inhibited additional DNA polymerization. Therefore, it was anticipated that each DNA polymerization reaction would contain a variety of incomplete DNA sense strands that would terminate in a dideoxynucleoside once the specified short reaction period had passed. All incomplete strands were expected to terminate in ddA when ddATP was used, ddGTP when ddCTP was used, ddTTP when strands ended in ddT, and all incomplete strands where ddATP was used to terminate in ddG. The famous DNA sequencing ladder was created by resolving the DNA sense strand products of all four reactions side by side on a polyacrylamide gel and visualizing the radioactivity using a photographic plate. Shorter strands run faster along the ladder, so each line that goes up in order from the bottom of the gel can be used to identify a subsequent deoxynucleoside residue in the sense strand DNA sequence beginning with the primer. These days, sequencing is carried out using a single DNA polymerization reaction that contains all four ddNTPs that have been differentially labeled with four distinct fluorescent label types that are unique to each ddNTP. Capillary electrophoresis can now resolve the entire set of individual DNA sense strand products. This means starting from the primer, each successive fluorescent label resolved and identified from the start of the elution run (shorter strands run faster) can correlate with and identify the deoxynucleoside residue sequence of sense strand DNA. These days, read lengths usually range from 400 to 800 primer deoxynucleoside residues.

3.3. CLONING AND IDENTIFICATION OF GENES IN DNA

A basic procedure in molecular biology is the cloning and identification of genes in DNA, which entails isolating and amplifying a particular gene to examine its structure and function.

There are two main approaches for this.

1. **Direct DNA Cloning:** This involves 'cutting' DNA from a source (such as genomic DNA), and then 'inserting' the desired DNA into plasmid DNA that can be replicated (in effect multiple copied), usually in *E. coli*.
2. **Polymerase Chain Reaction (PCR):** PCR is an enzyme-based method for amplifying (multiple copying) DNA, from either a DNA template or an RNA template (RT-PCR).

3.3.1. Direct DNA Cloning

Although it can also be used to isolate and clone eukaryotic genes, direct DNA cloning was initially developed for the isolation and cloning of prokaryotic genes. The isolation and multiple replications of DNA from a source of interest (i.e., genomic DNA primarily purified directly from a prokaryotic organism, or DNA established in a eukaryotic cDNA library) are referred to as direct DNA cloning. Once DNA has been isolated from prokaryotic organisms, it needs to be broken up into fragments, or "shot-gun" fragments, using restriction enzymes or chemicals. Recombinant pDNA constructs are then made by ligating restriction fragments of genomic DNA into pDNA vectors. To create alternative eukaryotic recombinant pDNA constructs, cDNA library components can be similarly inserted into pDNA with respect to eukaryotic organisms.

Recombinant DNA is made up of segments of DNA from two or more distinct organisms. These recombinant pDNA constructs are used to transform host cells, usually *Escherichia coli*. As the host grows, simple DNA replication allows for multiple copies of the recombinant pDNA constructs to be produced. The host cannot tell the difference between self- (host) DNA and recombinant DNA because the genetic code is universal. As a result, even the recombinant pDNA is replicated during host growth and division. When *Escherichia coli* cells are placed on a nutrient plate and given time to grow, colonies will form in which every cell is a clone of the original transformed cell and comes from a single parent cell. Recombinant pDNA (as a single plasmid or multiple copy/ clonal cell) containing a single, distinct restriction fragment that is absent from all other *Escherichia coli* colonies will be present in every cell of the colony. Since each clone contains a distinct restriction fragment from the original genomic DNA or cDNA of interest, the outcome is referred to as a clonal library.

After creating the library, it is necessary to determine which clone contains the desired gene. The traditional method involves using a synthetic, single-stranded DNA (ssDNA) molecule, known as a probe, which is complementary to a DNA sequence in the gene of interest. This probe will bind specifically to the target gene through base pairing in a process called hybridization. There are four general approaches to designing the probe in this scenario.

1. If the amino-acid sequence is known for the desired protein, then oligodeoxynucleotide hybridization probes may be designed with reference to the genetic code.
2. If the gene has previously been cloned from a related organism then a previously used probe (known as a heterologous probe) may be tried.
3. If a protein is known to be abundant in a particular cell/tissue, then this abundance may also be reflected in its statistical abundance within a clonal library.
4. If the protein of interest has been purified before, and an antibody has been raised against the protein, then the clonal library will have to be converted into an expression library (see later) and the presence of the desired protein is then screened using this antibody.

Clearly options (1) and (2) are much preferred since they can give unambiguous identification of a gene of interest within a very small number of clones (possibly one clone) within the clonal library.

3.3.2. Polymerase Chain Reaction

Since its development in 1986, the polymerase chain reaction (PCR) has completely changed biological research. The process of amplifying (multiply copying) DNA from DNA or RNA sources is known as PCR. While PCR has many uses, eukaryotic gene isolation and cloning from cDNA libraries is one of its most significant functions. The following is the PCR mechanism. For sense and complementary DNA strands to dissociate—a process known as DNA melting or denaturation—duplex template DNA (such as a cDNA library) is first identified and heated to a very high temperature (94°C). Then, two short oligodeoxynucleotides called primers are introduced to define the region to be amplified. These primers can anneal, or bind, antiparallel to mutually complementary sequences in either DNA strand of the split duplex template DNA when the temperature is lowered to 50–60°C, depending on the situation, using Watson–Crick base pairing. After that, the temperature is increased to about 74°C, which is the ideal range for a thermostable DNA polymerase to function at Taq polymerase gets its name from the fact that this enzyme was first isolated from the hyperthermophile Thermus aquaticus.

Beginning with primers that annealed to the complementary or sense strands of the original duplex DNA template, Taq polymerase reconstructs duplex DNA. Using the deoxynucleotide triphosphates dATP, dCTP, dGTP, and dTTP (commonly known as the dNTPs) as substrates for

DNA synthesis and duplex reconstruction, the enzyme works in a 5' → 3' pattern. We refer to this catalytic process as extension. PCR is a technique for semi-conservative replication because it creates two new duplex or double-stranded DNA (dsDNA) copies from the original duplex DNA template (parent original), each of which contains one DNA strand from the parent original template. After repeating the denaturing, annealing, and extension steps 20–30 times on average, the amount of dsDNA product grows exponentially, producing a sizable amount of dsDNA that has been amplified selectively in accordance with the two primers' designs.

PCR has its limitations despite its power. Initially, the base sequences of the regions in the duplex DNA template that border the target region of interest should be ascertained as possible in order to design the best primers. Second, amplifying very long DNA segments is challenging. Since Taq polymerase is not totally faithful (i.e., the fidelity rate is not 100%), the possibility of introducing mutations increases as the region of amplification increases further and further. In actuality, PCR amplifications of up to 3000 base pairs (3 kb) can be completed quickly and easily. With caution and expertise, this can be increased to 10 kb. The maximum size of a region that can be amplified by PCR is approximately 40 kb. These limitations don't, however, present a significant issue for the majority of applications. A protein with about 2000 amino acid residues, or 240 kDa, should be encoded by a 6 kb DNA coding sequence, assuming an average molecular weight of 120 Da per amino acid residue. Thus, direct cloning may be avoided if the gene is well-known and has already been cloned from a related source, or if aminoacid sequences are known. This is one of the primary benefits of PCR in the isolation and cloning of genes. Here, a gene of interest may be directly amplified by PCR using heterologous probes from prokaryotic genomic DNA or a eukaryotic cDNA library, negating the need for any of the previously mentioned direct cloning techniques. Then, regardless of the source of the duplex DNA template, PCR amplification proves to be a powerful method for obtaining quantities (μg–mg) of a desired gene. Naturally, there is no other option than to use direct cloning techniques to isolate, clone, and identify a gene of interest from either genomic DNA or cDNA libraries as appropriate if a gene of interest in prokaryotic genomic DNA or eukaryotic cDNA exceeds the limits of PCR amplification.

3.3.3. Gene Expression and Expression Vectors

Seeking high levels of gene expression and the purification of pure protein as a result come after the isolation, cloning, and identification of genes of interest from within genomic DNA or cDNA libraries by direct cloning or PCR amplification.

3.3.3.1. Expression Vectors

Usually, a pDNA expression vector designed specifically for high-level gene expression is used to express proteins in *Escherichia coli*. These pDNA expression vectors have the following properties and are usually in the range of 5000 base pairs in size. Regarding pDNA cloning vectors, both a selectable marker and an origin of replication are still necessary. In this instance, the palindromic sequence 5'-d(pGATC)-3,' which can be methylated on guanine to regulate replication, is also present in multiple copies at the origin of replication, which is a dA = dT base pair rich region. One well-known method of stopping replication and rendering a pDNA vector inactive with regard to replication is DNA methylation at the replication origins. The 468-base-pair non-coding region oriC, which supports a single copy per cell when incorporated into pDNAs, is the naturally occurring origin of replication in the *Escherichia coli* chromosome (there is only one origin of replication in *Escherichia coli* for the entire chromosome). A protein complex called the pre-replication complex attaches itself to oriC to facilitate pDNA replication.

The pre-replication complex has less control over replication when this oriC, also known as ori, origin of replication is diminished, and multiple copies of pDNA are found in each cell. We've already covered the significance and use of the selectable marker for a pDNA vector. In cases where the selectable marker confers resistance to antibiotics like tetracycline, kanamycin, and ampicillin, it is commonly referred to as Tet^R, Kan^R, and Amp^R, respectively. The most widely used marker, Amp^R, is associated with the β-lactamase gene. A pDNA expression vector's multiple cloning site (MCS) is another crucial component. In order to facilitate the ligation of the heterologous gene of interest that has been isolated, cloned, and identified from either genomic DNA or a cDNA library by direct cloning or direct PCR amplification, this region is prepared from a synthetic, heterologous DNA fragment engineered to contain as many unique (to the pDNA) restriction sites as possible.

To "complete the tour," additional significant regions found in a pDNA expression vector must be discussed. The promoter is foremost. Promoters are non-coding DNA segments that are found upstream of a heterologous gene of interest, that is, close to the 5 -side of the sense strand of pDNA. Essential to any pDNA expression vector, the promoter region regulates the transcription of the heterologous gene (5'→ 3') into mRNA. Early on in the field of molecular biology, several naturally occurring E. It was found that phage promoters like the λpL promoter of bacteriophage λ and coli promoters like the lac and trp promoters were present. In more recent times, sequence elements derived from original promoters have been fused to create artificial promoters, like the tac promoter. It's possible that promoters are inducible, meaning that transcription can be initiated by an outside agent. When IPTG (isopropyl-β-D-thiogalactopyranoside) is added to the growth medium, for example, the lac and tac promoters start to transcribe. Similar to this, adding indole acetic acid (IAA) activates the trp promoter, and briefly increasing the growth temperature to 42°C activates the λ_{pL} promoter by deactivating a transcription repressor encoded in phages. The T7 promoter is a potent, contemporary promoter element that must be used in conjunction with bacteriophage T7 RNA polymerase.

All of the enzymes that convert DNA to mRNA are known as RNA polymerase enzymes, or more precisely, DNA-dependent RNA polymerases. To start transcription, RNA polymerases bind at promoter sequences. An RNA polymerase's promoter structure is specific to that enzyme. Having said that, promoters share additional characteristics, such as TATA boxes (also called Hogness boxes in eukaryotes and Pribnow boxes in prokaryotes). A 10 bp deoxynucleotide region found within a promoter with a predominance of dA = dT base pairs makes up a TATA box. Consequently, transcription can start in this region because the duplex DNA structure is primed to split in response to RNA polymerase "invasion." In this process, transcription factors are also involved, particularly in eukaryotes. These factors are classified as proteins that either increase or decrease the amount of RNA polymerase that binds to promoters to control transcription. Up until a terminator sequence is reached, which indicates RNA polymerase dissociation, transcription proceeds in the 5'→ 3' direction. Many mRNA transcripts in prokaryotes are polycistronic, which means that multiple protein coding regions are sequentially incorporated into a single primary mRNA transcript. Since these types of proteins are frequently coded for by related functions, co-transcribed genes are referred to as belonging to an operon. Operator sites are typically located close to promoters. These are the binding sites for repressor proteins, which can bind to them to modify the transcription of genes, including operon genes. The process of RNA polymerase transcription is physically stopped as part of the repression mechanism. As a result, it is now known that small molecular transcription inducers like IPTG and IAA stimulate or induce transcription by directly binding to repressor proteins, preventing these proteins from binding to operator sites and suppressing transcription.

3.3.3.2. Protein Expression Strategy

Different methods exist for expressing proteins in *Escherichia coli*. Using specifically created pDNA expression vectors, like the pET vector family, which are freely accessible from commercial sources, is a contemporary method of expressing proteins. Since the T7 RNA polymerase promoter for transcription is present in this family of pDNA expression vectors, the corresponding *Escherichia coli* host should be a recombinant strain designed to express the heterologous T7 RNA polymerase constitutively. The promoterlacUV5 regulates this chromosomal T7 RNA polymerase. The lactose repressor, which regulates transcription from this lacUV5promoter, is the product of the LacI gene. When IPTG is added to recombinant *Escherichia coli* growth medium, transcription (and translation) of the T7 RNA polymerase is promoted. IPTG is an inducer of transcription from Lacfamily promoters by binding to the lactose repressor protein. Accordingly, the T7lac promoter, which is found in pET vectors, is a hybrid of the lac and wild-type T7 promoters. This allows T7 to function as an RNA polymerase. Because this is a lac family promoter as well, IPTG induces transcription (and subsequently translation) of the target heterologous gene by T7 RNA polymerase. A ribosome binding sequence (RBS) site found in the pET vector family enables post-transcriptional attachment of the primary mRNA transcript to *Escherichia coli* ribosomes. The heterologous gene of interest's RBS site is located seven to nine deoxynucleotides upstream, or before the fifth end of the gene, from the start codon. This is after the promoter. The MCS is used to insert this gene of interest into a pET family vector. Past the MCS is a transcriptional terminator site that is situated downstream (i.e., post 3 end of gene) from the heterologous gene of interest's stop codon. The location where the T7 RNA polymerase completes transcription is indicated by this terminator site.

3.3.3.3. Cloning for RNA Synthesis

A basic *in vitro* cell free transcription system can be used to synthesize large RNA molecules. Transcription of RNA from DNA is far simpler than translation of RNA into protein. Usually, a promoter like the T3, T7, or Sp6 promoter is used to insert the DNA coding for the desired RNA into pDNA. The most suitable DNA-dependent RNA polymerase for the promoter—such as T3 RNA polymerase, T7 RNA polymerase, or Sp6 RNA polymerase—is then used to synthesize RNA from the pDNA *in vitro*.

3.4. INTEGRATING CLONING AND EXPRESSION

For new chemical biology readers, the challenge lies in understanding the numerous tools and techniques used in molecular biology. The transition from chemistry to molecular biology can seem overwhelming at first, with its different tools, techniques, and language. However, once mastered, molecular biology can prove to be a valuable and efficient tool for chemical biology, easily integrated with minimal difficulty.

Large amounts of the iron-containing enzyme *Escherichia coli* iron-superoxide dismutase, or SodB, are needed in our example in order to remove superoxide radicals in order to detoxify. Cloning the relevant gene from the appropriate organism must be the first goal. Next, the gene must be expressed in fusion with an appropriate tag to aid in purification and boost the protein product's solubility. Basic details about the protein and the gene can be found using web-based *Escherichia coli* databases. For example, the protein has 192 amino acids (and is therefore coded by 576 bases), a molecular weight of about 21 kDa, and a theoretical pI of 5.9. For the purpose of identifying and cloning the gene, a PCR amplification approach is recommended due to the protein's size. Consequently, creating primers to amplify the gene from a suitable genomic source ought to be the first step. It is necessary to find specific restriction sites in the primers to facilitate the PCR product's subsequent easy insertion into a pDNA expression vector. Additionally, adding an N-terminal GST-tag in this instance will help the pDNA expression vector express a GST–sodB fusion gene rather than the sodB gene, which will result in a more soluble product upon translation and solve the SodB solubility issues during purification. Therefore, in order for the PCR product to be inserted by ligation, the pDNA expression vector must have a GST-tag sequence in frame with a downstream MCS region (i.e., post-3 end of the GST-tag deoxynucleotide sequence).

3.4.1. Designing Forward and Reverse Primers

In order to guarantee that a distinct Eco RI restriction site is active—as sites in terminal positions might not be usable restriction sites—the forward primer, which is initiated before the gene's 5 end, is constructed with three or four deoxynucleotide residues before it. The forward primer then ends with the first 20 deoxynucleotides (5'→ 3') of the sodB gene sequence that is known to exist. The Eco RI restriction site is exclusive to the chosen pDNA expression vector and is not present in the sequence of the sodB gene. The primer in reverse (i.e., likewise constructed with three or four deoxynucleotides before a distinct Xho I restriction site, and the first 20 deoxynucleotides (5'→3') of the known complementary sodB gene sequence. This construction is started before the 5 end of the complementary gene. It is evident that the Xho I restriction site is specific to the pDNA expression vector that is intended to be utilized and is not present in the corresponding sequence of the sodB gene.

As a result, the forward primer is needed to start the synthesis of the sense strand of DNA using the complementary (antisense) strand as a template. On the other hand, the sense strand of DNA serves as a template for the complementary strand DNA synthesis that is started by the presence of the reverse primer. When attempting primer design for the first time, it is usually easiest to sketch out the exact situation to ensure primer orientation. Additionally, by extending or shortening those deoxynucleotide stretches post the unique restriction sites as necessary, it is necessary to ensure that the melting temperatures of the forward and reverse primers are as similar to one another as possible. To maintain equal amounts of complementary strands and sense DNA synthesis during PCR amplification, this is crucial.

3.4.2. PCR Amplification and Product Isolation

The best method for achieving PCR amplification of a gene from prokaryotic DNA is to first isolate the genomic DNA from the target prokaryote in the manner described below. In this instance, amplifying the sodB gene sequence from E is necessary. Coli cells. In order to free the genomic DNA from DNAses without causing any degradation, these cells are first rinsed with PBS buffer and then heated to 95°C for a short while. The liberated genomic DNA, primers, and all other materials required for PCR amplification (Taq polymerase, four dNTPs, and buffer containing Mg^{2+}) can then be combined. It may be necessary to make some adjustments to the conditions and parameters, particularly the number of cycles and primer melting temperatures. The temperature at which the primer and template are 50% complexed is represented by this melting temperature. One can compute this using the primer sequence. Nonetheless, 5°C less than the melting temperature is the ideal annealing temperature. Moreover, excessively high melting temperatures will hinder sufficient primer binding to the template DNA. Primers will attach to complementary DNA sequences that are similar to or identical to one another, so if the temperature is too low, non-specific sequences will be amplified. 25 amplification cycles would be a typical number of cycles. This number is crucial because an excessive number of amplification cycles increases the risk of PCR-based mutations, whereas an insufficient number of amplification cycles may result in an insufficient amount of PCR amplification product to handle.

To verify that the PCR amplification was successful, part of the reaction mix might be run on an agarose gel. A clean band on a stained agarose gel that corresponds to the amplified sodB gene fragment, which is roughly 600 bp in length, should be visible if PCR was successful. The guanidinium isothiocyanate–silica method would then be the most straightforward way to purify this sodB gene fragment.

3.4.3. Ligation and Transformation

If necessary, additional pDNA expression vectors can be acquired through mini-prep preparation and digestion using Eco RI and Xho I restriction enzymes. The sodB gene fragment can also be digested with these enzymes. Complete digestion is crucial in both cases, and the resulting products should be purified through agarose gel electrophoresis. Once the main fragments are purified, they can be ligated together using T4 DNA ligase. In our specific example, we chose the pET pDNA expression vector to ligate the PCR-generated sodB gene fragment in frame with a GST tag, creating a GST– sodB fusion gene. The recombinant pDNA expression construct should be used to transform CaCl2 competent *Escherichia coli* cells, or other suitable host cells through electroporation. It is recommended to use an *Escherichia coli* DH10β strain that is proficient at taking up pDNA for transformation. After transformation, allow the cells to grow at 37°C without antibiotic for an hour, then spread them on agar plates for antibiotic selection and further incubation at 37°C for 16 hours.

3.4.4. Validation and Sequencing

A successful transformation using a recombinant pDNA expression construct should produce colonies on the antibiotic selection plate, with none on the control plate. The pDNA from these colonies should be validated and sequenced before further steps. To accomplish this, colony samples are selected and grown in selective media (5 ml). The pDNA is then isolated using the guanidinium isothiocyanate–silica method and digested with Eco RI and Xho I restriction enzymes to confirm the presence of two DNA fragments, one approximately 600 bp long. Subsequently, the recombinant 600 bp insert containing the sodB gene should be fully sequenced using designed sequencing primers, demonstrating successful cloning of the sodB gene from genomic DNA without PCR errors.

3.4.5. Protein Expression

To enable overexpression of the GST–sodB fusion gene, a host *E. coli* strain with T7 RNA polymerase is needed for optimal gene expression using the pDNA expression vector (pET family). The *E. coli* BL21 strain is recommended as it contains a chromosomal copy of the heterologous enzyme necessary for protein expression. After validating and sequencing the recombinant pDNA construct, competent *E. coli* BL21 cells are transformed and grown on antibiotic selection plates to select recombinant colonies. Once colonies are chosen, they are grown on a small scale until saturation (with antibiotics), and then on a larger scale until the A600 absorbance reaches approximately 0.6. At this point, IPTG is added to induce significant transcription of the GST–sodB fusion gene. Following several hours of further cell growth, samples are collected and analyzed by SDS-PAGE to confirm the presence of a protein overexpression band at the expected mass of 48 kDa for the GST–SodB fusion protein. If the band is visible, protein purification can proceed.

3.4.6. Cloning and Expressing From Eukaryotic Genes

To circumvent the intron issue, PCR amplification from cDNA libraries is the most widely used cloning method for higher eukaryotes. The fact that eukaryotic proteins usually do not fold correctly in E presents another difficulty when expressing them.

Coli and will not be processed using the appropriate post-translational modifications; consequently, it is frequently necessary to use alternative eukaryotic expression hosts. Yeast cells, insect cell lines, or mammalian cell lines are among the options.

3.5. SITE-DIRECTED MUTAGENESIS

A useful method for studying the structure and function of proteins is to intentionally alter or mutate the amino acid sequence, then compare the properties of the mutant proteins to those of the wild-type proteins. Prior to the development of recombinant DNA technology, various chemicals were employed to change particular amino acids, or rudimentary techniques like UV light were used to randomly alter the DNA coding sequence. However, with the use of recombinant tools and techniques, mutagenesis has never been simpler. As a result, site-directed mutagenesis can be carried out using PCR or without it.

One of the most significant lab methods for introducing mutations into DNA sequences and producing DNA libraries is site-directed mutagenesis. Although there are many ways to accomplish site-directed mutagenesis, artificial gene synthesis is now occasionally utilized as a substitute for site-directed mutagenesis due to the falling costs of oligonucleotide synthesis. Since 2013, the prokaryotic viral defense system that underpins the CRISPR/Cas9 technology has made it possible to edit the genome and carry out mutagenesis *in vivo* relatively easily.

3.5.1. History

Early attempts at mutagenesis produced random mutations because they were non-site-specific and used chemical or radiation mutagens. Later, localized point mutations were created using analogs of nucleotides and other chemicals, such as aminopurine, nitrosoguanidine, and bisulfite. In the lab of Charles Weissmann, site-directed mutagenesis was accomplished in 1974 through the use of the nucleotide analog N4-hydroxycytidine, which causes the transition of GC to AT. However, the types of mutations that can be produced by these mutagenesis techniques are restricted, and their specificity is not as great as that of later site-directed mutagenesis techniques.

Restrictions nucleases and small phage ϕX174 fragments can be used to create mutants, as demonstrated in 1971 by Clyde Hutchison and Marshall Edgell. Later, in 1978, Hutchison and his associate Michael Smith developed a more adaptable method of site-directed mutagenesis by employing oligonucleotides in a primer extension technique with DNA polymerase. Michael Smith later shared the Chemistry Nobel Prize in October 1993 with Kary Banks Mullis, the creator of the polymerase chain reaction, in recognition of his contribution to the process' development.

3.5.2. Basic Mechanism

The creation of a brief DNA primer is necessary for the fundamental process. In order to hybridize with the DNA in the target gene, this synthetic primer has the desired mutation and is complementary to the template DNA surrounding the mutation site. A point mutation is a single base change that can occur alone or in combination with other base changes, deletions, or insertions. After that, a DNA polymerase is used to extend the single-strand primer and copy the remaining portion of the gene. After being copied in this way, the gene with the mutated site is inserted using a vector into a host cell and cloned. In order to ensure that the mutants have the desired mutation, they are finally chosen through DNA sequencing.

3.5.3. Approaches

Due to a low yield of mutants, the original single-primer extension method was inefficient. A mixed population of mutant and non-mutant progenies is created by this mixture, which includes both the original unmutated template and the mutant strand. Additionally, the mutant strand is unmethylated whereas the template is methylated. Additionally, the presence of a mismatch repair system may cause the mutants to be counter-selected, leading to a reduction in the number of mutants, as the system favors the methylated template DNA. Since then, numerous strategies have been created to increase mutagenesis's effectiveness.

Many techniques exist for achieving site-directed mutagenesis, but since the early 2000s, most of them have hardly been employed in lab settings due to the availability of more straightforward methods for introducing site-specific mutation into genes.

3.5.3.1. Kunkel's Method

Thomas Kunkel presented a method in 1985 that lessens the requirement for mutant selection. After being inserted into a phagemid like M13mp18/19, the DNA fragment that needs to be mutated is changed into an *E. coli* strain lacking in the enzymes uracil deglycosidase (udg) and dUTPase (dut). Through the spontaneous deamination of dCTP to dUTP, both enzymes are a part of a DNA repair pathway that shields the bacterial chromosome from mutations. Because there is insufficient dUTPase, dUTP accumulates at a high level within the cell and cannot be broken down. The inability of uracil deglycosidase to function results in the uracil being left in newly synthesized DNA. Because of this, the enzymatic machinery of the double-mutant *Escherichia coli* may mistakenly incorporate dUTP rather than dTTP during the replication of the phage DNA, producing single-strand DNA that contains some uracils (ssUDNA). After the bacteriophage is released into the medium, its ssUDNA is extracted and used as a template for mutagenesis. For primer extension, an oligonucleotide with the intended mutation is utilized. One parental non-mutated strand with dUTP and one mutated strand with dTTP make up the heteroduplex DNA that forms. Next, an *Escherichia coli* strain with the wildtype dut and udg genes is created using the DNA. In this case, the parental DNA strand containing uracil is broken down, resulting in almost all of the resulting DNA being made up of the mutant strand.

3.5.3.2. Cassette Mutagenesis

Unlike other techniques, primer extension with DNA polymerase is not required for cassette mutagenesis. This process involves creating a DNA fragment and inserting it into a plasmid. It entails the ligation of two complementary oligonucleotides carrying the mutation in the target gene to the plasmid after a restriction enzyme cleaves a region in the plasmid. The sticky ends of the plasmid and insert can ligate to each other because restriction enzymes that cut at both the oligonucleotide and the plasmid are typically the same. The effectiveness of this method, which can produce mutants almost 100% of the time, is dependent on the availability of appropriate restriction sites that surround the desired mutation site.

3.5.3.3. PCR Site-Directed Mutagenesis

By employing polymerase chain reaction with oligonucleotide primers, it is possible to get around the restriction sites limitation in cassette mutagenesis and produce a larger fragment that spans two useful restriction sites. Through the process of exponential amplification in PCR, a fragment containing the desired mutation is generated in large enough quantities to be separated by gel electrophoresis from the original, unmutated plasmid. This fragment can then be reinserted into the original context using conventional recombinant molecular biology techniques. The same method exists in numerous variations. The most straightforward approach positions the mutation site near one of the fragment's ends, where it is present in one of the two oligonucleotides that were used to create the fragment. Even though only one PCR step is required, unless an extremely long primer is used, there is always the inherent issue of needing a suitable restriction site close to the mutation site. Therefore, other variations use three or four oligonucleotides. The mutagenic oligonucleotide may be complementary to a location within that fragment that is far away from any convenient restriction site, while the other two oligonucleotides may be non-mutagenic oligonucleotides that cover two convenient restriction sites and generate a fragment that can be digested and ligated into a plasmid. These techniques necessitate several PCR cycles in order for the final fragment to be ligated to carry the intended mutation. It can be difficult to design a fragment with the desired mutation and the appropriate restriction sites. SDM-Assist and other software tools can make the process easier.

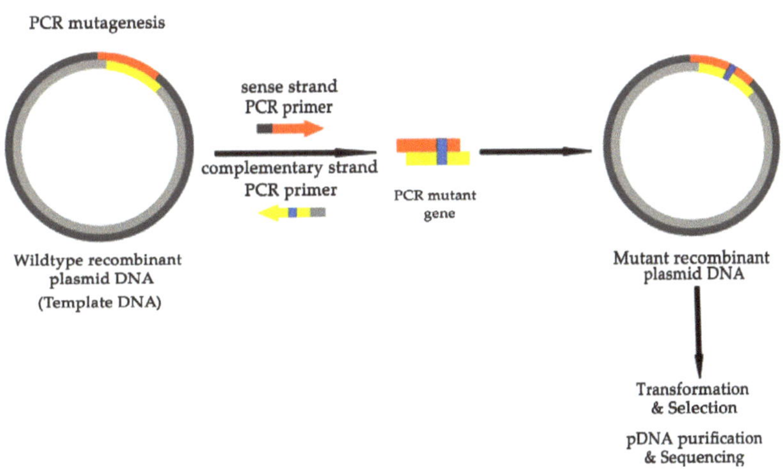

Figure 3.7. *PCR mutagenesis.*

Source: Andrew Miller and Julian Tanner, Essentials of Chemical Biology Structure and Dynamics of Biological Macromolecules, ISBN 978–0-470–84530–1.

3.5.3.4. Whole Plasmid Mutagenesis

Other site-directed mutagenesis methods for plasmid manipulations have mostly been replaced by methods that are commercially available as kits and are very simple, easy to use, and highly efficient. One of these methods is the Quik-change method, which uses a high-fidelity non-strand-displacing DNA polymerase like Pfu polymerase to amplify the entire plasmid in a thermocycling reaction using two complementary mutagenic primers. The result of the reaction is circular, nicked DNA. Enzymatic digestion using a restriction enzyme specific to methylated DNA, like DpnI, is required to remove the template DNA. Most *Escherichia coli* strains produce DNA that is methylated in its entirety; as a result, the template plasmid that is biosynthesized in the bacteria will be digested, but the mutated plasmid, which is generated *in vitro* and is therefore unmethylated, will remain undigested. It should be noted that although the thermocycling reaction can be employed in these double-strand plasmid mutagenesis techniques, if the two primers are formulated in a way that allows them to bind symmetrically to the same area surrounding the mutagenesis site, as per the original protocol, the DNA will not be amplified exponentially. Since there is no chain reaction in this instance of amplification, calling the process a PCR would be incorrect. Instead, it should be called linear amplification. However, the 3' region of the primers can bind to the amplified products as well if they are made to bind in an offset manner such that the mutagenesis site is near to the 5' end of both primers. This will result in the observation of exponential product formation. The term Quik-change comes from the registered trademark Quik-Change mutagenesis for site-directed mutagenesis kits owned by Stratagene, which is currently Agilent Technologies. The technique was created by Stratagene scientists (Papworth, Bauer, Braman, & Wright, 1996).

The extension reaction should be carried out at the suggested temperature of 68°C because Pfu polymerase can become strand-displacing at higher extension temperatures (≥70°C), which could lead to the experiment's failure. It has been noted that using this technique can result in the insertion of multiple primer copies in some applications. SPRINP, a variant of this technique, has been applied to various forms of site-directed mutagenesis and stops this artifact.

In plasmid mutagenesis, alternative methods such as scanning mutagenesis of oligo-directed targets (SMOOT) can combine mutagenic oligonucleotides in a semi-random manner. With this method, plasmid mutagenesis libraries containing single mutations or complete codon mutagenesis across a gene can be produced.

3.5.3.5. In Vivo Site-Directed Mutagenesis Methods

- Delitto perfetto
- Transplacement "pop-in pop-out"
- Direct gene deletion and site-specific mutagenesis with PCR and one recyclable marker
- Direct gene deletion and site-specific mutagenesis with PCR and one recyclable marker using long homologous regions
- *In vivo* site-directed mutagenesis with synthetic oligonucleotides

Since 2013, the efficient introduction of different mutations into the genomes of a wide range of organisms has been made possible by the development of CRISPR-Cas9 technology. The technique is the recommended method for genome editing because it leaves no marker and doesn't require a transposon insertion site.

3.5.4. Applications

Protein engineering, or site-directed mutagenesis, is a technique used to create mutations that could result in a well-designed protein with enhanced or unique features.

3.5.4.1. Tools for Investigation

Particular mutations in DNA enable logical research into the characteristics and operation of a protein or DNA sequence. Moreover, site-directed mutagenesis in proteins that results in single amino acid alterations can shed light on the significance of post-translational modifications. For example, phosphorylation can be studied when a specific serine (phosphoacceptor) in a substrate protein is changed to an alanine (phospho-non-acceptor), which prevents a phosphate group from attaching. This method has been used to identify the kinase HIPK-induced phosphorylation of the protein CBP. Site saturation mutagenesis is an additional thorough method in which a single codon or a group of codons can have all possible amino acids substituted at the designated positions.

3.5.4.2. Commercial Applications

It is possible to engineer proteins to create mutant forms that are suited for a particular use. Common laundry detergents, for instance, might contain subtilisin, a protein whose wild-type form contains a methionine that bleach can oxidize and greatly reduce the protein's activity. The protein will remain active in the presence of bleach if this methionine is swapped out for alanine or another residue that is resistant to oxidation.

3.5.5. Gene Synthesis

Artificial synthesis of an entire gene is now a practical technique for introducing mutations into genes, as the cost of synthesizing DNA oligonucleotides continues to decline. With this technique, extensive mutagenesis across multiple sites is possible, including a complete redesign of a gene's codon usage to optimize it for a specific organism.

CASE STUDY

THE ORDER OF NUCLEOTIDES IN A GENE REVEALED BY DNA SEQUENCING

All of the information needed to build and maintain an organism — whether it's a human, a dog, or a bacterial cell — is contained in its DNA. DNA molecules are composed of four nucleotides, and these nucleotides are linked together much like the words in a sentence. Together, all of the DNA "sentences" within a cell contain the instructions for building the proteins and other molecules that the cell needs to carry out its daily work.

- How Do Researchers "Read" Gene Sequences?

Determining the order of the nucleotides within a gene is known as DNA sequencing. The earliest DNA sequencing methods were time-consuming, but a breakthrough came in 1975 with the development of the process called Sanger sequencing. Sanger sequencing is named after English biochemist Frederick Sanger, and it is sometimes also referred to as chain-termination sequencing or dideoxy sequencing. Some 25 years after its creation, the Sanger method was used to sequence the human genome, and, with the addition of many technological improvements and modifications, it remains an important method in laboratories across the world today.

- How Does Sanger Sequencing Work?

Sanger sequencing is modeled after the natural process of DNA replication, and it uses dummy nucleotides to stop replication whenever a specific nucleotide is encountered. Because this truncated replication occurs over and over again, nucleic acids of varying lengths accumulate and can be used to determine the position of each nucleotide in the sequence.

- **Understanding DNA Replication:** In order to understand how Sanger sequencing works, its first necessary to understand the process of DNA replication as it exists in nature. DNA is a double-stranded, helical molecule composed of nucleotides, each of which contains a phosphate group, a sugar molecule, and a nitrogenous base. Because there are four naturally occurring nitrogenous bases, there are four different types of DNA nucleotides: adenine (A), thymine (T), guanine (G), and cytosine (C). Within double-stranded DNA, the nitrogenous bases on one strand pair with complementary bases along the other strand; in particular, A always pairs with T, and C always pairs with G. Then, during DNA replication, the two strands in the double helix separate. This allows an enzyme called DNA polymerase to access each strand individually. As the DNA polymerase moves down the single-stranded DNA, it uses the sequence of nucleotides in that strand as a template for replication. Thus, whenever the DNA polymerase recognizes a T in the template strand, it adds an A to the complementary daughter strand it is building; similarly, whenever it encounters a C in the original strand, it adds a G to the daughter strand. This process happens along both strands simultaneously, resulting in the eventual production of two double-helical molecules, each of which contains one "old" strand and one "new" strand of DNA.

- **Setting up the Sequencing Experiment:** The Sanger method relies upon a variation of the replication process described above in order to determine the sequence of nucleotides in a segment of DNA. Before Sanger sequencing can begin, however, researchers must first make many copies of, or amplify, the DNA segment they wish to sequence. This is done either by cloning the DNA or by triggering the polymerase chain reaction (PCR). Once the DNA has been amplified, it is heated so that the two strands separate, and a synthetic primer is added to the mixture. The primer's sequence is complementary to the first piece of target DNA, which means that the primer and the DNA target bind with each other. At this point, the target sequence is exposed to a solution that contains DNA polymerase and all of the nucleotides required for synthesis of the complementary DNA strand — along with one special ingredient.
- **Adding ddNTPs:** As described above, the next major step in the Sanger process is to expose the target sequence to DNA polymerase and significant amounts of all four nucleotides. In their unbound form, nucleotides have three phosphate groups and are formally called deoxynucleotide triphosphates, or dNTPs (where the "N" is a placeholder for A, T, G, or C). During the construction of a new DNA strand, a molecule called a hydroxyl group (which contains an oxygen atom and a hydrogen atom) attaches to the sugar of the last dNTP in the strand and chemically binds to the phosphate group on the next dNTP. This binding causes the DNA chain to grow. In Sanger sequencing, however, a special type of "dummy" nucleotide is included with the regular dNTPs that surround the growing DNA strand. These special nucleotides are known as dideoxynucleotide triphosphates, or ddNTPs, and they lack the crucial hydroxyl group that is attached to the sugar of dNTPs. Therefore, whenever a ddNTP is added to a growing DNA strand, it is unable to chemically bind with the next nucleotide in the chain, and the DNA strand stops growing.

When researchers carry out the Sanger process, they are manipulating many copies of the template strand at once, so an overabundance of dNTPs is required in order for DNA synthesis to proceed unimpeded on these copies until a ddNTP is added. Then, after the supply of dNTPs has been exhausted, the final result of the sequencing experiment is a group of new DNA strands of varying lengths. These strands all have a terminal ddNTP that indicates whether an A, T, G, or C occurs in that position on the template strand.

- **Reading the Sequence: Now and Then:** When Sanger sequencing was first introduced, four separate reagents were used, one for each type of ddNTP. The four reaction products were then separated by gel electrophoresis, a process that organizes DNA fragments in order of size. This enabled researchers to assess the lengths of the truncated strands in each sample. This was important, because the end of each truncated strand was used to determine the position at which a ddNTP was added to the strand, thereby halting DNA elongation.

More recently, automation of the Sanger technique has made this process more efficient by combining all four ddNTP reactions in a single test tube. Each of the four ddNTPs in the tube is labeled with a different fluorescent color. Rather than being run on a gel and read manually, the reaction products are passed through a small tube containing a gel-like matrix. As the different-sized DNA fragments pass through the tube, a sequencing machine reads the fluorescent label at each position. Sequencing machines have vastly increased the speed and efficiency of DNA sequencing, and this technology continues to evolve at an astonishing rate.

- How Is DNA Sequencing Used by Scientists?

In recent years, DNA sequencing technology has advanced many areas of science. For example, the field of functional genomics is concerned with figuring out what certain DNA sequences do, as well as which pieces of DNA code for proteins and which have important regulatory functions. An invaluable first step in making these determinations is learning the nucleotide sequences of the DNA segments under study. Another area of science that relies heavily on DNA sequencing is comparative genomics, in which researchers compare the genetic material of different organisms in order to learn about their evolutionary history and degree of relatedness. DNA sequencing has also aided complex disease research by allowing scientists to catalog certain genetic variations between individuals that may influence their susceptibility to different conditions.

- How Can All People Benefit From DNA Sequencing?

At the individual level, biomedical research into the cause and course of common human diseases is primed to greatly improve health care. The application of DNA sequencing to the identification of disease-causing genetic variants will lead to improvements and expansion in genetic testing, as well as development of more targeted, personalized drug therapies in the years to come. Already today, the benefits of DNA sequencing can be seen in agriculture thanks to the production of disease-resistant plants and animals. In addition, microbial genome sequencing projects may someday lead to the development of new biofuels and pollutant-monitoring systems. DNA sequencing techniques are also used in forensic science, providing crucial evidence in criminal cases. In the United States, for instance, the Federal Bureau of Investigation (FBI) funds and operates a national database containing the genetic profiles of known offenders that can be searched whenever DNA evidence is obtained at a crime scene. According to the FBI, as of 2008, this database had profiles of over 6.5 million offenders and had assisted in almost 81,000 investigations.

CLASS ACTIVITY

Write the complementary sequence to following DNA strand:

A A T T C G C C G G T A T T A G A C G T T
| | | | | | | | | | | | | | | | | | | | |

SUMMARY

- Molecular biology and chemical biology are interdisciplinary fields that have significantly advanced our understanding of biological systems.
- Molecular biology is a powerful tool for the directed synthesis of biological macromolecules, particularly nucleic acids and proteins. It is also a highly diverse and forward-looking discipline in its own right. The key to manipulation in molecular biology is the isolation, cloning and identification of genomic information in an appropriately useful DNA form.
- The plasmid is central to almost all molecular biology manipulations of genomic DNA or cDNA. The term plasmid refers to DNA that can replicate independently of chromosomal DNA. PlasmidDNA(pDNA) systems were first found in prokaryotes and were quickly realized to be ideal potential vectors.
- Restriction enzymes are enzymes that cut double-stranded DNA (dsDNA) by hydrolyzing two phosphodiester links (one per strand) without altering attached deoxynucleotides.
- DNA ligases are needed for the covalent ligation of heterologous duplex DNA fragments. DNA fragments that have been restricted (generated by restriction endonucleases) are rejoined with the aid of ligase enzymes.
- Exogenous pDNA is introduced into host cells by a process known as transformation Cells are rendered competent to take up added exogenous pDNA by pre-treatment with calcium chloride solution on ice.
- Selectable markers are typically genes that code for resistance to antibiotics such as ampicillin, tetracycline and kanamycin.
- Electrophoresis is the movement of charged molecules through a solid-phase medium under the influence of an electrical potential difference.
- Circular DNA are more strongly affected by ethidium bromide concentration than linear DNA if ethidium bromide is present in the gel during electrophoresis.

- The most common dye used to make DNA or RNA bands visible for agarose gel electrophoresis is ethidium bromide, usually abbreviated as EtBr. It fluoresces under UV light when intercalated into the major groove of DNA (or RNA).
- The standard method of sequencing DNA is the chain-termination sequencing method. In this method DNA polymerase is used for short time periods to rebuild duplex DNA using the complementary DNA strand as template.
- Cloning and identification of genes in DNA is a fundamental process in molecular biology that involves isolating and amplifying a specific gene to study its function and structure.
- Site-directed mutagenesis is one of the most important laboratory techniques for creating DNA libraries by introducing mutations into DNA sequences.

REVIEW QUESTIONS

1. Discuss the central dogma of molecular biology.
2. Differentiate between prokaryotic and eukaryotic genes.
3. Explain the role of DNA ligases.
4. Discuss the factors affecting the migration of nucleic acid.
5. What do you understand by DNA sequencing? Discuss.
6. What is the difference between DNA cloning and PCR?
7. Describe PCR amplification and product isolation.

REFERENCES

1. Avery, O. T., Macleod, C. M., & McCarty, M. (1944). Studies on the chemical nature of the substance inducing transformation of pneumococcal types: Induction of transformation by a desoxyribonucleic acid fraction isolated from pneumococcus type III. *Journal of Experimental Medicine, 79*(2), 137–158. doi:10.1084/jem.79.2.137.
2. Beadle, G. W., & Tatum, E. L. (1941). Genetic control of biochemical reactions in neurospora. *Proceedings of the National Academy of Sciences, 27*, 499–506. doi:10.1073/pnas.27.11.499.
3. Braig, M., & Schmitt, C. (2006). Oncogene-induced senescence: Putting the brakes on tumor development. *Cancer Research, 66*(6), 2881–2884. doi: 10.1158/0008-5472.CAN-05-4006.
4. Estell, D. A., Graycar, T. P., & Wells, J. A. (1985). Engineering an enzyme by site-directed mutagenesis to be resistant to chemical oxidation. *Journal of Biological Chemistry, 260*(11), 6518–6521. doi:10.1016/S0021-9258(18)88811-1.
5. Gerstein, M. B., Bruce, C., Rozowsky, J. S., Zheng, D., Du, J., Korbel, J. O., Emanuelsson, O., Zhang, Z. D., Weissman, S., & Snyder, M. (2007). What is a gene, post-ENCODE history and updated definition. *Genome Research, 17*(6), 669–681. doi:10.1101/gr.6339607.

6. McCarthy, B. J., & Holland, J. J. (1965). Denatured DNA as a direct template for in vitro protein synthesis. *Proceedings of the National Academy of Sciences of the United States of America, 54*(3), 880–886.

7. Min Jou, W., Haegeman, G., Ysebaert, M., & Fiers, W. (1972). Nucleotide sequence of the gene coding for the bacteriophage MS2 coat protein. *Nature, 237*(5350), 82–88. doi:10.1038/237082a0.

8. Papworth, C., Bauer, J. C., Braman, J., & Wright, D. A. (1996). Site-directed mutagenesis in one day with >80% efficiency. *Strategies, 9*(3), 3, 4.

9. Saiki, R. K., Gelfand, D. H., Stoffel, S., Scharf, S. J., Higuchi, R., Horn, G. T., Mullis, K. B., & Erlich, H. A. (1988). Primer-directed enzymatic amplification of DNA with a thermostable DNA polymerase. *Science, 239*, 487–491. doi:10.1126/science.2448875.

10. Saiki, R. K., Scharf, S., Faloona, F., Mullis, K. B., Horn, G. T., Erlich, H. A., & Arnheim, N. (1985). Enzymatic amplification of beta-globin genomic sequences and restriction site analysis for diagnosis of sickle cell anemia. *Science, 230*(4732), 1350–1354. doi:10.1126/science.2999980.

11. Thomas, P. S. (1980). Hybridization of denatured RNA and small DNA fragments transferred to nitrocellulose. *Proceedings of the National Academy of Sciences, 77*(9), 5201–5205. doi:10.1073/pnas.77.9.5201.

CHAPTER 4

Electronic and Vibrational Spectroscopy

LEARNING OBJECTIVES

After studying this chapter, you will be able to:

- Understand electronic and vibrational spectroscopy in chemical biology
- Explain ultraviolet-visible spectroscopy
- Describe circular dichroism spectroscopy
- Focus on vibrational spectroscopy
- Identify fluorescence spectroscopy
- Know probing metal centers by absorption spectroscopy

KEY TERMS FROM THIS CHAPTER

Atomic layers
Circular dichroism
Discrete wavelengths
Electron spectroscopy
Fluorescence spectroscopy
Infrared spectroscopy
Molar absorptivity
Photoelastic Modulator (PEM)
Radiation
Beer-lambert's Law
Cuvette
Double-beam spectrophotometer
Electronic structure
Franck–Condon principle
Magnetic resonance
Molecule
Photoelectrons
Single-beam spectrophotometer

INTRODUCTION

Electron spectroscopy is a series of techniques that analyze the energies of emitted electrons, such as photoelectrons and Auger electrons. Some common techniques in this group include X-ray photoelectron spectroscopy (XPS), also known as Electron Spectroscopy for Chemical Analysis (ESCA), Electron energy loss spectroscopy (EELS), Ultraviolet photoelectron spectroscopy (UPS), and Auger electron spectroscopy (AES). These methods are utilized to identify and characterize the elements and electronic structures present on the surface of a sample, which can be solid, gas, or liquid. Due to the low energies of Auger electrons and photoelectrons (typically 20–2,000 eV), chemical information is only obtained from the top atomic layers of the sample with a depth of 10 nm or less. As a result, electron spectroscopy techniques are primarily used for surface chemical analysis.

Vibronic spectroscopy focuses on the simultaneous changes in electronic and vibrational energy levels of a molecule during the absorption or emission of a photon. In the gas phase, these transitions also involve changes in rotational energy. Analysis of vibronic spectra of diatomic molecules reveals that emission spectra are more complex than absorption spectra. The intensity of allowed vibronic transitions follows the Franck–Condon principle and can provide valuable information on the electronic excited states of stable molecules, including bond length. Vibronic spectroscopy has also been used to study unstable molecules like dicarbon (C_2) in various environments, such as discharges, flames, and astronomical objects.

4.1. ELECTRONIC AND VIBRATIONAL SPECTROSCOPY IN CHEMICAL BIOLOGY

In comparison to techniques like magnetic resonance, diffraction, and microscopy, electronic and vibrational spectroscopy are classified as lower-resolution methods. This means that achieving atomic-level descriptions of biological macromolecular structures with these techniques is nearly impossible. However, electronic spectroscopy has the advantage of being able to provide valuable data from small quantities of biological macromolecules.

This allows for rough characterizations of biological macromolecular structures in the range of nanometers to be done before more advanced and resource-intensive atomic-level structural characterizations like X-ray diffraction. Additionally, electronic and vibrational spectroscopy can also show changes in the structures of biological macromolecules in response to environmental changes or molecular interactions, giving a broader view of biological function.

This perspective can even be applied to the real-time observations of molecular binding interactions occurring in living cells when using fluorescence spectroscopy. Therefore, before conducting more in-depth research, electronic and vibrational spectroscopies serve as "first-pass" structural and even functional characterization methods that can help provide a general, if not always precise, understanding of biological macromolecular structure and function. A lot of the time, the lack of sufficient amounts of biological macromolecules for research means that the data from vibrational and electronic spectroscopies may be all that is currently possible to obtain. For a very long time to come, such realities guarantee that electronic and vibrational spectroscopies will play a crucial role in connecting early structural to early functional characterization.

4.2. ULTRAVIOLET-VISIBLE SPECTROSCOPY

UV-Vis spectroscopy is a method used to analyze the absorption or transmission of specific wavelengths of UV or visible light by a sample compared to a reference sample. The composition of the sample can affect this property, offering insight into its contents and concentration. As this technique utilizes light, it is important to understand the properties of light first.

The wavelength of light and its energy content are inversely correlated. As a result, light has a greater energy at shorter wavelengths and less at longer wavelengths. To detect absorption, a substance's electrons must be promoted to a higher energy state with a certain amount of energy. To move electrons to a higher energy state, a substance's electrons in various bonding environments need distinct amounts of energy. This explains why different substances absorb light at different wavelengths. Visible light has a wavelength range that is visible to humans, ranging from about 380 nm, or violet, to 780 nm, or red. Up to about 100 nm, the wavelengths of UV light are shorter than those of visible light. Consequently, the wavelength of light can be used to characterize it. This can be helpful in UV-Vis spectroscopy to identify or analyze various substances by identifying the precise wavelengths that correspond to maximum absorbance.

4.2.1. UV Visible Spectroscopy Principle

UV-visible spectroscopy is a precise method utilized in analytical chemistry to quantify the light absorbed by a substance. When light hits a substance, it both absorbs and reflects a specific amount of radiation. As the light traverses through the sample, the quantity of radiation absorbed by the substance is determined by the variance between the incoming radiation (I_o) and the outgoing radiation (I). The absorbed radiation is referred to as the absorbance (A) and transmittance (T), which is a ratio (I/ I_o) signifying the proportion of light that has traveled through the sample.

Transmittance, $T = I/ I_o$

Absorbance, $A = \log_{10}(I_o/I) = \log_{10} (1/T) = -\log_{10} (T)$

Beer-Lambert's Law states that the concentration of the absorbing species (the compound) and the path length are directly correlated with the absorbance of a solution containing the compound. This means that the extent of absorption increases as the number of molecules (concentration) able to absorb the radiation at a given wavelength increases. More absorption is also influenced by the molecule's efficiency (measured by its molar absorptivity) in absorbing the radiation.

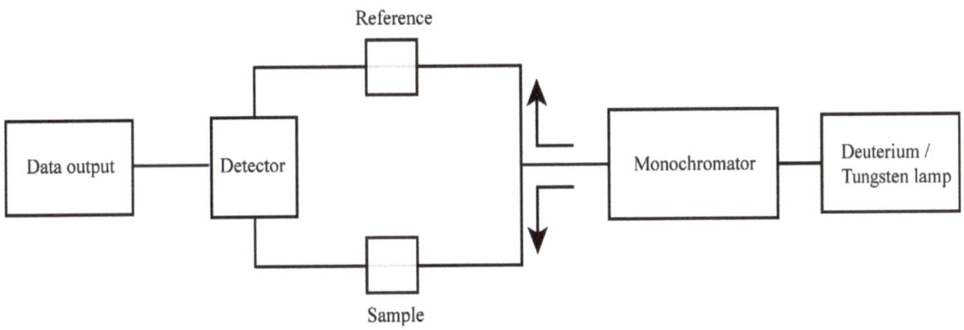

Figure 4.1. *Simplified schematic of a double beam UV–visible spectrophotometer.*

Source: By Sobarwiki – Own work, Public Domain, https://commons.wikimedia.org/w/index.php?curid=48800666.

The formulation for Beer-Lambert's law is given by:

A = ε**cl**; for a given wavelength.

where; ε is the molar absorptivity (also known as molar extinction coefficient); c is the molar concentration of the absorber (solute); l is the path length (length of the sample cell or cuvette; in cm).

The Beer-Lambert law enables the direct determination of the concentration of a substance in a solution based on its absorbance for a fixed path length.

The sample is put inside the sample compartment to complete the measurement. A cuvette is a rectangular holder for liquid samples that is constructed of glass, quartz, or plastic. Standard cuvettes facilitate easy UV radiation transmittance and have a 10 mm path length. When the monochromator beam travels through the sample and onto the detector, the sample is positioned in the radiation's path. The spectrophotometer displays the absorbance value (A or Abs) by comparing the light intensity of the transmitted radiation (I) that has already passed through the sample with the intensity of the incident radiation (I_o) prior to passing through the sample.

4.2.2. Instrumentation of UV-Vis Spectroscopy

The main components of UV-Vis spectrophotometer are:

1. Light Source
2. Wavelength selector
3. Sample container
4. Detectors

4.2.2.1. Light Source

It is needed for a UV-Vis spectrometer to function in order to emit light in a broad range of wavelengths. Xenon lamps are frequently used as high-intensity light sources in the UV and visible spectrums. It is more expensive and less stable than tungsten and halogen lamps. Thus, the two light sources for this instrument are a tungsten or halogen lamp for visible light and a deuterium lamp for UV light. Good intensity is provided by the two lamps. The spectrometer should switch during the light intensity measurement. When the transition happens between 300 and 350 nm, it can happen more smoothly because both visible and UV light sources emit the same amount of light at that wavelength.

4.2.2.2. Wavelength Selector

To enable sample examination with the light source's emitted wavelengths, wavelength selection is utilized to determine the most suitable wavelength for the analyte and sample. The monochromator is the typical wavelength selector in UV-Vis spectrometers, as it divides light into a focused wavelength band. Radiation of varying wavelengths enters the monochromator through the entrance slit. The beam then collides at a specific angle with the dispersing element within the device. A prism within the monochromator separates the different wavelengths of light in the beam, bending the monochromatic light and causing non-linear dispersion. Only radiation of a specific wavelength or color is able to pass through the monochromator and exit through the ultimate slit.

4.2.2.3. Sample Container

All of the light from the light source passes through the sample in a single beam spectrophotometer. By measuring the differences in light source intensity before and after a sample is inserted, single-beam spectrophotometers are able to determine color. The wavelength range is measured to be between 190 and 750 nm, but it might reach 1,100 nm.

All of the light from the light source splits into two beams in a double-beam spectrophotometer: one beam travels through the sample and the other beam only passes through the reference. In a similar vein, the wavelength range of double-beam spectrophotometers is 190–1,100 nm. In addition, the absorbance versus wavelength or sample and reference beam ratio is measured by the double-beam spectrophotometer.

For every measurement, lamp brightness fluctuations are corrected using the reference detector. The sample detector is measured in the sample position and subtracted from the sample spectrum after the sample is collected. It has a sample chamber as well as a reference chamber. A cuvette, also known as a sample chamber, is a transparent, flat container used to hold samples. The reference chamber, sometimes referred to as the blank, contains the solvent that dissolves the sample. The selection of the sample cell is contingent upon the desired wavelength, path length, shape, size, and transmission characteristics, as well as the relative cost.

The light intensity travels through the beam separator to the reference chamber (I_o) and sample chamber (I) for each wavelength. The number of photons per second is represented by the symbol for light intensity, or Io. The blank solution does not absorb light when it travels through it; this is known as (1). Sample I has absorbed some light if it is less than I_o.

The absorbance (A) of the sample is related to I and Io according to the following equation:

Absorbance (A)= -log(T)= -log (I/I_o)

The relationships between transmittance and absorbance are displayed in this equation.

Additionally, transmittance (T), which indicates how much light has passed through a sample, is defined as the fraction I divided by I_o.

$T= I/I_o$

$T= I/I_o$ = e–kbc

where; – I_o is the incident intensity; I is the transmitted intensity; e is the base of natural logarithms; – k is a constant; and b is the path length (usually in cms).

Transmittance increases with refracted lightness. The transmittance increases with decreasing absorbance. Quartz cuvettes or fused silica cuvettes are frequently used in UV and visible regions.

4.2.2.4. Detector

Semiconductors or photoelectric coatings are used in detectors. It transforms the sample's incoming light into an electric signal, or current. The intensity increases with current level. It responds linearly because of its high sensitivity and low noise characteristics. Every detector has a unique sensitivity range and range of wavelengths. Finally, in the UV and visible regions of the electromagnetic spectrum, the data recorder typically plots the absorbance against wavelength (nm).

4.2.3. Strengths and Limitations of UV-Vis Spectroscopy

While no technique is flawless, UV-Vis spectroscopy has several key strengths that contribute to its popularity.

- Firstly, it is non-destructive, allowing samples to be reused or further analyzed.
- Secondly, measurements can be taken quickly, seamlessly integrating into experimental protocols.
- Additionally, the instruments are user-friendly, requiring minimal training.
- Data analysis is also straightforward, requiring minimal processing.
- Furthermore, UV-Vis spectroscopy is cost-effective, making it accessible for many laboratories.

While the advantages of this technique are significant, there are also some drawbacks to consider:

- **Stray Light:** Because wavelength selectors in real instruments aren't flawless, a tiny amount of light from a broad wavelength range may still be transmitted from the light source. This could lead to significant measurement errors. Other potential sources of stray light include the surroundings and an instrument's loosely fitting compartment.
- **Light Scattering:** Scattered solids in liquid samples are frequently the source of light scattering, which can lead to significant measurement errors. Results are not repeatable if there are bubbles in the cuvette or sample because they scatter light.

- Multiple absorbing species can cause interference. For instance, a sample may contain different kinds of the green pigment chlorophyll. When analyzed simultaneously in the same sample, the spectra of the various chlorophylls will overlap. Every chemical species in the sample needs to be taken out and analyzed separately in order to perform an accurate quantitative analysis.
- **Geometric Considerations:** Inaccurate and non-reproducible results may result from misaligned positioning of any one of the instrument's components, particularly the cuvette containing the sample. As a result, it's critical that each part of the instrument be positioned consistently and aligned in the same direction for each measurement. To prevent misuse, some basic user training is thus usually advised.

4.2.4. Applications of UV-Vis Spectroscopy

UVVis has found itself applied to many uses and situations including but not limited to:

4.2.4.1. Pharmaceutical Analysis

The pharmaceutical industry is among the industries that utilize UV-Vis spectroscopy the most frequently. Specifically, overlapping absorbance peaks in the original spectra can be resolved to identify specific pharmaceutical compounds by processing UV-Vis spectra using mathematical derivatives. For example, by using the first mathematical derivative on the absorbance spectra, it is possible to identify the antibiotic chlortetracycline and the local anesthetic benzocaine simultaneously in commercial veterinary powder formulations. By developing a calibration function for each compound, simultaneous quantification of both substances was achievable on a microgram per milliliter concentration range.

4.2.4.2. DNA and RNA Analysis

The main goal is to confirm the concentration and purity of DNA and RNA, which is essential for subsequent processes such as sequencing. This ensures that the DNA or RNA samples being prepared for sequencing are free from contaminants. A pure DNA sample typically has an absorbance ratio of 1.8, while pure RNA has a ratio of 2. The 260 nm/280 nm absorbance ratio is important for detecting protein contamination in nucleic acids. Additionally, the 260nm/230nm absorbance ratio can vary for RNA and DNA samples, ranging from 2.15 to 2.50.

4.2.4.3. Bacterial Culture

When culturing bacteria, UV-Vis spectroscopy is frequently utilized. At a wavelength of 600 nm, OD measurements are routinely and rapidly taken to estimate the concentration of cells and monitor growth. Because of the optical characteristics of the bacterial culture media in which they are cultivated and to prevent cell damage in situations where the cells are necessary for further experimentation, 600 nm is frequently used and preferred.

4.2.4.4. Beverage Analysis

UV-Vis spectroscopy is frequently used to identify specific compounds in beverages, including monitoring caffeine levels to ensure they comply with legal regulations. Additionally, UV light is utilized to quantify the concentration of colored compounds like anthocyanin, found in fruits like blueberries, raspberries, blackberries, and cherries, by comparing their absorbance wavelengths in wine for quality control purposes.

4.2.4.5. Other Applications

Numerous other industries may also make use of this technique. To guarantee that electric power is delivered safely, for instance, measuring a color index can be helpful in keeping an eye on transformer oil. Research on cancer may benefit from measuring hemoglobin absorbance to ascertain hemoglobin concentrations. In wastewater treatments, UV-Vis spectroscopy can be used in kinetic and monitoring studies to ensure certain dyes or dye byproducts have been removed properly by comparing their spectra over time. It also finds great utility in food authenticity analysis and air quality monitoring.

UV-Vis spectroscopy is also beneficial for qualitative analysis in specialized research. It can be used to detect changes in the wavelength of peak absorbance, making it valuable for studying structural protein alterations and identifying battery components. Additionally, shifts in peak absorbance wavelengths are helpful for characterizing ultra-small nanoparticles in modern applications. The potential uses of this technique are diverse and limitless.

4.3. CIRCULAR DICHROISM SPECTROSCOPY

The dichroism involving circularly polarized light is known as circular dichroism (CD), i.e., the difference in how left- and right-handed light is absorbed [Edward I. Solomon; A. B. P. Lever (February 3, 2006)]. Circular dichroism is also known as dichroism for spin angular momentum because left-hand circular (LHC) and right-hand circular (RHC) polarized light represent two possible spin angular momentum states for a photon. In the first part of the 1800s, Jean-Baptiste Biot, Augustin Fresnel, and Aimé Cotton made the discovery of this phenomenon. Optical activity manifests as circular dichroism and circular birefringence. It is visible in the absorption bands of chiral molecules that are optically active. Applications for CD spectroscopy are numerous and span a wide range of industries. UV CD is most famously used to study the secondary structure of proteins. A technique for studying charge-transfer transitions is UV/Vis CD. By probing metal d→d transitions, near-infrared CD is used to study geometric and electronic structure. Utilizing light from the infrared energy region, vibrational circular dichroism is used to study the structures of small organic molecules, most recently proteins and DNA.

Chirality, or the asymmetry in a molecule's structure, is a characteristic that characterizes optically active molecules. Circular birefringence, also known as optical rotation, is a property of chiral molecules. It states that the refractive index of an optically active medium affects the difference in light velocity that passes through it. Consequently, the propagation speed of left-handed CPL in a chiral sample will differ from that of its right-handed counterpart. The electric field vector that results from a chiral molecule absorbing left- and right-handed CPL to varying degrees traces an ellipse. As the circularly polarized light passes through the birefringent material, the phase relationship between the CPL waves changes, rotating the linearly polarized wave and tilting the elliptically polarized light. This is because linearly polarized light can be thought of as a superposition of equal intensities of left- (E_l) and right-handed (E_r) circularly polarized light (Figure 4.2). This wavelength-dependent ellipticity (q) characterizes a circular dichroism spectrum.

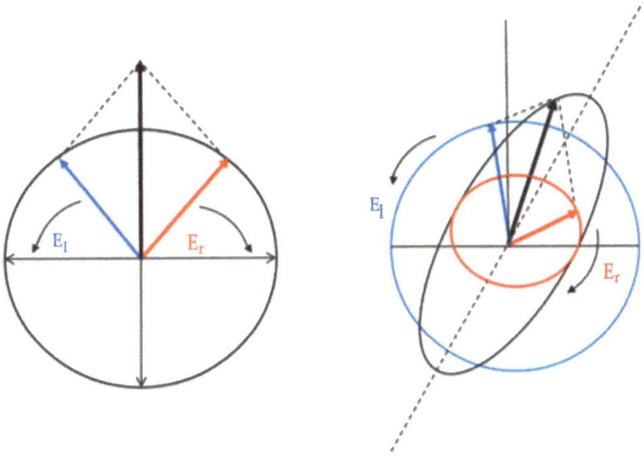

Figure 4.2. *El and Er components when they are absorbed equally (left), resulting in linearly polarized light, and unequally (right), resulting in elliptically polarized light.*

Source: https://jascoinc.com/learning-center/theory/spectroscopy/circular-dichroism-spectroscopy/.

4.3.1. Instrumentation Used in Circular Dichroism Spectroscopy

4.3.1.1. Photoelastic Modulator (PEM)

The circular dichroism spectrophotometer generates circularly polarized light by converting linearly polarized light. Initially, non-polarized light is filtered through a polarizer aligned with the crystal axes and molecular orientation. Next, a photo-elastic modulator (PEM) transforms the linearly polarized light into circularly polarized light.

4.3.1.2. Nitrogen Purge

A xenon arc lamp has a significant UV radiation output. Ozon is produced when intense UV radiation interacts with oxygen molecules. The instrument's mirrors become oxidized by ozone, which lowers their reflectivity and reduces their ability to focus light through the monochromator and lowers the S/N. Furthermore, oxygen molecules absorb far UV light. In addition to preserving the mirrors' longevity and reflectivity, nitrogen purging of the monochromator also reduces oxygen absorption, enabling the collection of sample data at higher UV wavelengths.

Circular dichroism (CD) is a technique that provides information on the molecular and electronic structure of chiral chromophores, either through quantitative or empirical measurements. Chromophores that are not inherently chiral can be placed in an asymmetric environment or paired with chiral chromophores to induce CD in their achiral electronic transitions. CD experiments usually require minimal sample preparation, are simple to execute, and are non-destructive, allowing for sample recovery. Sample measurements can be obtained in liquid form or as solids, films, or gels, typically at lower concentrations compared to other structural analysis methods.

4.3.2. The Use of CD Spectroscopy to Study Protein Folding, Form, and Function

Typically, this phenomenon will appear in absorption bands of all optically active molecules. This means that biological molecules exhibit circular dichroism due to their dextrorotary and levorotary components. Furthermore, different secondary structures like the alpha helix of proteins and the double helix of nucleic acids will also show unique CD spectral signatures. This ability of CD to provide a structural signature makes it a valuable tool in contemporary biochemistry, with applications across various fields of study.

Though generally regarded as more sophisticated, CD is closely related to the optical rotatory dispersion (ORD) method.

While ORD can be measured far from these bands, CD can only be measured in or close to the molecule of interest's absorption bands. The data analysis makes CD's advantage clear. Because of the less extensive overlap of their recorded bands at specific wavelengths compared to ORD, structural elements can be distinguished more easily. If all of the absorptions are included in the measurements, then these two spectral measurements can theoretically be interconverted using an integral transform (Kramers–Kronig relation).

The CD spectrum of proteins in the far-UV range can provide valuable information about their secondary structure. This information can help determine the percentages of alpha-helix, beta-sheet, beta-turn, and other conformations (such as random coil) in the molecule [Hall V, Nash A, Hines E, Rodger A (2013)]. The protein's potential secondary conformations are significantly limited by these fractional assignments. Generally speaking, CD is unable to pinpoint the exact location of detected alpha helices within a molecule or even estimate their total number. Nevertheless, CD remains a useful tool, particularly when demonstrating conformational changes. It can, for instance, be used to study how the secondary structure of a molecule changes as a function of temperature or of the concentration of denaturing agents, e.g., Guanidinium chloride or urea.

In this way, it can provide crucial thermodynamic details about the molecule that would otherwise be difficult to obtain, like the enthalpy and Gibbs free energy of denaturation. In order to ensure that a protein is in its native conformation before conducting lengthy and/or costly experiments with it, CD is an invaluable tool for anyone studying proteins. Furthermore, CD spectroscopy has several other applications in protein chemistry besides the estimation of the alpha-helix fraction. Additionally, studies on the bioinorganic interface have made use of CD spectroscopy. It has been specifically used to examine the variations in an engineered protein's secondary structure between its pre- and post-reagent titration states.

The near-UV CD spectrum of proteins, which ranges from greater than 250 nm, offers insight into the tertiary structure. The signals observed in the 250–300 nm range are a result of the absorption, dipole orientation, and surrounding environment of amino acids like phenylalanine, tyrosine, cysteine (or S-S disulfide bridges), and tryptophan. In contrast to far-UV CD, the near-UV CD spectrum does not correspond to a specific 3D structure but instead provides information on the nature of prosthetic groups in proteins, such as heme groups found in hemoglobin and cytochrome c.

Visible CD spectroscopy can resolve individual d-d electronic transitions as distinct bands, making it an extremely powerful tool for studying metal-protein interactions. Free metal ions in solution are not detected because CD spectra in the visible light spectrum are only generated when a metal ion is in a chiral environment. The benefit of this is that pH dependence and stoichiometries can be easily obtained because it only looks at the metal that is bound to proteins. Transition metal ion complex optical activity has been ascribed to conformational, configurational, and vicinal effects. A set of empirical guidelines for anticipating the emergence of visible CD spectra for Cu^{2+} and Ni^{2+} square-planar complexes involving histidine and main-chain coordination was developed by Klewpatinond and Viles (2007).

Compared to other methods that provide atomic resolution data, such as protein NMR spectroscopy and X-ray crystallography, CD provides less precise structural information. On the other hand, CD spectroscopy is a fast technique that doesn't require a lot of proteins or sophisticated data processing. Consequently, a wide range of solvent conditions, including changes in salinity, pH, temperature, and the presence of different cofactors, can be surveyed using CD.

Since CD spectroscopy is typically used to study proteins in solution, it is a useful addition to techniques used to study solid states. Another drawback is that a lot of proteins are naturally embedded in membranes, and solutions with membrane structures frequently scatter strongly. Thin films are sometimes used to measure CD. In order to obtain large signals in the UV range of wavelengths, where the electronic transitions for biomolecules frequently occur, CD spectroscopy has also been performed using semiconducting materials like TiO_2.

4.3.3. Advantages and Limitations of CD

A CD has its advantages and disadvantages, just like any other instrument. Table 4.1's comparison of CD and NMR provides a good idea of CD's capabilities.

Table 4.1. A Comparison of CD Spectroscopy to NMR Spectroscopy

CD	NMR
Molecules of any size can be studied	There is size limitation
The experiments are quick to perform; single wavelength measurements require milliseconds	This is not always the case.
Unique sensitivity to asymmetry in sample's structure	Special conditions are required to differentiate between enantiomers.
Can work with very small concentrations, by lengthening the cuvette width until discernible absorption is achieved	There is a limit to sensitivity of instrument
Timescale is much shorter (UV) thus allowing to study dynamic systems and kinetics	Timescale is long, use of radio waves gives average of all dynamic system
Only qualitative analysis of data is possible	Quantitative data analysis can be performed to estimate chemical composition
Does not provide atomic level structure analysis	Very powerful for atomic level analysis, providing essential information about chemical bonds in system
The observed spectrum is not enough for claiming one and only possible structure	The NMR spectrum is key information for assigning a unique structure.

Source: https://chem.libretexts.org/Bookshelves/Analytical_Chemistry/Physical_Methods_in_Chemistry_and_Nano_Science_(Barron)/07%3A_Molecular_and_Solid_State_Structure/7.07%3A_Circular_Dichroism_Spectroscopy_and_its_Application_for_Determination_of_Secondary_Structure_of_Optically_Active_Species.

4.4. VIBRATIONAL SPECTROSCOPY

Infrared spectroscopy, also known as IR spectroscopy or vibrational spectroscopy, involves measuring how infrared radiation interacts with different substances through absorption, emission, or reflection. This technique is used to analyze and identify chemical compounds or functional groups in solid, liquid, or gas states. It can also be used to characterize new materials or confirm the identity of known and unknown samples. The process of infrared spectroscopy is carried out using a specialized instrument called an infrared spectrometer or spectrophotometer, which generates an infrared spectrum. The spectrum is typically shown as a graph with infrared light absorbance or transmittance on the vertical axis and frequency, wavenumber, or wavelength on the horizontal axis. Wavenumber is commonly expressed in reciprocal centimeters and wavelength in micrometers (or "microns"), with the symbol μm, which are inversely related to each other. A popular type of infrared spectrometer used in laboratories is the Fourier transform infrared (FTIR) spectrometer. Additionally, two-dimensional IR spectroscopy is also achievable, as described further below.

The electromagnetic spectrum's infrared region is commonly separated into three areas: the near-, mid-, and far-infrared, so named because of their proximity to the visible spectrum. Higher-energy near-IR radiation, with a wavelength of roughly 14,000–4,000 cm^{-1} (0.7–2.5 μm), can stimulate combination or overtone modes of molecular vibrations. The fundamental vibrations and related rotational-vibrational structure are typically studied in the mid-infrared, or roughly 4,000–400 cm−1 (2.5–25 μm). With its low energy, the far-infrared range of 400–10 cm^{-1} (25–1,000 μm) can be utilized for low frequency vibrations and rotational spectroscopy. The terahertz region, which borders the microwave region and spans 2–130 cm^{-1}, is thought to be able to investigate intermolecular vibrations. These subregions' names and classifications are arbitrary and only partially determined by the relative electromagnetic or molecular characteristics.

4.4.1. Uses and Applications

Inorganic and organic chemistry, as well as research and industry, use infrared spectroscopy extensively because it is a straightforward and trustworthy method. It is a very helpful tool in catalysis research to identify products and intermediates during the catalytic reaction, as well as to characterize the catalyst. Applications such as the long-term, unattended measurement of CO_2 concentrations in greenhouses and growth chambers by infrared gas analyzers use it for quality control, dynamic measurement, and monitoring. Additionally, it is employed in forensic analysis for both civil and criminal cases, such as when determining the degree of polymer degradation. It can be applied to ascertain a suspected drunk driver's blood alcohol content. The analysis and identification of pigments in paintings and other works of art, such as illuminated manuscripts, have been accomplished with success using infrared spectroscopy. When making polymers, infrared spectroscopy can also be used to measure the degree of polymerization. Measuring at a particular frequency over time allows one to evaluate changes in the quantity or character of a given bond. As many as 32 infrared measurements per second across the range of interest can be made with current research instruments. This can be carried out while other methods are used to make simultaneous measurements. As a result, observations of chemical processes and reactions can be made more quickly and precisely. The field of semiconductor microelectronics has also successfully employed infrared spectroscopy. For instance, semiconductors such as silicon, gallium arsenide, gallium nitride, zinc selenide, amorphous silicon, silicon nitride, etc., can all be studied using infrared spectroscopy.

The food industry uses infrared spectroscopy extensively to measure the concentration of different compounds in various food products. These days, the instruments are portable and small enough to be used in field trials. Devices like the DP-IR and EyeCGAs that detect gas leaks also use infrared spectroscopy. These instruments find hydrocarbon gas leaks in natural gas and crude oil transportation. NASA unveiled a significantly enhanced database for tracking polycyclic aromatic hydrocarbons (PAHs) in space in February 2014. The database is based on infrared spectroscopy. Scientists believe that PAHs, which could have served as precursors to life, may account for over 20% of the carbon in the universe. PAHs are widely distributed throughout the universe, appear to have formed soon after the Big Bang, and are linked to newly discovered stars and exoplanets.

In addition to being a useful stand-off technique for sorting plastics of various polymers (such as PET, HDPE), infrared spectroscopy is an essential analysis tool in the recycling process of household waste plastics. Other advancements include NIR-spectroscopic chips that can be incorporated into smartphones and other devices, as well as a tiny IR spectrometer that can be used for daily personal use and is connected to a cloud-based database.

The combination of infrared spectroscopy, artificial intelligence, and machine learning holds promise for quick, precise, and non-invasive bacterial detection. Bacteria have a complex chemical makeup that includes nucleic acids, proteins, carbohydrates, and fatty acids. This complexity produces high-dimensional datasets in which the key characteristics are essentially hidden beneath the entire spectrum. Therefore, advanced statistical techniques like machine learning and deep neural networks are needed to extract the essential features. This technique has been shown to have potential for classifying bacteria at the genus, species, and serotype taxonomic levels. It has also shown promise for antimicrobial susceptibility testing, which is crucial in many clinical settings where a reduction in needless blind treatment with broad-spectrum antibiotics due to faster susceptibility testing is desired. Large-scale database construction is challenging due to the technique's high sensitivity to technical equipment and sample preparation methods, which is its primary limitation for clinical applications. However, Bruker has made attempts in this direction with their food microbiology IR Biotyper.

4.5. FLUORESCENCE SPECTROSCOPY

Using the sample's fluorescence characteristics as a basis, fluorescence spectroscopy is an analytical technique used to quantify pharmaceuticals, metabolites, and other chemical products.

4.5.1. Use of Fluorescence

This technique employs fluorescence for sample analysis. Fluorescence involves the absorption of light by a fluorophore at a specific wavelength, causing an electron to move from a lower energy state to a higher one. As the electron returns to its original state, it releases energy in the form of fluorescence, resulting in light of a distinct wavelength. These specific wavelengths are known as the "excitation wavelength" and "emission wavelength."

The light that is released has a longer wavelength than the light that is absorbed because energy is lost during this transition. The "stokes shift" is the difference between the excitation and emission wavelengths; if the absorbed radiation wavelength is "a" and the emitted radiation wavelength is "b," the stokes shift is computed as "b-a." For fluorophores, a larger Stokes shift is desired because it lessens the interference between the emission and excitation spectra.

4.5.2. Intrinsic and Extrinsic Fluorophores

Molecules that are naturally fluorescent because of aromatic groups found in the side chains of amino acids are referred to as intrinsic fluorophores. Tyrosine, tryptophan, phenylalanine, and co-factors like FMN, FAD, and NAD are a few examples of these substances. Extrinsic fluorophores are substances that are not naturally fluorescent. By affixing a fluorescent probe to non-fluorescent compounds, fluorescence spectroscopy can be carried out on these compounds. Fluorescein, ethidium bromide, and acridine orange are a few instances of these.

4.5.3. Quantum Efficiency

The ratio of incident photons absorbed to converted electrons, or photon vs. converted electrons, is measured by quantum efficiency, an essential parameter in fluorescent measurements. When samples are concentrated at low levels, this parameter is high.

4.5.4. Factors That Affect Fluorescence Spectroscopy

- **Molecular Rigidity:** For fluorescence spectroscopy, fluorophores with stiff structures are favored because they have fewer vibrations and a lower chance of going into the triplet state. Phenolphthalein is non-rigid and non-fluorescent, whereas fluorescein and eosin have stiff structures and are highly fluorescent.

- **Solvent Polarity:** The amount of fluorescence can also be controlled by the solvent's polarity. If there are heavy atoms in the solvent, the structures' fluorescence may diminish.
- **Dissolved Oxygen:** By lessening the emission intensity, dissolved oxygen in the solvent can also have an impact on the fluorescence. The fluorophore is photochemically oxidized to accomplish this. The quenching of fluorescence can also result from oxygen's paramagnetic characteristics.
- **pH:** The fluorescent properties can be impacted by pH. Aniline is one such example, which can be an anion at higher pH values or a cation at lower pH values. The fluorescence disappears in each case.
- **Quenching:** The term "quenching" describes a drop in fluorescence intensity. Either the solution's absorption of fluorescence or the fluorescent material itself may be to blame for this. We refer to the latter result as self-quenching.

4.5.5. Instrumentation

The following tools are used in fluorescence spectroscopy: a monochromator, a detector, and a light source. The typical light source is a mercury arc lamp or xenon lamp (300–800 nm). A particular wavelength from the incident beam is chosen using a monochromator before it is incident on the sample. The emission wavelength is set by the emission monochromator. A sensitive photomultiplier that detects and records the emission wavelength is referred to as a detector.

4.5.6. Applications of Fluorescence Spectroscopy

- **Fluorescent Probes:** This technique finds biological compounds in low concentrations both qualitatively and quantitatively.
- **Protein Structure:** Tryptophan, tyrosine, and phenylalanine are just a few of the intrinsic fluorophores found in proteins. All of a protein's amino acids, the characteristics of the solvent, and the existence of co-factors like NAD and FAD all affect how fluorescent the protein is. Thus, the protein structure can be ascertained with the aid of this method.
- A mixture of cells can be sorted using the fluorescence-activated cell sorting (FACS) method according to the variations in their fluorescent properties.

4.5.7. Disadvantages of Fluorescence Spectroscopy

The above-mentioned parameters, including pH, temperature, and solvent, can affect the fluorescence. The fluorescence may also be altered and impacted by additional interfering materials, such as tissue, filter paper, and detergents. Interactions with other molecules can also cause fluorescence to be lost, particularly in concentrated samples.

4.6. PROBING METAL CENTERS BY ABSORPTION SPECTROSCOPY

Investigating chemical bonding dynamics quantitatively at the molecular level is one of chemistry's main objectives. That is, to watch the creation and breaking of chemical bonds, track nuclear motions and valence-charge distributions in real time, and comprehend how these processes interact with atomic rearrangement and the emergence of new molecular structures. The growing field of ultrafast x-ray spectroscopy offers crucial new insights into molecular dynamics because x-ray transitions are selective for certain elements (C, N, O, etc.), selective for certain symmetry and spin states, and useful for probing lighter elements. A quantitative fingerprint of the local chemical bonding geometry (orbital hybridization, etc.) can be obtained using the x-ray near-edge spectral region (XANES), while extended x-ray absorption fine structure (EXAFS) provides quantitative structural information with sub-Å accuracy close to the target atom or targets. This is especially useful for comprehending molecular dynamics in solutions, where significant chemistry takes place and the solvent environment significantly affects reaction dynamics, frequently making it impossible to quantitatively interpret transient absorption spectra in the visible range.

Transition metal complexes solvated in a solution are important because of the significant relationship between the electronic and molecular structure. Specifically, in Fe^{2+} complexes, the octahedral ligands efficiently connect optical charge-transfer excitations to minor alterations in the molecular structure, resulting in quick spin-state interconversion. The rapid spin crossover event in an Fe^{2+} complex, switching between low and high spin states, is a key reaction for studying electron transfer mechanisms in chemistry and biology. Additionally, light-induced excited spin-state trapping (LIESST) in Fe^{2+} molecular crystals has technological implications.

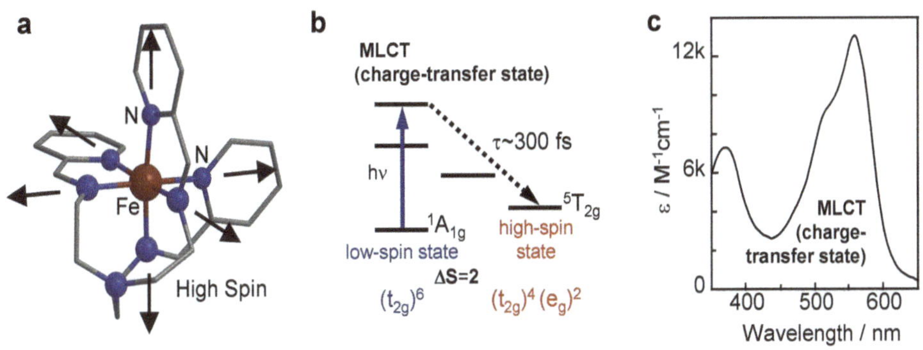

Figure 4.3. *(a) Schematic of [Fe(tren(py)$_3$)]$^{2+}$, arrows indicate the metal-ligand bond change associated with the spin transition. (b) Simplified energy levels of intersystem crossing after optical excitation to the charge-transfer state. (c) Absorption spectrum of the charge transfer band centered at 560 nm.*

Source: https://www.osti.gov/servlets/purl/953772.

The complex $[Fe(tren(py)_3)]^{2+}$ (Figure 4.3a) has a central Fe atom surrounded by an octahedral ligand geometry. The degeneracy of the Fe-3d orbitals is lifted by hybridization with the closest ligand N-2p orbitals, forming t_{2g} and e_g (Figure 4.3b). The related charge-transfer transition Fe-N is depicted in Fig. The maximum absorption of 4.3c occurs at 560 nm. The 300 fs time constant that optical spectroscopy has extracted has been interpreted as a signature of intersystem crossing to a high-spin state with a spin change of ΔS=2.

4.6.1. K-Edge Spectroscopy

The relationship between molecular structure and ultrashort spin cross-over transitions in the low-spin complex $[Fe(tren(py)_3)](PF_6)_2$ dissolved in acetonitrile is directly probed by time-resolved EXAFS measurements at the Fe K-edge. X-ray absorption spectra are obtained using tunable 70 ps x-ray pulses after excitation with a 100 fs optical pulse, and compared with model calculations and the static x-ray absorption spectrum of a high-spin analog. The results show that structural and electronic changes are complete within 70 ps in the photo-excited high-spin state, with the Fe-N bond length increasing by 0.21 Å.

Figure 2a shows the static spectra of the low-spin complex and the chemically-stabilized high-spin analog. Analysis of the static EXAFS spectra reveals that the average Fe-N bond length is approximately 0.23 Å longer in the high-spin analog compared to the low-spin complex. This bond elongation is a result of the difference in electronic configuration between the two compounds and explains all features observed in the XANES region. In Figure 4.4b, the transient difference XAS spectrum of $[Fe(tren(py)_3)]^{2+}$ (photo-excited high-spin state compared to the low-spin ground state) is displayed above the Fe K-edge, 330 ps after laser excitation. The difference of the steady-state spectra from Figure 4.4a is also shown below the transient data, scaled by a factor of 0.1 to adjust for 10% laser photolysis.

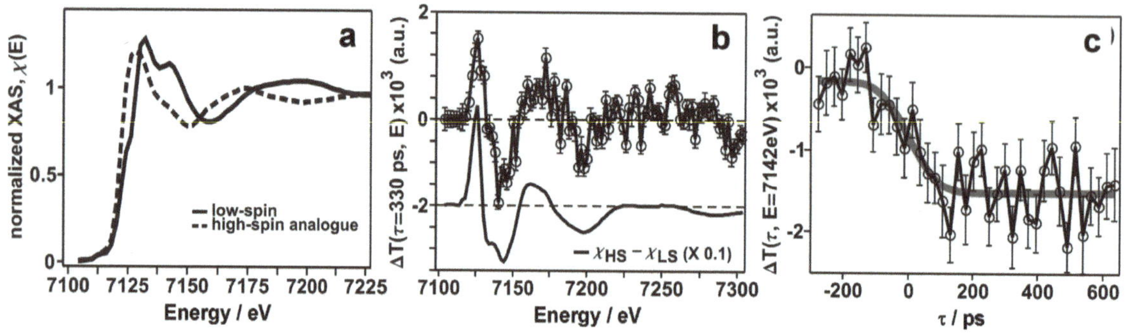

Figure 4.4. *(a) XANES spectra at the Fe K-edge of the low-spin complex* $[Fe(tren(py)_3)](PF_6)_2$ *and the highspin analog* $[Fe(tren(6\text{-}Me\text{-}py)_3)](PF_6)_2$, *both dissolved in acetonitrile. (b) Transient difference XAS spectrum at* $\tau = 330$ *ps* ($HS_{photo\text{-}excited}\text{-}LS_{ground\ state}$)*. The difference between steady-state XAS spectra of the high-spin and low-spin compounds is plotted below. (c) Pump-probe x-ray absorption signal at 7142 eV.* $\Delta T = I(\tau) - I_{ref}$, *where* $I(\tau)$ *and* I_{ref} *refer to the XAS signals with and without the laser pulse, respectively.*

Source: https://www.osti.gov/servlets/purl/953772.

The spectral feature at 7142 eV's time-evolution is displayed in Figure 4.4c in relation to the interval between the optical pump and x-ray probe pulses. The amplitude of the spectral feature at 7142 eV decreases after photoexcitation of the low-spin compound, according to the results. A thick solid curve representing the 70 ps x-ray pulse serves as the fit for the temporal evolution of the difference x-ray signal. It suggests that, within the time resolution of our experiment, the structural evolution of the low-spin compound after photoexcitation is finished. The structure of the transient high-spin state was determined by fitting the k space differential XAS spectrum obtained at τ = 330 ps. This fit shows good agreement and indicates that the excited high-spin state of $[Fe(tren(py)_3)](PF_6)_2$ is characterized by a lengthening of the Fe–N bond by 0.21Å, an interpretation corroborated by further experiments.

4.6.2. L-Edge Spectroscopy

The complex interaction of electronic, atomic, and magnetic structures is revealed through ultrafast intersystem crossing and structural changes. To gain further understanding, time-resolved soft x-ray spectroscopy is utilized at the Fe L3-edge, with Beamline 6.0.2 at the ALS being employed. Liquid-phase studies have been facilitated by the creation of a sample cell (see Figure 4.5, left) that securely holds a liquid film between two 100 nm-thick silicon nitride membranes, allowing for precise control of the film thickness with sub-200 nm accuracy. Additionally, a third silicon-nitride membrane isolates the low-pressure sample chamber from the ultrahigh vacuum of the femtosecond soft x-ray beamline at the ALS. $[Fe(tren(py)_3)](PF_6)_2$ is dissolved in acetonitrile at 100 mM concentration and stimulated with ultrashort 560 nm pulses at 1 kHz. Transient absorption changes at the Fe L_2- and L_3-edges are then examined using tunable 70 ps synchrotron pulses at 2 kHz, considering x-ray probe energy and the time delay between excitation and probe pulses. Through spectral measurements of both the ground and photoexcited state, transient XAS spectra of the high-spin state at the Fe L_2- and L_3-edges are reconstructed.

The high- and low-spin states' absorption following the photo-induced metal-to-ligand charge transfer is depicted in Figure 4.5. To the best of our knowledge, these are the first time-resolved solution-phase transmission spectra in the soft x-ray region to be captured on record. Regarding the low-spin state, the L_3-edge of the high-spin state exhibits a strong red-shift of 1.6 eV and no appreciable change in linewidth. On the blue side of the maximum, there are, nonetheless, more intricate absorption characteristics. When excited by a laser, the L_2-edge only shows a slight redshift of about 0.4 eV, but the absorption cross-section is reduced by almost a factor of 2 (data not shown). Consistent with the spectral scans, the time delay scans in the inset display variations in sample absorption at particular probe photon energies. After excitation, absorption decreases at the low-spin state's spectral absorption maxima but increases at 707 eV, the high-spin state's maximal absorption.

Regarding the K-edge measurements, the delay scans show that the Fe^{2+} compound achieves its meta-stable high-spin state within the experiment's time resolution and reflect the 70 ps width of the x-ray probe pulses. The high-spin electronic structure evolves completely in 70 ps, as shown by transient XAS spectra at 5 ns delay.

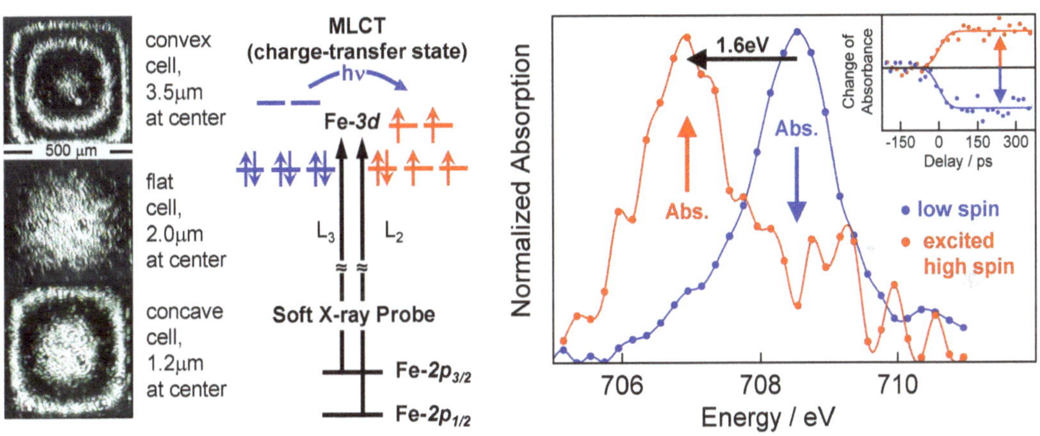

Figure 4.5. Transmission interferograms of the sample cell that indicate liquid film thickness are shown on the left. In the center is a simplified level scheme of the soft x-ray probe transition. To the right, absorption spectra of the Fe L3-edge for the low-spin ground state (blue) and excited high-spin state (red) at 5ns delay. The inset shows time delay scans at the absorption maxima of the low and high-spin state.

Source: https://www.osti.gov/servlets/purl/953772.

The photo-excited high-spin state of $[Fe(tren(py)_3)](PF_6)_2$is typified by a 0.21 Å F–N bond lengthening that takes place in 70 ps. Our time-resolved L-edge measurements in liquid phase show that the changes in ligand-bonding, which mediate the ligand-field splitting, cause a red-shift of the Fe^{2+} L-edges by 1.6 eV. The L-edge measurements provide an important opportunity to study the influence of the solvent on the ultrafast dynamics of the electronic and spin states. These measurements are consistent with static x-ray measurements of a similar compound in crystalline phase (where the high-spin state is trapped at low temperature). The theoretical calculations of the high- and low-spin electronic structure agree well with our experimental findings. Specifically, it reproduces the size of the red-shift of the L_2- and L_3-edges. Our observations are explained by variations in 3D occupancy according to crystal field multiplet calculations.

CASE STUDY

FLUORESCENCE SPECTROSCOPY IDENTIFIES PHENOLIC CONTENT OF HEALTHY OLIVE OILS

After all, consumers know that olive oil has many health benefits and contains "good" fat, so what would a salad be without a drizzle of the stuff? Scientists believe that phenolic compounds found in olive oil may help reduce the incidence of colon, prostate, and coronary heart disease. Phenolic compounds influence olive oils' oxidative stability as well as their sensory qualities. Researchers can quantify those phenolic compounds and ascertain the composition of olive oil by using fluorescence, near-infrared, and mid-infrared spectroscopy. Fluorescence spectroscopy, however, proves to be a quicker method.

Phenolic acids and other components are phenolic compounds found in dietary plants and medicinal herbs. The chemopreventive qualities of phenolic compounds, such as their anti-inflammatory, anticarcinogenic, and antioxidant effects, are caused by a variety of bioactivities. By stopping the cell cycle, controlling the metabolism of carcinogens and the expression of ontogenesis, preventing DNA binding and cell adhesion, migration, proliferation, or differentiation, and obstructing signaling pathways, these actions also aid in the phenolic compounds' ability to induce a cell's natural death.

A popular vegetarian diet, the Mediterranean diet calls for a lot of olive oil consumption. There are at least 30 phenolic compounds found in olive oil. These substances are both beneficial and potent antioxidants and radical scavengers. Compounds called antioxidants are present in food and work to prevent or postpone cell damage. Radical scavengers aid in shielding cells from the harm that free radicals can do. Term "free radicals" describes substances with high reactivity. It has the ability to bind to cells and is frequently linked to cell damage that results in the emergence of cancer.

However, the phenolic benefits of unrefined extra-virgin olive oil are absent from refined oils due to the absence of antioxidants and anti-inflammatories. Regular olive oil has been refined and has lost valuable antioxidants and nutrients. On the other hand, extra virgin olive oil is made by a natural extraction method that guarantees the fruit's antioxidants and nutrients are all retained. Scientists have discovered that olive oils possess distinct fluorescent fingerprints. Fluorescence emissions in fresh, extra-virgin olive oil are caused by phenols, tocopherols, and chlorophylls. Fresh fluorescence arises from oxidation products during the degradation of oil and the consequent loss of compounds that promote health.

Thus, fluorescence spectroscopy can be used by researchers to screen for fluorescent components in olive oil while it is being stored. This enables you to keep an eye on the deterioration of extra virgin olive oil. Furthermore, the amount of refined and deodorized oil that is mixed with extra-virgin olive oils can be measured using fluorescence spectroscopy.

Food spectra can give distinct sample fingerprints that can be used to evaluate food quality. In one study, total phenolic content concentration was determined by utilizing front-face fluorescence measurements in conjunction with chemometrics to develop models that discriminate between virgin olive oils with low and high total phenolic contents using fluorescence spectra. A set of statistical and mathematical techniques known as "chemometrics" are used to enhance comprehension of chemical data. Fluorescence spectroscopy was used in the study to demonstrate its ability to quickly screen for olive oils' total phenolic content (Figure 4.6).

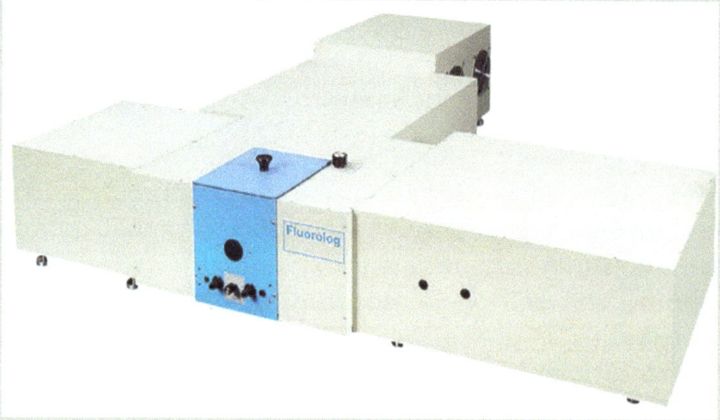

Figure 4.6. *Fluorolog steady-state and lifetime modular spectrofluorometer.*

Source: https://www.horiba.com/bra/scientific/resources/spectroscopy-matters/fluorescence-spectroscopy-identifies-phenolic-content-of-healthy-olive-oils/.

A HORIBA Fluorolog® 3–11 spectrofluorometer—a sensitive and adaptable modular spectrofluorometer with TCSPC lifetimes and interchangeable accessories—was used by researchers to record fluorescence spectra for the study. The instrument's basic setup consisted of a red-sensitive photomultiplier and single-grating monochromators in the emission and excitation positions. Excitation-emission matrices, EEMs, and the synchronous and total synchronous spectra of food samples were all measured by scientists using the Fluorolog.

The team has recently obtained a HORIBA Aqualog® spectrofluorometer that is capable of simultaneously acquiring Absorbance, Transmittance, and fluorescence Excitation-Emission Matrices (A-TEEM). This instrument can collect EEMs up to 100 times faster than traditional scanning fluorescence instruments. The team plans to use the Aqualog for rapid recording of EEMs in food samples.

CLASS ACTIVITY

ULTRAVIOLET-VISIBLE SPECTROSCOPY OF TEXTILE FIBERS

- Scope

Any fiber color comparison must include a quantitative, objective method of color analysis and comparison. For this, visible spectroscopy can be applied. Furthermore, dyes and additives may exhibit distinctive ultraviolet (UV) absorbance that proves advantageous in terms of discrimination. It is not necessary to compute complementary chromaticity coordinates, or colorimetry, in order to compare the colors of fibers used in forensic cases. If the sample size and dye concentration allow, thin layer chromatography or high performance liquid chromatography (HPLC) should be used in addition as a complementary technique for dye analysis when only the visible wavelength range is being used.

- Summary of Activity

This activity utilizes quantitative and qualitative UV-visible (UV-VIS) microscopical spectroscopy or microspectrophotometry in the range of 240nm to 760nm. It is used to analyze single questioned fibers and sets of known fibers during forensic investigations. The method has limitations, such as not being suitable for opaque fibers that have not been reduced in cross section, fibers with low colorant levels that are not visible, and fibers colored with similar compounds like synthetic indigo dyes. Additionally, some fiber polymers can cause interference in the UV range.

- Significance and Use

This activity aims to assist individuals and laboratories conducting forensic fiber examinations by offering guidance on the use of visible or UV-VIS spectroscopy in analyzing fiber evidence. It is designed to be suitable for use with a variety of visible and UV-VIS spectrometers.

- Sample Handling

The samples must adhere to or surpass the specifications outlined in ASTM 1492–05 and the pertinent sections of SWGMAT Trace Evidence Quality Assurance Guidelines in terms of their overall management and monitoring.

- Analysis

Operators must be familiar with the operating manuals of manufacturers due to variations in software and hardware configurations between instrumentation. For microspectrophotometric analysis, the specimen mounting medium must have minimal fluorescence, with options including XAM, fluorescent free glycerol, Phytohistol, Fluoro Mount, Permount, and Norland Optical Adhesive 65. Some of these media may have slight fluorescence but should not interfere with the analysis. To analyze fibers in the full UV-VIS range, they must be mounted on quartz slides with quartz cover slips as glass and many mountants absorb below 300nm.

Sometimes, a scented solvent-based mounting media like XAM can harm certain fiber dyes and fluorescent brighteners by dissolving them and causing them to spread from the fiber. This typically occurs soon after mounting. If any "bleeding" is observed when mounting the known sample, it is advised to use a different mountant for both the known and questioned fibers. It is crucial to use only a small amount of mountant to create a thin, flat, and void-free preparation. Additionally, make sure that the fiber's long axis stays as parallel as possible to the microscope slide surface.

The variety of colors and dyeing depths found in the known fabric, yarn, or other fiber source should be fully reflected in the known fiber sample selection. It is important to make sure that the sample accurately depicts the degree of wear, biological deterioration, thermal, mechanical, bleaching, and laundering artifacts that the item has displayed. To ensure that the fibers are mounted in a single layer, known fibers should be carefully separated (microscopically) and mounted in the same way as the fibers under question.

Spectrophotometer Performance Check: Depending on the equipment being used, all spectroscopy components should warm up before each use. This time will vary. Since absorption spectrophotometry is a quantitative process by nature, it needs proper wavelength and photometric response verification.

Before starting any casework, it is crucial to conduct a wavelength and photometric check to ensure the system's functionality and to keep a record of its performance. This can be achieved by using appropriate standard filters like Holmium Oxide and Didymium Oxide for wavelength and photometric standard filters with documented absorption values and compared against a known standard. The instrument operating parameters for the performance check should match those used for normal casework. To ensure daily data is comparable, the optical path setup must be consistent, including maintaining the objective at a consistent focal position, keeping the measuring and luminous field diameters at similar sizes, and placing the standard filters at a consistent point in the optical path. Ideally, the system's wavelength accuracy should be within +/– one resolution unit under optimized conditions. The instrument's photometric accuracy should be within +/– 5% transmittance (%T) or +/– 0.02 absorbance units (A) for true values above 0.1A, and photometric stability should be within half the allowed accuracy variation or +/– 0.005A for true values above 0.1A. These guidelines should be followed according to the instrument manufacturer's specifications.

Performance records must be kept and should contain the date, system parameters, original instrument output data, including system background scans and single beam "object" or "sample" scans. Other system parameters like dark current, 100% line stability vs. time, and scattered light interference can also be measured and recorded. These measurements are necessary to ensure quality assurance on the instrumentation.

Apertures with fixed or variable sizes and circular or rectangular shapes are used to control the areas (fields) of sample illumination or detector measurement. It is advised to align the fibers in the same direction (N/S or E/W) in systems with rectangular apertures because orientation may affect the instrument response. While all samples should be oriented in the same direction for comparison, circular or square apertures are less sensitive to sample orientation.

Samples must be precisely aligned along the optical axis of the system, with the focus adjusted as close to the center of the sample volume as possible. The setup should prioritize Köhler illumination, with the sample preparation centered on the microscope stage. The luminous field should also be centered on the optical axis, and the detector measuring aperture must be selected and positioned over the luminous field aperture consistently throughout scans. The sizes and positions of the apertures must remain constant between sample and background scans for accurate comparisons.

The absorbance contributions of all system components—aside from the sample of interest—are included in the background reference absorption spectrum, which is referred to as the "blank" or "background" of the system. Sample slide, mountant, and cover slip are all components of the "system," which starts with the data output device and ends with the lamp power supply. The parameters used for the sample (object) scans should be the same as those used for "blank" scans.

In the scan region of interest, the detector sensitivity (voltage or gain) should be adjusted to the system's maximum blank energy transmission wavelength. To ensure the identification of absorbance curve inflection points, the monochromator resolution should be set to 5 nm or higher. Non-scanning systems should be configured to acquire at least two data points per desired resolution unit. The monochromator driver should be set to advance at least two steps per resolution unit (roughly 2 nm steps in this example). Much of this might be automatically optimized or set as a preset value in newer instruments.

Although the visible spectrum is typically thought to lie between 380 and 760 nm, different systems and operator preferences can cause detection to vary by as much as 20 nm. The scanning range is expanded in UV-VIS spectroscopy to encompass the area below 380 nm. To increase accuracy and lower signal noise, the photometric value at each scan step can be calculated using an average of 2 to 50 measurements. Generally, unless the sample shows extremely high absorbance values or a small cross section, a nominal value of 10 measurements per step is sufficient.

Due to their microscopic heterogeneity, most fibers may need to have multiple absorbance spectral scanning locations on one or more fibers in order to produce representative spectra for the entire sample. It's possible that a single fiber won't dye evenly. The cross sections of natural fibers are typically not uniform throughout their length. Variations in color depth can be produced by these circumstances at various points along a fiber. Selecting a measuring location should prevent glaring inhomogeneities from existing in the region being measured. It might be necessary to scan several spots along a single fiber or fibers. If a representative mean absorbance curve and standard deviation curve for a single fiber are required, then more scans might be required. Compared to natural fibers, synthetic fibers may produce good results with fewer scan locations. It's possible for known fiber samples to show dye variations. The goal of sampling these sets of fibers is to ensure that each one displays the broadest range of dyeing depths visually. It is advised to choose at least five fibers from a synthetic fiber sample or 10 fibers from a natural fiber sample for analysis when working with known samples. The scans should show both the midranges and the extremes of the apparent dyeing depths. Make sure to sample a range of cross sections and fiber thicknesses.

Several things may need to be taken into account when comparing UV-VIS to visible spectroscopy in terms of usefulness. For UV-VIS spectroscopy, the fibers mounted on glass slides must be taken out and remounted on quartz slides with quartz cover slips. Certain fiber polymers (e.g., polyester) absorb UV light moderately to strongly, making analysis in this range challenging or useless. The UV range may provide useful discriminating information for other types of fibers.

- Documentation

Depending on the preference of the operator, spectra can be recorded in transmittance or absorbance. However, when recording spectra from very dark fibers, absorbance is advised. Other options include spectral derivatives and averaging. It is possible to compare known and questioned spectra by plotting them sequentially on the same graph or by superimposing them. To ascertain whether there is a positive correlation, each fiber spectrum under investigation needs to be compared to the known fiber spectra. It is necessary to take into account the peak width, peak intensity, and peak maxima's position (nm). A spectral inclusion occurs when, after taking into account the absorbance values and curve shape, the questioned spectrum falls within the range of the known spectra. A spectral exclusion occurs when the absorbance value or curve shape of the questioned spectrum is outside the range of the known spectra. When neither the questioned nor the known spectra have any meaningful points of comparison, the result is inconclusive (e.g., spectra from extremely pale or microscopically black fibers that fall outside of the instrument's dynamic range).

SUMMARY

- Compared with magnetic resonance, or diffraction and microscopy, electronic and vibrational spectroscopy are decidedly low-resolution techniques, in that atomic-level descriptions of biological macromolecular structures are all but impossible to achieve with these techniques.
- UV-Vis spectroscopy is an analytical technique that measures the amount of discrete wavelengths of UV or visible light that are absorbed by or transmitted through a sample in comparison to a reference or blank sample.
- UV-visible spectroscopy is a quantitative technique used in analytical chemistry to measure the amount of light absorbed by a substance. When light falls upon a substance it absorbs and reflects a certain amount of radiation.
- Circular dichroism (CD) is dichroism involving circularly polarized light, i.e., the differential absorption of left- and right-handed light. Left-hand circular (LHC) and right-hand circular (RHC) polarized light represent two possible spin angular momentum states for a photon, and so circular dichroism is also referred to as dichroism for spin angular momentum.
- Infrared spectroscopy (IR spectroscopy or vibrational spectroscopy) is the measurement of the interaction of infrared radiation with matter by absorption, emission, or reflection. It is used to study and identify chemical substances or functional groups in solid, liquid, or gaseous forms. It can be used to characterize new materials or identify and verify known and unknown samples.
- Infrared spectroscopy is a simple and reliable technique widely used in both organic and inorganic chemistry, in research and industry. In catalysis research it is a very useful tool to characterize the catalyst, as well as to detect intermediates and products during the catalytic reaction.
- Fluorescence spectroscopy is an analytical method based on the fluorescence properties of the sample, and is used for quantitative measurements of drugs, metabolites, and other chemical products.
- That is: to follow nuclear motions and valence-charge distributions in real time, to observe the formation and dissolution of chemical bonds, and to understand their cooperative relationship with atomic rearrangement and the formation of new molecular structures. The emerging technique of ultrafast x-ray spectroscopy provides important new insight to molecular dynamics because x-ray transitions are element specific, symmetry and spin state selective, and effective for probing lighter elements (C, N, O, etc.).

REVIEW QUESTIONS

1. What is the limit of detection for a UV-Vis Spectrophotometer?
2. What applications is UV-Vis spectroscopy used for?
3. What is the difference between UV and visible spectrophotometry?
4. What is the CD spectroscopy method?

5. What is the difference between UV and CD spectroscopy?
6. How does a CD spectroscope work?
7. What is the application of vibrational spectroscopy?
8. What are the principles of vibrational spectroscopy?
9. Where is fluorescence spectroscopy used?

REFERENCES

1. Begum, R., Farooqi, Z. H., Naseem, K., et al. (2018). Applications of UV/Vis spectroscopy in characterization and catalytic activity of noble metal nanoparticles fabricated in responsive polymer microgels: A review. *Critical Reviews in Analytical Chemistry, 48*(6), 503–516. doi:10.1080/10408347.2018.1451299.
2. Behzadi, S., Ghasemi, F., Ghalkhani, M., et al. (2015). Determination of nanoparticles using UV-Vis spectra. *Nanoscale, 7*(12), 5134–5139. doi:10.1039/C4NR00580E.
3. Bosch Ojeda, C., & Sanchez Rojas, F. (2013). Recent applications in derivative ultraviolet/visible absorption spectrophotometry: 2009–2011. *Microchemical Journal, 106*, 1–16. doi: 10.1016/j.microc.2012.05.012.
4. Brown, J. Q., Vishwanath, K., Palmer, G. M., & Ramanujam, N. (2009). Advances in quantitative UV–visible spectroscopy for clinical and pre-clinical application in cancer. *Current Opinion in Biotechnology, 20*(1), 119–131. doi: 10.1016/j.copbio.2009.02.004.
5. Diffey, B. L. (2002). Sources and measurement of ultraviolet radiation. *Methods, 28*(1), 4–13. doi:10.1016/S1046-2023(02)00204-9.
6. Domingo, C., & Saurina, J. (2012). An overview of the analytical characterization of nanostructured drug delivery systems: Towards green and sustainable pharmaceuticals: A review. *Analytica Chimica Acta, 744*, 8–22. doi: 10.1016/j.aca.2012.07.010.
7. Gaikwad, J., Sharma, S., & Hatware, K. V. (2020). Review on characteristics and analytical methods of tazarotene: An update. *Critical Reviews in Analytical Chemistry, 50*(1), 90–96. doi:10.1080/10408347.2019.1586519.
8. Gendrin, C., Roggo, Y., & Collet, C. (2008). Pharmaceutical applications of vibrational chemical imaging and chemometrics: A review. *Journal of Pharmaceutical and Biomedical Analysis, 48*(3), 533–553. doi: 10.1016/j.jpba.2008.08.014.
9. Hall, V., Nash, A., Hines, E., & Rodger, A. (2013). Elucidating protein secondary structure with circular dichroism and a neural network. *Journal of Computational Chemistry, 34*(32), 2774–2786. doi:10.1002/jcc.23456.
10. Kalb, V., & Bernlohr, R. (1977). A new spectrophotometric assay for protein in cell extracts. *Analytical Biochemistry, 82*, 362–371. doi:10.1016/0003-2697(77)90173-7.
11. Kristo, E., Hazizaj, A., & Corredig, M. (2012). Structural changes imposed on whey proteins by UV irradiation in a continuous UV light reactor. *Journal of Agricultural and Food Chemistry, 60*(24), 6204–6209. doi:10.1021/jf300278k.

12. Lange, R., & Balny, C. (2002). UV-visible derivative spectroscopy under high pressure. *Biochimica et Biophysica Acta (BBA)–Protein Structure and Molecular Enzymology, 1595*(1–2), 80–93. doi:10.1016/S0167-4838(01)00336-3.

13. Leong, Y., Ker, P., Jamaludin, M., et al. (2018). UV-Vis Spectroscopy: A New Approach for Assessing the Color Index of Transformer Insulating Oil. *Sensors, 18*(7), 2175. doi:10.3390/s18072175.

14. Liu, P.-F., Avramova, L. V., & Park, C. (2009). Revisiting absorbance at 230nm as a protein unfolding probe. *Analytical Biochemistry, 389*(2), 165–170. doi:10.1016/j.ab.2009.03.028.

15. Lourenço, N. D., Lopes, J. A., Almeida, C. F., Sarraguça, M. C., & Pinheiro, H. M. (2012). Bioreactor monitoring with spectroscopy and chemometrics: A review. *Analytical and Bioanalytical Chemistry, 404*(4), 1211–1237. doi:10.1007/s00216-012-6073-9.

16. Namioka, T. (2000). Diffraction gratings. In *Vacuum Ultraviolet Spectroscopy* (Vol. 1, pp. 347–377). Experimental Methods in Physical Sciences. Elsevier. doi:10.1016/B978-012617560-8/50018-9.

17. Patel, M. U. M., Demir-Cakan, R., Morcrette, M., et al. (2013). Li-S Battery Analyzed by UV/Vis in Operando Mode. *ChemSusChem, 6*(7), 1177–1181. doi:10.1002/cssc.201300142.

18. Picollo, M., Aceto, M., & Vitorino, T. (2019). UV-Vi's spectroscopy. *Physics Reports, 4*(4). doi:10.1515/psr-2018-0008.

19. Pinheiro, H. M., Touraud, E., & Thomas, O. (2004). Aromatic amines from azo dye reduction: Status review with emphasis on direct UV spectrophotometric detection in textile industry wastewaters. *Dyes and Pigments, 61*(2), 121–139. doi: 10.1016/j.dyepig.2003.10.009.

20. Sánchez Rojas, F., & Bosch Ojeda, C. (2009). Recent development in derivative ultraviolet/visible absorption spectrophotometry: 2004–2008. *Analytica Chimica Acta, 635*(1), 22–44. doi: 10.1016/j.aca.2008.12.039.

21. Sharpe, M. R. (1984). Stray light in UV-VIS spectrophotometers. *Analytical Chemistry, 56*(2), 339A–356A. doi:10.1021/ac00266a003.

22. Solomon, E. I., & Lever, A. B. P. (2006). *Inorganic Electronic Structure and S.* Wiley-Interscience.

23. Stevenson, K., McVey, A. F., Clark, I. B. N., Swain, P. S., & Pilizota, T. (2016). General calibration of microbial growth in microplate readers. *Scientific Reports, 6*(1), 38828. doi:10.1038/srep38828.

24. Tadesse Wondimkun, Z. (2016). The determination of caffeine level of wolaita zone, Ethiopia coffee using UV-visible spectrophotometer. *American Journal of Applied Chemistry, 4*(2), 59. doi: 10.11648/j.ajac.20160402.14.

25. Tom, J., Jakubec, P. J., & Andreas, H. A. (2018). Mechanisms of enhanced hemoglobin electroactivity on carbon electrodes upon exposure to a water-miscible primary alcohol. *Analytical Chemistry, 90*(9), 5764–5772. doi: 10.1021/acs.analchem.8b00117.

26. Yu, J., Wang, H., Zhan, J., & Huang, W. (2018). Review of recent UV–Vis and infrared spectroscopy researches on wine detection and discrimination. *Applied Spectroscopy Reviews, 53*(1), 65–86. doi:10.1080/05704928.2017.1352511.

CHAPTER 5 Chemical Biology of Cells

LEARNING OBJECTIVES

After studying this chapter, you will be able to:

- Evaluate the need for chemical biology: beyond genetics and biochemistry
- Elaborate the chemical components of a cell
- Determine the chemistry of cells

KEY TERMS FROM THIS CHAPTER

Cell differentiation
Cell proliferation
Cellular phenotypes
Chemical biology
Environmental changes
Gene expression
Inhibitors
Miniaturization
Tissue engineering
Cell migration
Cellular metabolome
Chemical biologists
Chemical genetic screen
Fluorescent probes
High-throughput screening (HTS)
Microfluidics
Molecules

INTRODUCTION

Chemical biology of cells is an interdisciplinary field that explores the interface between chemistry and biology, aiming to understand and manipulate cellular processes at the molecular level. This area employs chemical techniques, tools, and principles to investigate biological systems, offering insights into cellular function and regulation. Central to chemical biology is the synthesis and application of small molecules, which can be designed to interact with specific cellular components, thereby elucidating their roles and mechanisms within the complex milieu of the cell. These small molecules can serve as probes to study protein function, enzyme activity, and signal transduction pathways, providing a powerful means to dissect and manipulate biological processes in a controlled and precise manner.

One of the key focuses in the chemical biology of cells is the study of proteins, which are fundamental to virtually all cellular functions. Chemical biologists develop inhibitors, activators, and modulators that target specific proteins, allowing for the investigation of their roles in health and disease. For instance, kinase inhibitors have been instrumental in understanding signaling pathways involved in cancer and other diseases. The ability to precisely target proteins with small molecules has also paved the way for drug discovery and development, leading to novel therapeutic strategies.

Another significant aspect of chemical biology is the exploration of the cellular metabolome, the complete set of metabolites within a cell. By employing metabolomics techniques, researchers can gain insights into metabolic pathways and their alterations in response to genetic or environmental changes. This has profound implications for understanding diseases such as cancer, where metabolic reprogramming is a hallmark. Chemical tools, such as metabolic tracers and enzyme inhibitors, are used to map and manipulate metabolic networks, thereby uncovering new therapeutic targets and strategies.

Chemical biology also extends to the study of nucleic acids, including DNA and RNA. Techniques such as CRISPR-Cas9 genome editing have revolutionized the ability to precisely alter genetic sequences, providing a deeper understanding of gene function and regulation. Additionally, small molecules that target specific RNA structures or modulate the function of RNA-binding proteins are being developed, offering new avenues for therapeutic intervention, particularly in diseases caused by genetic mutations or dysregulated gene expression.

The integration of chemical biology with advanced imaging techniques has further enhanced our understanding of cellular processes. Fluorescent probes and labels enable the visualization of molecular interactions and dynamics in real time within living cells. These tools provide a window into the spatial and temporal organization of cellular components, revealing intricate details about how cells operate and respond to various stimuli.

5.1. THE NEED FOR CHEMICAL BIOLOGY: BEYOND GENETICS AND BIOCHEMISTRY

Traditional biological sciences fall short in filling the niche that chemical biology fills. Proteins in isolation or in lysates can be understood through the study of biochemistry and enzymology. These techniques have been used to determine the functions of individual proteins, clarify the rates of particular steps in enzyme-catalyzed reactions, and develop tools to modify a particular enzymatic process (Knowles & Albery, 1977). For example, because inhibitors can quickly downregulate particular enzymatic function, inhibition experiments have a direct bearing on physiological research.

As can happen during genetics experiments, these events can have a swift effect on cellular pathways and processes without providing the system with an opportunity to make up for the signaling changes caused by activity loss (see infra). Time-dependent effects allow one to see how the system reacts when the function of the target protein is lost. Naturally, drug design and discovery can also directly benefit from these inhibition studies. The greatest example of the application of inhibitors for research and business purposes may be Daniel Bovet's 1957 Nobel Prize in physiology or medicine. Daniel Bovet won the 1957 Nobel Prize in Physiology or Medicine for his research on artificial compounds that block the effects of specific body substances, particularly those that affect the skeletal muscles and vascular system. Inhibitors are still employed to better model diseases or investigate age- or context-specific loss of protein function in comparison to genetic knockout studies (Ogasawara et al. 2018). Equally important, these chemical-biology experiments necessitate careful consideration and data interpretation because knockout and chemical inhibition are not always equivalent in terms of percentage of loss of function and effect on global protein function (Hedstrom, 2017).

Strong complexes or the function of individual proteins are also frequently studied by biochemistry and enzymology, using techniques like immunoprecipitation and activity assays. However, lysates or purified proteins are usually used in these experiments. In these circumstances, cellular compartmentalization is lost, and a number of additional contextual factors are also thrown off. The nature of triplet codons, for example, and other fundamental processes were revealed by means of such biochemistry experiments (Matthaei et al., 1962). However, lysis or isolation-induced loss of context frequently results in artifacts or a loss of activity or structure because of things like improper preparation methods or expression conditions (Osz-Papai et al., 2015). Large amounts of protein are also needed for these assays, which means that data on, say, cell-to-cell variation is lost. When studying cell-surface behaviors, small-molecule bulk exposure regimens typically lack resolution at the sub-cellular and organ/tissue scales unless they employ cell-impermeable molecules or, as we'll see below, chemical-biological tricks. It is important to remember that a small molecule's intracellular concentration does not always correspond to its media concentration. Throughout the cell, there is frequently a gradient in concentration, particularly for small molecules that are reactive.

However, genetic tools have been shown to be effective in examining the requirements and roles of genes, pathways, and specific protein functions. These studies can frequently be conducted in the preferred organ or tissue, or in the subcellular location. High-throughput screening methods have been used in the global analysis of protein stability, cell-to-cell variation, and pathway intersections to find specific genetic associations that are challenging to find in other ways. Genetics provides methods to control steady-state information flow. These methods are frequently guided by biochemical/structural studies, random mutation of putatively-important mutants, genomic sequencing of specific mutants demonstrating inhibitor resistance, or rescue of a particular phenotype. These studies can frequently be conducted at the single-cell level. However, in all but a few cases, genetic techniques lack precise temporal control and dynamic range, especially in higher eukaryotes. Therefore, new techniques were needed to provide high-resolution data on fleeting associations, to facilitate studies of quick gain of function, to cause localized disruption, or to hone in on the signaling behavior of specific proteins or pathways.

Chemical biology has successfully addressed various questions by providing the ability to create reactive chemical signals and manipulate specific proteins in targeted locations and time frames. By generating highly-reactive small molecules on demand with short half-lives and diffusion distances, chemical biology methods offer valuable insights. These methods often incorporate genetically-encodable elements to control localization and duration of effects, allowing for biocompatible processes. The overarching goals of chemical biology are to apply *in vitro* analysis to cells and organisms and to expand the functions encoded by genetics beyond canonical amino acids. Researchers in chemical biology work at the intersection of multiple fields, such as chemistry, enzymology, and cell signaling, to drive advances in the field. While chemical biology spans various areas of life sciences, there are inherent challenges that must be addressed through thoughtful experimental design.

5.1.1. Array Technologies, Microfluidics, and Miniaturization

Biology, chemistry, and medicine have greatly advanced research and applications due to the transformative fields of array technologies, microfluidics, and miniaturization. The convergence and combination of these fields allow for high-throughput analysis, accurate regulation of chemical and biological processes, and the creation of small and effective devices.

5.1.1.1. Array Technologies

The methodical arrangement of several samples or analytes in a structured format, usually on a flat surface like a microarray chip, is known as array technologies. Thousands of chemical or biological interactions can be analyzed simultaneously thanks to these arrays. For example, DNA microarrays allow scientists to study patterns of gene expression throughout the whole genome, providing insights into disease mechanisms, genetic regulation, and personalized medicine. Enzyme activities, protein-protein interactions, and biomarker identification are all studied using protein and antibody arrays. Because arrays are high-throughput, large-scale screening and analysis are made possible, which speeds up research and makes them essential for fields like proteomics, genomics, and drug discovery.

5.1.1.2. Microfluidics

The science and technology of working with fluids at the microscale, usually in channels with sizes ranging from tens to hundreds of micrometers, is known as microfluidics. This technology makes it possible to precisely control and manipulate small volumes of liquids, which is essential for applications that call for minimal sample usage and fine-scale precision. Lab-on-a-chip systems, also known as microfluidic devices, combine several laboratory operations onto a single chip to enable automated and high-throughput processing. These tools are employed in many different fields, such as medical diagnostics, where their ability to carry out intricate assays with great sensitivity and specificity using small sample volumes is invaluable. Microfluidics is used in research to study cellular behaviors in controlled microenvironments, perform drug screening, and analyze single cells. Tissue engineering and organ-on-a-chip models are further supported by microfluidic devices' capacity to precisely create gradients and environments.

5.1.1.3. Miniaturization

The development of more compact and effective systems and gadgets is known as miniaturization. Miniaturization in the context of array technologies and microfluidics facilitates the integration of multiple functionalities into compact platforms, improves portability, and lowers material and reagent costs. Smaller systems are advantageous for point-of-care diagnostics because they allow for quick on-site testing without requiring bulky, sophisticated laboratory equipment. Miniaturized analytical instruments lower experimental setup footprints and boost throughput in both industrial and research contexts. Advances in nanotechnology, microfabrication methods, and materials science are what are driving the trend toward miniaturization. Examples of these innovations include nanopore sequencing, which enables quick and thorough genetic analysis at the nanoscale.

5.1.1.4. Synergies and Applications

There is great promise for scientific and technological advancement from the combination of array technologies, microfluidics, and miniaturization. For instance, high-throughput platforms for carefully regulated screening of sizable libraries of compounds or genetic variants can be developed by fusing array technologies and microfluidics. These platforms' miniaturization improves their usability and portability, enabling them to be used in a variety of contexts, from field diagnostics to research labs. By providing quick and precise diagnosis, customized treatment regimens, and real-time therapeutic response monitoring, these integrated systems have the potential to completely transform personalized medicine. They also make it easier to test food safety, monitor the environment, and create new materials and sensors.

5.1.2. Chemical Genetics and Potential New Therapeutics

The study of biological systems with the aid of small molecules is known as "chemical genetics." It falls into two categories: chemical forward and chemical reverse. Like forward genetics, forward chemical genetics uses small molecules to evaluate possible effects on gene activity, cellular mechanisms, or organismal behavior. The tiny compounds that result in a desired phenotype (e.g., Following the selection of a compound that causes either gene repression or activation, their structure can be changed to improve the biological response (a process known as structure-activity relationship, or SAR) studies. The identification of the biological target accountable for the desired phenotype is the final and most important stage in a forward chemical genetic approach. Forward chemical genetics has shown to be a useful tool in the field of regenerative medicine. It has been used, for instance, to identify novel factors with cardiogenic potential and to decipher the pathways involved in tissue regeneration. On the other hand, reverse chemical genetics functions similarly to reverse genetics, working from "genotype to phenotype," with an emphasis on a particular protein and the identification of substances from a library that bind to and modify its function. One common instance is the discovery of small molecule inhibitors that specifically target protein kinases. For instance, 2,6,9-trisubstituted purines are known to target the human cyclin-dependent kinase 2's adenosine triphosphate-binding site, which is a crucial cell cycle regulator. Another example of a medication that has been approved by the US Food and Drug Administration (FDA) for the treatment of cutaneous T cell lymphoma is suberoylanilide hydroxamic acid, a histone deacetylase inhibitor that targets all 11 human known class I and class II histone deacetylases. GPCRs are a common class of reverse chemical genetic targets. One of the main downstream targets of activated GPCRs, regulator of G-protein signaling protein 4 (RGS4), is specifically inhibited by methyl N-[(4-chlorophenyl)sulfonyl]-4-nitrobenzenesulfinimidoate (CCG-4986).

Chemical genetic screens require three main components: (i) a large and diverse library of small molecules with both known and unknown biological targets; (ii) an assay to screen the components of the library; and (iii) a way to link an active compound to its target. Small molecules extracted from naturally occurring bioactive products, synthetic natural product-like compounds, or a specially created commercial collection could all be found in the library. Notably, the vast majority of FDA-approved medications in recent years have been made from natural sources. A well-known example of a medication made from naturally occurring bioactive substances is paclitaxel, which was initially extracted from the bark of the Pacific yew tree Taxus brevifolia. It is approved for the treatment of lung, breast, and ovarian cancer. When utilizing a library of naturally occurring bioactive substances in a screen, it's crucial to keep in mind that the clarification of the chemical structure. The study of the target compound through mass spectrometry, nuclear magnetic resonance spectroscopy, and x-ray crystallography is essential to future drug development. The fact that bespoke commercial collections are made up of highly pure substances with well-established structures and functions is advantageous in this sense. Moreover, multi-well plates are commonly used in commercial libraries, allowing for high-throughput screening (HTS) (Lokey, 2003). Commercially accessible libraries typically contain FDA-approved therapeutic medications, but they may also contain substances that were toxic in clinical trials and did not pass.

Any study on chemical genetics must also carefully consider which small molecule screening technique to use. Screening assays typically use living cells or organisms and are based on phenotypic changes. Small molecules may have an impact on cellular phenotypes such as viability, proliferation, migration, and differentiation. Cellular screens can also be used to assign molecular phenotypes, such as variations in gene expression. Whole organism phenotypes can include variations in body size, weight, and tissue remodeling and regeneration. Because zebrafish can regenerate their heart, spinal cord, or fins, they are often used as models for studying tissue regeneration. Chemical genetics has proven to be a valuable tool in the investigation of the pathways involved in zebrafish tissue regeneration. Mathew and colleagues, for instance, screened 2000 biologically active small molecules and found 17 compounds, including beclomethasone, a medication often prescribed for asthma prophylaxis, which specifically inhibited tissue regeneration by activating glucocorticoid receptor signaling. Notably, the process of identifying phenotypic changes in cells or organisms usually takes a while because it frequently entails evaluating biological parameters like body mass measurements. It is crucial to take into account the automation of phenotypic alterations and the high-throughput recording that follows. High throughput cell imaging screening can be used to evaluate changes in mitotic disruption, protein trafficking, cell size, shape, and proliferation brought on by small molecules. A widely employed substitute method relies on the alteration of reporter gene activity through small molecule-mediated expression (e.g., luciferase, eGFP, or mCherry) under the direction of a desired promoter. This method's primary benefit is that it makes automated screening based on fluorescence and cell imaging possible. The following considerations must be made, though, when relying on reporter activity to evaluate the impact on gene expression: (i) selecting the promoter region to use in the reporter construct to accurately recapitulate the expression of the gene of interest; (ii) choosing the cell type to use in the screening—primary tissue versus cell line; and (iii) the origin of the cells used (e.g., Rat versus. human).

After the initial screening of the library, compounds exhibiting a desired phenotype are chosen, and structural antigen recognition (SAR) studies are conducted to ascertain whether structural alterations could augment compound activity and whether particular chemical substructures could be linked to a particular biological effect. Additionally, the "drug-likeness" of the chosen compounds will be enhanced by SAR studies in terms of their pharmaceutical and pharmacological profile, which includes toxicology, excretion, metabolism, distribution, and absorption (ADMET properties). The quality of a compound with respect to its ADMET properties can be evaluated using a variety of parameters, as suggested by Lipinski's rule of five (Lipinski et al. 2001). These parameters include calculated values like molecular weight, polar surface, ionization (pKa), solubility, and cLogP to estimate lipophilicity and drug absorption, respectively.

The identical primary assay is then used for a secondary screen. These screens' primary goals are to locate lead "hits" and identify any possible toxicity problems. In conducting secondary screens, the chosen compounds can be mixed together to evaluate potential synergistic effects; similarly, growth factors that are known to regulate or modulate the target phenotype can be included as controls in the secondary screens and/or combined with the hits to evaluate potential modifier functions. The biological target must be identified after the phenotypic action of the "hits" has been verified. It is vital to ascertain whether a known or previously unidentified interaction was the cause of the phenotypic event, even if the small molecules chosen have known function(s) and target(s). There are numerous ways to link an active substance to the biological target or targets. The most popular methods are affinity-based ones, which typically entail immobilizing the small molecule on a glass slide, microarray slide, magnetic bead, or affinity chromatography bead. The immobilized form of the small molecule is then used to purify the biological target from a cellular extract or tissue homogenate, and mass spectrometry is used to identify it. Alternatively, biotin can be radiolabeled or tagged for non-covalent immobilization of small molecules. Biotin binds strongly to streptavidin-coated beads. It is necessary to test the modified small molecule in the primary screen because immobilizing or tagging small molecules carries the risk of influencing their binding to the molecular target.

Numerous biological discoveries have been made possible by small molecules, which offer a high degree of temporal control over protein function. They can cause rapid activation or inhibition of protein function, and their effects are frequently reversible. One compound has the ability to disrupt multiple targets within or across protein families and signaling pathways, in addition to a single function of a multifunction protein. The effects of small molecules on biological systems can be precisely controlled by adjusting their concentration. Furthermore, small molecules have a broad potential for therapeutic development because they can interfere with protein-protein interactions and generally function in a variety of cell types and species (Kawasumi & Nghiem, 2007). Ultimately, the discovery that a number of artificial and natural substances, such as flavonoids and dimethyl sulfoxide (DMSO), have the ability to induce cardiogenesis has raised the profile of small molecules and chemical genetics as feasible methods for drug development in the field of cardiovascular regenerative medicine.

5.1.3. Chemical Cellular Dynamics

The complex mechanisms influencing the movement and interaction of molecules within cells are studied in the field of chemical cellular dynamics. This field focuses on comprehending how molecular interactions and chemical reactions affect cellular responses, dynamics, and overall functions. It includes a number of topics, including metabolic networks, signaling pathways, and the temporal and spatial organization of cellular constituents. Through the application of a blend of chemical, biological, and computational methodologies, scientists endeavor to elucidate the intricacies of cellular behavior and the fundamental mechanisms that govern cellular reactions to both internal and external stimuli. The study of signaling pathways, which are networks of proteins and other molecules that carry signals from the cell surface into the interior, is one of the main topics of chemical cellular dynamics. Critical cellular processes like growth, division, differentiation, and apoptosis are regulated by these pathways. Researchers can investigate the functions of individual components by manipulating these pathways with high specificity thanks to chemical biology's tools. For example, kinases, which are essential enzymes in signal transduction and add phosphate groups to proteins, can be specifically inhibited by small molecule inhibitors. Through the observation of these inhibitors' effects, researchers can determine which kinases contribute to which cellular signaling networks. An additional essential component of chemical cellular dynamics is metabolic networks, which include the chain of chemical reactions that take place inside a cell to sustain life. Metabolism is the process by which nutrients are transformed into energy and macromolecule building blocks and waste products are eliminated. Analysis of metabolic changes in response to different conditions is made possible by metabolomics, the thorough study of metabolites in a biological system. The flow of metabolites through various pathways is studied using chemical probes and metabolic tracers, which show how cells modify their metabolism in response to environmental changes like stress or nutrient availability. Understanding cellular dynamics requires an understanding of the temporal and spatial organization of cellular components. Real-time visualization of molecular interactions and processes within living cells is made possible by the combination of chemical probes and advanced imaging techniques. Researchers can follow the location and movement of fluorescent markers within a cell by attaching them to specific proteins or nucleic acids. This method has provided insight into dynamic processes like the trafficking of proteins, the dynamics of organelles, and the formation and dissolution of molecular complexes. Chemical cellular dynamics encompasses the investigation of the interactions that cells have with one another and with their surroundings. Cellular behavior is influenced by the extracellular matrix, signaling molecules, and cell adhesion molecules in the microenvironment. Chemical instruments that mimic and modify the extracellular environment, like surface-modified materials and biosensors, can reveal how cells react to chemical and physical stimuli. This is especially crucial to comprehending processes such as wound healing, tissue growth, and cancer metastasis.

5.2. THE CHEMICAL COMPONENTS OF A CELL

Combinations of elements, such as carbon or hydrogen, that are incapable of chemically dissolving or changing into other substances, make up matter. An atom is the smallest known particle of an element that nevertheless possesses its unique chemical characteristics. However, the way atoms are arranged in groups to form molecules determines the properties of substances other than pure elements, such as the building blocks of living cells. Therefore, understanding the formation of all the chemical bonds that hold atoms in molecules together is essential to understanding how living organisms are constructed from inanimate matter.

5.2.1. Cells Are Made From a Few Types of Atoms

A cloud of negatively charged electrons, held in a series of orbitals by electrostatic attraction to the nucleus, surrounds the positively charged nucleus at the center of every atom. Conversely, protons, which have a positive charge, and neutrons, which have an electrical neutrality, make up the nucleus. The atomic number is determined by the quantity of protons in the atomic nucleus. Since a hydrogen atom's nucleus is made up of just one proton, hydrogen is the lightest element with an atomic number of 1. An atom of carbon has an atomic number of six and a nucleus made up of six protons (Figure 5.1). Every proton carries an electric charge that is precisely equal to the charge borne by an individual electron. The number of negatively charged electrons surrounding the nucleus is equal to the number of positively charged protons the nucleus contains because an atom is electrically neutral overall. As a result, the number of electrons in an atom also equals the atomic number. Every atom in a given element has the same atomic number, and it is these electrons that determine how an atom behaves chemically.

Even though the electrons are depicted here as separate particles, the laws of quantum mechanics control their behavior, making it impossible to predict with certainty where an electron will be at any given time. All atoms, except hydrogen, have neutrons that are electrically neutral and positively charged protons in their nuclei. An atom has no net charge because the number of electrons in it equals the number of protons or the atomic number. The size of the protons, neutrons, and electrons is greatly exaggerated here; in actuality, they are tiny in comparison to the atom as a whole. Furthermore, the nucleus' diameter is only roughly 10^{-4} that of the electron cloud.

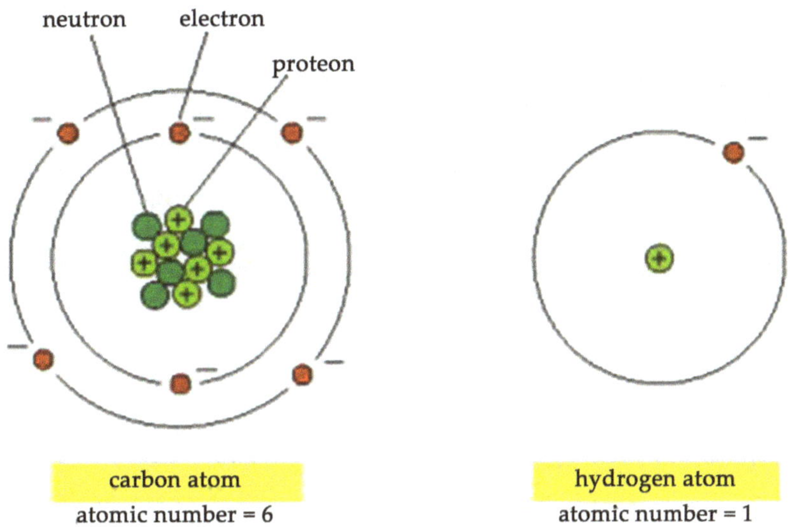

Figure 5.1. *Highly schematic representations of an atom of carbon and an atom of hydrogen.*

Source: https://www.ncbi.nlm.nih.gov/books/NBK26883/bin/ch2f1.jpg.

Uncharged subatomic particles, neutrons and protons have nearly equal masses. They do not change the chemical makeup of the atom, but they do help keep the nucleus structurally stable. If there are too few or too many, radioactive decay could cause the nucleus to break apart. As a result, an element can exist in multiple isotopes—physically distinct but chemically identical forms—each of which has an equal number of protons but a different number of neutrons.

A mole is X grams of a substance, where X is its realative molecular mass (molecular weight). A mole will contain 6x 10^{23} molecules of the substance.

1 mole of carbon weighs 12 g
1 mole of glucose weighs 180 g
1 mole of sodium chloride weighs 58 g

Molar solutions have a concentration of 1 mole of the substance in 1 liter of solution. A molar solution (denoted as 1 M) of glucose, for example, has 180 g/l, while a millimolar solution (1mM) has 180 mg/l.

The standard abbreviation for gram is g; the abbreviation for liter is l.

Figure 5.2. *Moles and molar solutions.*

Source: https://www.ncbi.nlm.nih.gov/books/NBK26883/bin/ch2f2.jpg.

Nearly every element has multiple isotopes in nature, including some unstable ones. As an illustration, the majority of carbon on Earth is found in the stable isotope carbon 12, which has six protons and six neutrons. However, trace amounts of the unstable isotope, radioactive carbon 14, which has six protons and eight neutrons in each atom, are also present. Radioactive decay of carbon 14 occurs gradually but steadily. This serves as the foundation for the carbon 14 dating method, which is employed in archaeology to establish the chronological age of organic material origins.

The mass of an atom in relation to the mass of a hydrogen atom is known as its atomic weight, also known as its molecular weight. Because electrons are much lighter than protons and add very little to the total, this is essentially equal to the number of protons plus neutrons that the atom or molecule contains. As a result, the major isotope of carbon, designated ^{12}C, has an atomic weight of 12, while the recently discussed unstable isotope, designated ^{14}C, has an atomic weight of 14. A Dalton is an atomic mass unit that is roughly equivalent to the mass of an atom of hydrogen. The mass of an atom or molecule is commonly expressed in Daltons. It is difficult to comprehend the tiny size of an atom. Since the diameter of a single carbon atom is approximately 0.2 nm, a millimeter would require around 5 million of them to be spread out in a straight line. A proton or neutron weighs roughly $1/(6 \times 10^{23})$ gram, meaning that 6×10^{23} atoms make up one gram of hydrogen. The key scale factor that describes the relationship between everyday quantities and quantities measured in terms of individual atoms or molecules is this enormous number, also known as Avogadro's number (6×10^{23}). A substance with a molecular weight of X will weigh X grams when 6×10^{23} molecules of it are present. One mole of the substance is this amount (Figure 5.2).

There are 92 naturally occurring elements, each differing from the others in the number of protons and electrons in its atoms. Living organisms, however, are made of only a small selection of these elements, four of which—carbon (C), hydrogen (H), nitrogen (N), and oxygen (O)—make up 96.5% of an organism's weight. This composition differs markedly from that of the nonliving inorganic environment (Figure 5.3) and is evidence of a distinctive type of chemistry. The most common elements in living organisms are listed in Table 5.1 with some of their atomic characteristics.

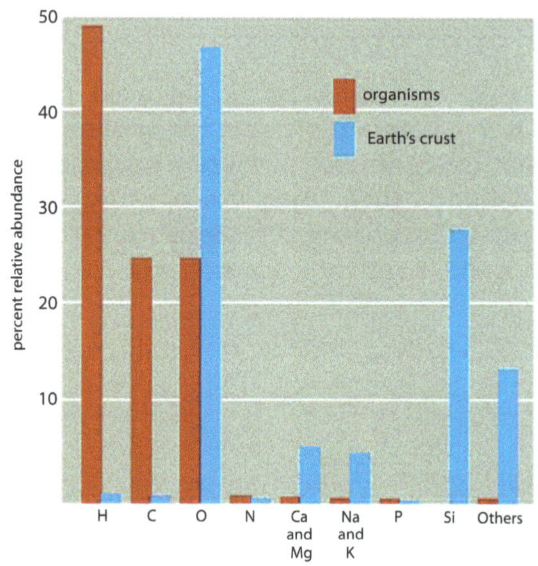

Figure 5.3. *The abundances of some chemical elements in the nonliving world (the Earth's crust) compared with their abundances in the tissues of an animal.*

Source: https://www.ncbi.nlm.nih.gov/books/NBK26883/bin/ch2f3.jpg.

The abundance of each element is expressed as a percentage of the total number of atoms present in the sample. Thus, for example, nearly 50% of the atoms in a living organism are hydrogen atoms.

The survey here excludes mineralized tissues such as bone and teeth, as they contain large amounts of inorganic salts of calcium and phosphorus. The relative abundance of elements is similar in all living organisms.

Table 5.1. Atomic Characteristics of the Most Abundant Elements in Living Tissues

Common Elements in Living Organisms						
Element		**Protons**	**Neutrons**	**Electrons**	**Atomic Number**	**Atomic Weight**
Hydrogen	H	1	0	1	1	1
Carbon	C	6	6	6	6	12
Nitrogen	N	7	7	7	7	14
Oxygen	O	8	8	8	8	16
Less Common Elements						
Element		**Protons**	**Neutrons**	**Electrons**	**Atomic Number**	**Atomic Weight**
Sodium	Na	11	12	11	11	23
Magnesium	Mg	12	12	12	12	24
Phosphorus	P	15	16	15	15	31
Sulfur	S	16	16	16	16	32
Chlorine	Cl	17	18	17	17	35
Potassium	K	19	20	19	19	39
Calcium	Ca	20	20	20	20	40

Source: https://www.ncbi.nlm.nih.gov/books/NBK26883/table/A170/?report=objectonly.

5.2.2. The Outermost Electrons Determine How Atoms Interact

We need to focus on an atom's electrons in order to comprehend how atoms combine to form molecules, which are the building blocks of living things. In the nucleus, protons and neutrons are fused together tightly and only undergo partner switching in extreme circumstances, such as in radioactive decay, the sun's interior, or a nuclear reactor. Only an atom's electrons can rearrange themselves in living tissues. They define the chemistry that unites atoms to form molecules and form the outer shell of an atom. Although motions on this submicroscopic scale follow different laws from those we are accustomed to in daily life, electrons are always in motion around the nucleus. According to these laws, an atom's electrons can only exist in specific discrete states known as orbitals, and the number of electrons that can fit in an orbital of a particular type—referred to as an electron shell—is strictly limited. The electrons in the innermost, most tightly bound shell are drawn to the positive nucleus the strongest because they are, on average, the closest to it. There can be no more than two electrons in this shell. The electrons in the second shell are less firmly bonded and are located farther from the nucleus. Up to eight electrons can be contained in this second shell. Even loosely bound electrons can be found in the third shell, which has the capacity to accommodate up to eight electrons. Each of the fourth and fifth shells has a capacity of 18 electrons. In biological molecules, atoms with more than four shells are extremely uncommon.

An atom's electron configuration is most stable when all of its electrons are in the tightest bound states that are feasible for them, or when they are all occupied by the innermost shells. Therefore, the electrons of an atom fill the orbitals in the following order: the first shell fills before the second, the second shell fills before the third, and so forth, with some exceptions in the larger atoms. A completely electron-filled outermost shell indicates that an atom is exceptionally stable and hence chemically inert. Inert gases such as argon with 2 + 8 + 8, neon with 2 + 8, and helium with 2 electrons are examples. In contrast, hydrogen is extremely reactive because it only has one electron and a partially filled shell. The other atoms present in living tissues can also give, take, or share electrons with one another to form molecules and ions because they all have incomplete outer electron shells (Figure 5.4).

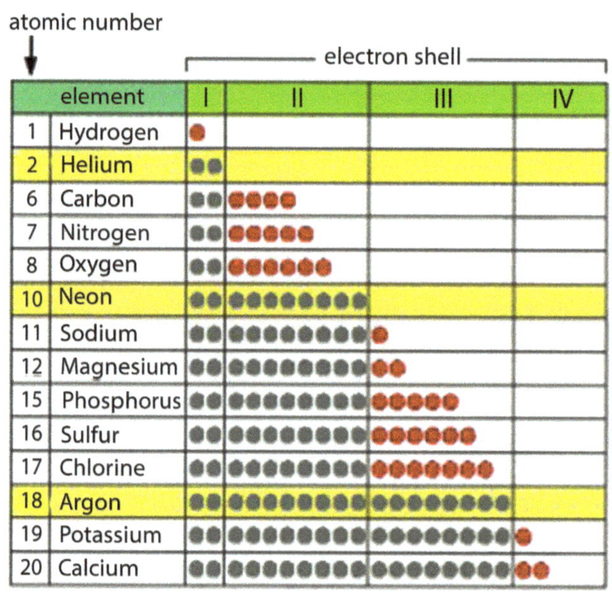

Figure 5.4. *Filled and unfilled electron shells in some common elements.*

Source: https://www.ncbi.nlm.nih.gov/books/NBK26883/bin/ch2f4.jpg.

Due to their empty outer shells (red), all elements that are frequently found in living things are capable of reacting chemically with other atoms. Some elements that are chemically unreactive and only have filled shells (yellow) are displayed for comparison. Atoms lacking an electron shell are more likely to interact with other atoms in a way that results in them gaining or losing enough electrons to form a completed outermost shell, as an unfilled electron shell is less stable than a filled one. Two ways to accomplish this electron exchange are by sharing electrons between two atoms or by moving electrons from one atom to another. Two types of chemical bonds are produced between atoms by these two strategies: covalent bonds are created when two atoms share a pair of electrons, while ionic bonds are created when one atom donates electrons to another (Figure 5.5). A polar covalent bond is the result of an intermediate strategy wherein the pair of electrons is shared unequally, with a partial transfer between the atoms, as we shall discuss later.

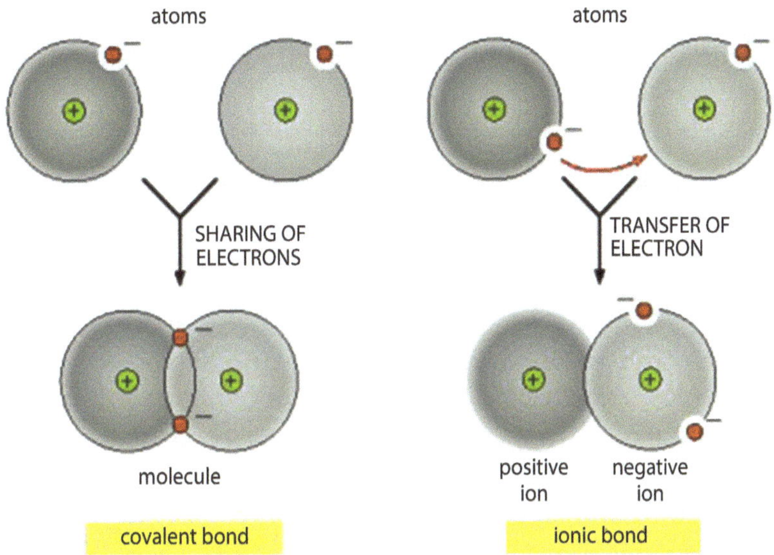

Figure 5.5. *Comparison of covalent and ionic bonds. Atoms can attain a more stable arrangement of electrons in their outermost shell by interacting with one another.*

Source: https://www.ncbi.nlm.nih.gov/books/NBK26883/bin/ch2f5.jpg.

An H atom can obtain an additional electron by electron sharing, forming a covalent bond with another atom that is frequently polar. This allows the H atom to reach its shell with just one extra electron. The remaining most prevalent elements in living cells—C, N, and O, which have an incomplete second shell, and P and S, which have an incomplete third shell (see Figure 5.4)—usually share electrons in order to form multiple covalent bonds that fill the outer shell with eight electrons. The term "valence" refers to the quantity of electrons that an atom needs to gain or to lose (by sharing or by transfer) to reach a full outer shell. Because the outer electron shell plays such a significant role in determining an element's chemical properties, elements with similar properties can be found periodically when elements are listed in order of atomic number. For example, an element with one electron in its incomplete second shell will behave similarly to an element with one electron in its incomplete third shell. For instance, the outer shells of metals are incomplete, containing only one or a few electrons, while the outer shells of inert gases are complete, as we have just seen.

5.2.3. Ionic Bonds Form by the Gain and Loss of Electrons

Atoms with one or two extra electrons on top of a filled outer shell, or those that are just one or two electrons away from obtaining a filled outer shell, are most likely to form ionic bonds. Transferring electrons to or from another atom makes it easier for them to achieve a fully filled outer electron shell than sharing electrons. For instance, atomic number 11 sodium (Na) atoms can decompose into a filled shell by relinquishing the one electron that separates them from their second shell (Figure 5.4). On the other hand, an atom of chlorine (Cl), which has atomic number 17, can finish its outer shell with just one additional electron. As a result, an electron can move from a Na atom to a Cl atom when they come into contact, leaving both atoms with filled outer shells. Table salt is the result of this union of sodium, a soft and highly reactive metal, and chlorine, a poisonous green gas (NaCl).

These two atoms become electrically charged ions when an electron jumps from Na to Cl. The Na atom that experienced an electron loss now has one fewer electron in its nucleus than protons, resulting in a single positive charge (Na^+). With one extra electron than protons, the Cl atom with the gained electron now has a single negative charge (Cl^-). Negative ions are referred to as anions, and positive ions as cations. Ions can be further divided based on the number of electrons they gain or lose. This means that while magnesium and calcium have two electrons to lose and form cations with two positive charges (Mg^{2+} and Ca^{2+}), sodium and potassium (K) only have one electron to lose and form cations with a single positive charge (Na^+ and K^+). Na^+ and Cl^- are attracted to one another due to their opposing charges, which keeps them together in an ionic bond. A single 1 mm across crystal of salt contains astronomical amounts of Na+ and Cl^- (approximately 2×10^{19} ions of each type), precisely arranged in three dimensions with their opposing charges perfectly balanced (Figure 5.6).

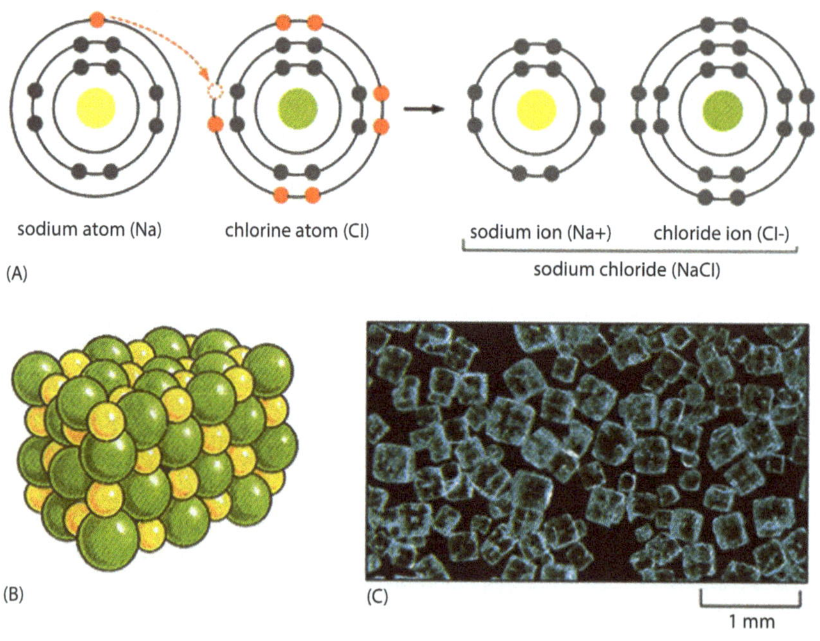

Figure 5.6. (A) An atom of sodium (Na) reacts with an atom of chlorine (Cl). Electrons of each atom are shown schematically in their different energy levels; electrons in the chemically reactive (incompletely filled) shells are red. The reaction takes place with transfer of a single electron from sodium to chlorine, forming two electrically charged atoms, or ions, each with complete sets of electrons in their outermost levels. The two ions with opposite charge are held together by electrostatic attraction. (B) The product of the reaction between sodium and chlorine, crystalline sodium chloride, consists of sodium and chloride ions packed closely together in a regular array

in which the charges are exactly balanced. (C) Color photograph of crystals of sodium chloride.

Source: https://www.ncbi.nlm.nih.gov/books/NBK26883/bin/ch2f6.jpg.

Generally speaking, substances such as NaCl—which are only held together by ionic bonds—are referred to as salts rather than molecules. We will discuss other examples of the various kinds of noncovalent bonds that can form between atoms, of which ionic bonds are just one.

A sodium (Na) atom and a chlorine (Cl) atom interact in the above figure (A). Each atom's electrons are represented schematically in relation to their various energy levels; electrons in chemically reactive (partially filled) shells are indicated by red. One electron is transferred from sodium to chlorine during the reaction, resulting in the formation of two electrically charged atoms, or ions, each of which has a full set of electrons at the outermost level. Electrostatic attraction holds the two oppositely charged ions together. (B) The result of the reaction between sodium and chlorine is crystalline sodium chloride, which is made up of closely spaced sodium and chloride ions arranged in a regular array with perfectly balanced charges. (C) Color photograph showing sodium chloride crystals. Water significantly weakens ionic bonds because it interacts favorably with ions. As a result, many salts, including NaCl, dissolve into individual ions like Na^+ and Cl^-, which are encircled by a collective of water molecules. On the other hand, covalent bond strengths are not impacted in this manner.

5.2.4. Covalent Bonds Form by the Sharing of Electrons

A cell's molecules determine every one of its properties. A molecule is characterized as a group of atoms bound together by covalent bonds; in these cases, electrons are shared rather than transferred between atoms to complete the outer shells. The number of electrons needed to fill the first shell is shared by two H atoms, each of which has one electron, in the most basic possible molecule: a hydrogen (H_2) molecule. In contrast to the mutual repulsion between like charges that would otherwise drive them apart, these shared electrons create a cloud of negative charge that is densest between the two positively charged nuclei and aids in holding them together. When there is a characteristic distance, known as the bond length, between the nuclei, the attractive and repulsive forces are in balance.

Strength is an additional essential characteristic of any bond, whether covalent or noncovalent. The amount of energy required to break a bond is a measure of bond strength. Kilocalories per mole, or kcal/mole, are commonly used to express this. One kilocalorie is the amount of energy required to raise the temperature of one liter of water by one degree Celsius. Thus, the strength of a bond is 1 kcal/mole if 1 kilocalorie is required to break 6×10^{23} bonds of a particular type, or 1 mole of these bonds. Equivalent to 0.239 kilocalories, the kilojoule is a commonly used unit of measurement for energy. It is useful to compare bond strengths with the average energies of the impacts that molecules constantly experience from collisions with other molecules in their environment (their thermal, or heat, energy), as well as with other biological energy sources like light and glucose oxide fermentation (Figure 5.7) to gain an understanding of what bond strengths mean. Covalent bonds are typically only broken during particular chemical reactions with other atoms and molecules because they are typically 100 times stronger than the thermal energies.

As a result, they are resistant to being pulled apart by thermal motions. Covalent bond formation and breakage are violent processes that are meticulously regulated by highly specialized catalysts known as enzymes in living cells. Noncovalent bonds are generally much weaker, but as we will see later, they play a crucial role in the cell in the various circumstances in which molecules must associate and dissociate quickly to perform their tasks.

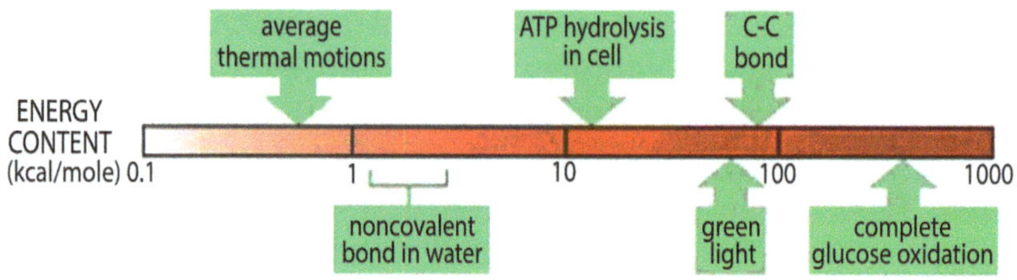

Figure 5.7. *Some energies important for cells. Note that these energies are compared on a logarithmic scale.*

Source: https://www.ncbi.nlm.nih.gov/books/NBK26883/bin/ch2f7.jpg.

The other common atoms that form covalent bonds in cells—O, N, S, and P, as well as the crucial C atom—can form multiple bonds, whereas an H atom can only form one. As we've seen, these atoms can have up to eight electrons in their outermost shell, and they can form covalent bonds with as many other atoms as needed to get there.

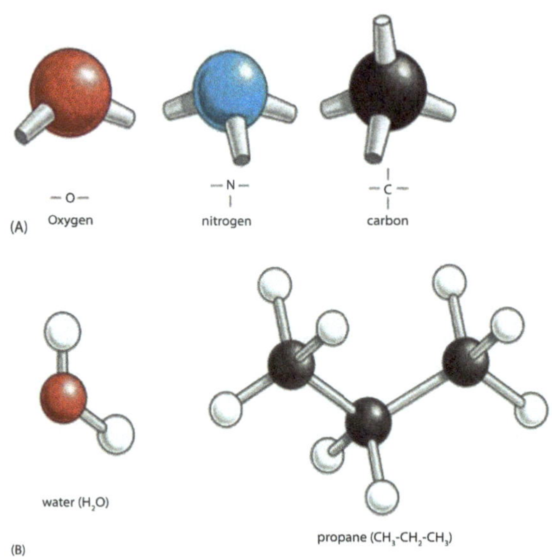

Figure 5.8. *The geometry of covalent bonds.*

Note: (A) The spatial arrangement of the covalent bonds that can be formed by oxygen, nitrogen, and carbon. (B) Molecules formed from these atoms have a precise three-dimensional structure, as shown here by ball and stick models for water and propane. A structure can be specified by the bond angles and bond lengths for each covalent linkage.

Source: https://www.ncbi.nlm.nih.gov/books/NBK26883/bin/ch2f8.jpg.

With its outer shell containing six electrons, oxygen is most stable when it shares two additional electrons with other atoms to create up to two covalent bonds. As can be seen in Figure 5.4, nitrogen, having five outer electrons, can form up to three covalent bonds, whereas carbon, having four outer electrons, can form up to four covalent bonds, sharing four pairs of electrons in the process. When an atom forms a covalent bond with multiple other atoms, the orbits of the shared electrons are reflected in the distinct spatial orientations of these multiple bonds. As a result, particular bond angles, bond lengths, and bond energies define covalent bonds involving multiple atoms (Figure 5.8). The arrangement of the four covalent bonds that can form, for instance, around a carbon atom is likened to the four corners of a regular tetrahedron. The three-dimensional geometry of organic molecules is based on the exact orientation of covalent bonds.

5.2.5. There Are Different Types of Covalent Bonds

The majority of covalent bonds, also known as single bonds, are characterized by the sharing of two electrons, one from each participating atom. On the other hand, sharing multiple pairs of electrons occurs in some covalent bonds.

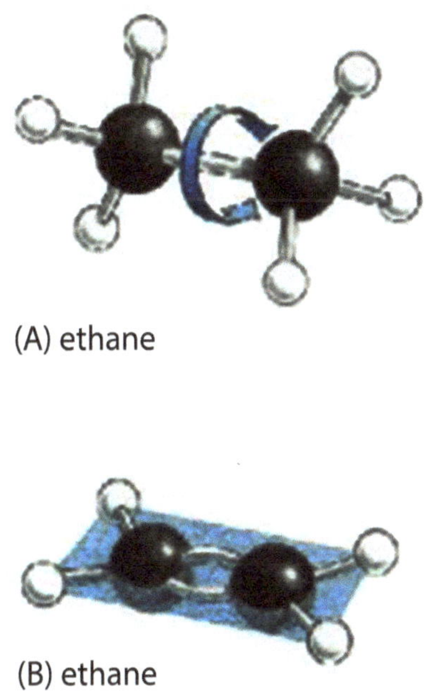

Figure 5.9. *Carbon-carbon double bonds and single bonds compared.*

Note: (A) The ethane molecule, with a single covalent bond between the two carbon atoms, illustrates the tetrahedral arrangement of single covalent bonds formed by carbon. One of the CH_3 groups joined by the covalent bond can rotate relative to the other around the bond axis. (B) The double bond between the two carbon atoms in a molecule of ethene (ethylene) alters the bond geometry of the carbon atoms and brings all the atoms into the same plane (blue); the double bond prevents the rotation of one CH_2 group relative to the other.

Source: https://www.ncbi.nlm.nih.gov/books/NBK26883/bin/ch2f9.jpg.

A double bond is created when four electrons are shared, for instance, with two coming from each participating atom. Double bonds have a distinctive impact on the three-dimensional geometry of molecules containing them because they are stronger and shorter than single bonds. One atom can often rotate one portion of a molecule relative to the other around the bond axis when two atoms form a single covalent bond. A double bond stops this kind of rotation, resulting in an atom arrangement that is less flexible and more rigid (Figure 5.9).

Certain molecules exhibit a hybrid nature that lies in between single and double bonds, as a result of three or more atoms sharing electrons. For instance, the extremely stable benzene molecule consists of a ring of six carbon atoms with evenly distributed bonding electrons (typically represented as an alternating series of single and double bonds).

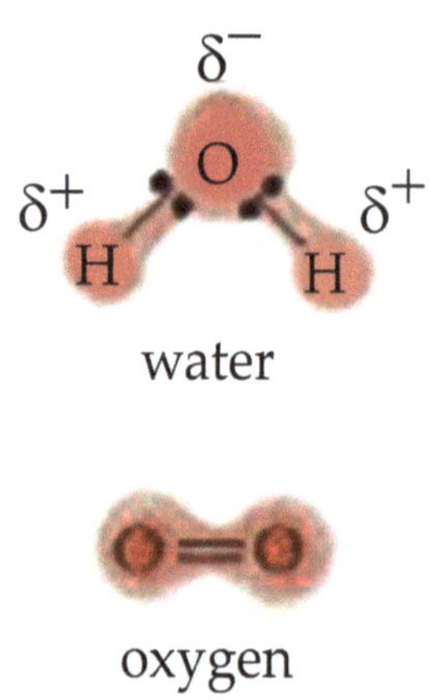

Figure 5.10. *Polar and nonpolar covalent bonds. The electron distributions in the polar water molecule (H2O) and the nonpolar oxygen molecule (O2) are compared (δ+, partial positive charge; δ-, partial negative charge).*

Source: https://www.ncbi.nlm.nih.gov/books/NBK26883/bin/ch2f10.jpg.

The two atoms typically attract the shared electrons to varying degrees when they are joined by a single covalent bond but are composed of different elements. For instance, O and N atoms draw electrons more strongly than C atoms do, but H atoms draw electrons less strongly. In the context of electricity, a polar structure is by definition one in which the positive pole is the center of positive charge, while the negative pole is the center of negative charge. Polar covalent bonds are thus defined as those in which the electrons are distributed unevenly (Figure 5.10). Examples of polar covalent bonds are those between oxygen and hydrogen (-O-H) and nitrogen and hydrogen (-N-H). On the other hand, the covalent bond between carbon and hydrogen (-C-H) is comparatively nonpolar, with the electrons being attracted to both atoms much more equally.

In order to enable molecules to interact through electrical forces, polar covalent bonds form permanent dipoles, which are crucial for biological processes. There will be a pattern of partial positive and negative charges on the surface of any large molecule with numerous polar groups. A permanent dipole interaction, similar to but weaker than the ionic bonds previously discussed for NaCl, will draw such a molecule and any other molecule with a complementary set of charges together.

5.2.6. An Atom Often Behaves as if It Has a Fixed Radius

The sharing of electrons that occurs during the formation of a covalent bond between two atoms causes their nuclei to become unusually close. However, most of the atoms in cells that are jostling with each other so quickly are found in different molecules. For the sake of simplicity and clarity, atoms and molecules are typically depicted in a highly schematic fashion, such as a ball and stick model or a line drawing of the structural formula. What happens when two of these atoms come into contact? But using what are known as space-filling models, a more accurate representation can be achieved. Here, the so-called van der Waals radius for an atom—the radius of the electron cloud at which strong repulsive forces prevent any second, non-bonded atom from approaching closer—is represented by a solid envelope. Because of the extremely sharp increase in repulsion that occurs when two of these atoms get close to one another, this is made possible. Van der Waals attraction is the term for the weak attractive force that occurs between any two atoms at slightly larger distances. Therefore, in each atom's interaction with an atom of a second, non-bonded element, there exists a distance at which repulsive and attractive forces precisely balance to produce an energy minimum (Figure 5.11).

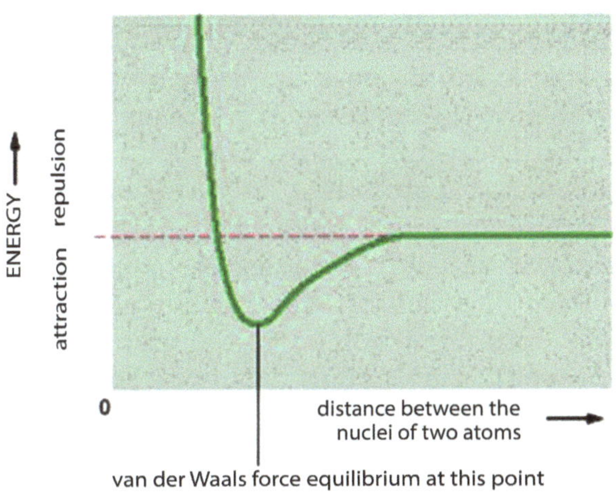

Figure 5.11. *The balance of van der Waals forces between two atoms.*

Source: https://www.ncbi.nlm.nih.gov/books/NBK26883/bin/ch2f11.jpg.

When two atoms' nuclei come close together, they initially have a weak bonding interaction due to their fluctuating electric charges. However, if they get too close, they will strongly repel each other. The balance of these attractive and repulsive forces occurs at the energy minimum. Depending on the purpose, small molecules will be represented as line drawings, ball and stick models, or space-filling models in this book. For example, the water molecule is shown in all three ways in Figure 5.12. When dealing with large molecules like proteins, a simplified representation will often be used.

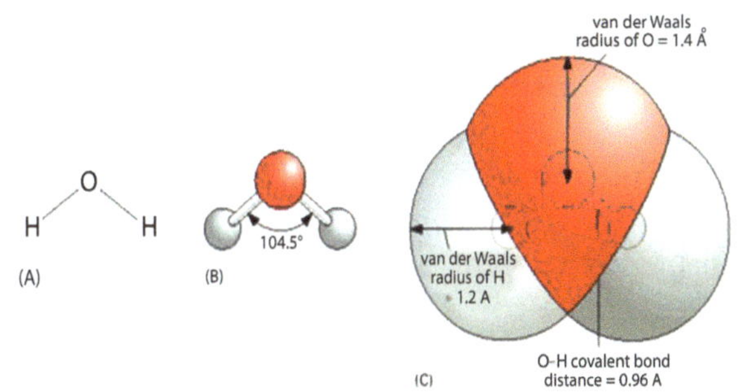

Figure 5.12. *Three representations of a water molecule.*

Note: (A) The usual line drawing of the structural formula, in which each atom is indicated by its standard symbol, and each line represents a covalent bond joining two atoms. (B) A ball and stick model, in which atoms are represented by spheres of arbitrary diameter, connected by sticks representing covalent bonds. Unlike (A), bond angles are accurately represented in this type of model (see also Figure 5.8). (C) A space-filling model, in which both bond geometry and van der Waals radii are accurately represented.

Source: https://www.ncbi.nlm.nih.gov/books/NBK26883/bin/ch2f12.jpg.

5.2.7. Water Is the Most Abundant Substance in Cells

A cell's weight is primarily composed of water, and the majority of intracellular reactions take place in an aqueous environment. The conditions in the ocean during the beginning of life on Earth left a lasting impression on the chemistry of living things. Thus, the qualities of water are essential to life. Covalent bonds bind the two H atoms in each water molecule (H_2O) to the O atom (see Figure 5.12). Due to the O's strong attraction to electrons and the H's weak attraction, the two bonds are extremely polar. As a result, a water molecule has an uneven electron distribution, with a preponderance of positive charge on the two H atoms and negative charge on the O (refer to Figure 5.10). A weak bond known as a hydrogen bond can form when the positively charged region of one water molecule (one of its H atoms) approaches the negatively charged region (the O) of another water molecule due to electrical attraction. Each bond is only formed for a very brief period of time because these bonds are far weaker than covalent bonds and are easily broken by the random thermal motions caused by the heat energy of the molecules. However, the sum effect of numerous weak bonds is not insignificant. Every water molecule has the ability to create hydrogen bonds with two other water molecules via its two H atoms, resulting in a network where hydrogen bonds are constantly being formed and broken. Water is not a gas at room temperature; rather, it is a liquid with a high boiling point and high surface tension due to the hydrogen bonds that hold its molecules together.

Molecules with polar bonds that can form hydrogen bonds with water, like alcohols, dissolve easily in water. As was previously mentioned, molecules with ions—which have positive or negative charges—also interact favorably with water. These molecules are known as hydrophilic molecules, denoting their affinity for water. This group inherently includes sugars, DNA, RNA, and most proteins, as well as a significant portion of other molecules found in a cell's aqueous environment. Water does not dissolve hydrophobic (or water-hating) molecules because they are uncharged and form few or no hydrogen bonds. One noteworthy example is hydrocarbons. The H and C atoms in these molecules are covalently bound by a mostly nonpolar bond. The H atoms can't create strong hydrogen bonds with other molecules because they essentially have no net positive charge. Since cells' membranes are made of molecules with long hydrocarbon tails, this renders the hydrocarbon as a whole hydrophobic—a property that is utilized by them.

5.2.8. Some Polar Molecules Form Acids and Bases in Water

When a molecule with a highly polar covalent bond between a hydrogen atom and a second atom dissolves in water, one of the most basic types of chemical reactions occurs, one that has significant biological implications. In a molecule of this type, the hydrogen atom has essentially given up its electron to the companion atom, leaving it as a nearly naked positively charged hydrogen nucleus, or proton (H^+). As water molecules encircle the polar molecule, the proton is drawn to the partial negative charge on the O atom of a nearby water molecule. It can then split off from its initial partner and bind with the oxygen atoms of the water molecule to form a hydronium ion (H_3O^+) (Figure 5.13A). Millions of protons must be continuously darting from one solution molecule to another in order to imagine an equilibrium state where the opposite reaction occurs with great ease.

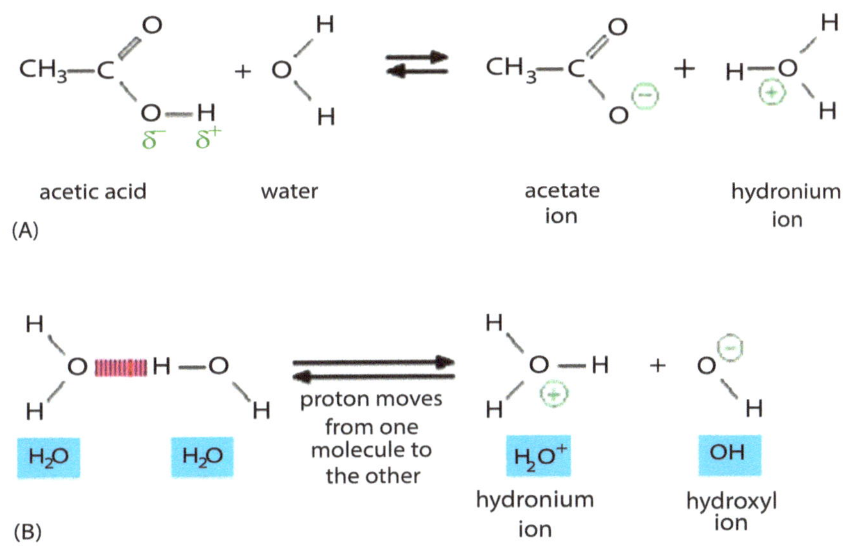

Figure 5.13. *Acids in water.*

Note: (A) The reaction that takes place when a molecule of acetic acid dissolves in water. (B) Water molecules are continuously exchanging protons with each other to form hydronium and hydroxyl ions. These ions in turn rapidly recombine to form water molecules.

Source: https://www.ncbi.nlm.nih.gov/books/NBK26883/bin/ch2f13.jpg.

Acids are substances that, when dissolved in water, release protons to form H_3O^+. The solution becomes more acidic as the concentration of H_3O^+ increases. Because protons transfer from one water molecule to another, H_3O^+ can be found in pure water as well, at a concentration of 10^{-7} M (Figure 5.13B). Even though the majority of H^+ in an aqueous solution is present as H_3O^+, the H_3O^+ concentration is typically referred to as the H^+ concentration due to tradition. The pH scale is a logarithmic scale used to express the concentration of H^+ instead of using unwieldy numbers. The pH of pure water is seven zero. Because a hydronium ion's proton can easily transfer to a variety of molecules within a cell, changing their nature, the concentration of H_3O^+ (or acidity) within a cell needs to be carefully controlled. If the concentration of H_3O^+ in the solution is low, molecules that are capable of giving up protons will do so more easily, and if the concentration is high, they will typically receive them back.

A base is the opposite of an acid. Similar to how an acid is defined by giving protons to a water molecule to increase the concentration of H_3O^+ ions, a base is defined by increasing the concentration of hydroxyl (OH^-) ions, which are created when a proton is taken out of a water molecule. Because sodium hydroxide (NaOH) dissociates to form Na^+ and OH^-ions in aqueous solution, it is therefore considered basic (although the term "alkaline" is also used). Another class of bases is those with NH_2 groups, which are particularly significant in living cells. By removing a proton from water, these groups can produce;

$$-NH_2 + H_2O \rightarrow -NH_3{}^+ + OH^-.$$

An increase in OH-concentration results in a decrease in H_3O^+ concentration and vice versa because an OH^-ion combines with an H_3O^+ ion to form two water molecules. A pure solution of water has a pH of 7.0, meaning that it is neither acidic nor basic due to the equally low concentration of both ions (10^{-7} M). Cell interiors are maintained near neutrality.

5.2.9. Four Types of Noncovalent Interactions Help Bring Molecules Together in Cells

Covalent bonds can distinguish one molecule from another in aqueous solutions because they are 10–100 times more powerful than other atomic attractive forces. However, the precise way that various molecules bind to one another is essential to much of biology. A collection of weak noncovalent attractions, whose bond energies add up to form an effective force between two distinct molecules, mediate this binding. Three of these noncovalent forces—ionic bonds, hydrogen bonds, and van der Waals attractions—have already been covered. The strengths of these three bond types are contrasted with the typical covalent bond in Table 5.2, both with and without water. We will discuss their characteristics here, as they are crucial to all biological systems.

Table 5.2. Covalent and Noncovalent Chemical Bonds

		Strength (kcal/mole)	
	Length (nm)	**In Vacuum**	**In Water**
	0.15	90	90
	0.25	80	3
hydrogen	0.30	4	1
van der Waals attraction (per atom)	0.35	0.1	0.1

Source: https://www.ncbi.nlm.nih.gov/books/NBK26883/table/A191/?report=objectonly.

- **Ionic Bonds:** These are just the effects of electrostatic attraction between atoms with opposing charges. These forces are fairly strong in the absence of water, as we observed for NaCl. On the other hand, the polar water molecules group around polar molecules with permanent dipoles as well as fully charged ions (Figure 5.14). This significantly lowers the likelihood that these charged species will find each other attractive (see Table 5.2).
- **Hydrogen Bonds:** A typical hydrogen bond's structure is shown in Figure 5.15. This bond is an example of a unique type of polar interaction where two electronegative atoms share a portion of an electropositive hydrogen atom. Its hydrogen can be thought of as a proton that has partially split off from a donor atom so that another acceptor atom can share it. This bond is highly directional, unlike a typical electrostatic interaction, and is strongest when a straight line can be drawn between all three of the involved atoms. As was previously mentioned, water forms competing hydrogen-bond interactions with

the molecules involved, weakening these bonds (see Table 5.2).

Hydrogen bonds form when hydrogen atoms are covalently bonded to highly electronegative atoms such as fluorine (F), oxygen (O), or nitrogen (N). Here are some sequences where hydrogen bonding can occur, following the H-FON pattern:

- **H-F-H**: Hydrogen fluoride dimer, where hydrogen bonds form between the hydrogen of one HF molecule and the fluorine of another.
- **H-O-H**: Water (H_2O), where hydrogen bonds form between the hydrogen atoms of one water molecule and the oxygen atom of another.
- **H-N-H**: Ammonia (NH_3), where hydrogen bonds form between the hydrogen atoms of one ammonia molecule and the nitrogen atom of another.
- **O-H-O**: Hydrogen peroxide (H_2O_2) or in a hydronium ion cluster (H_3O^+), where hydrogen bonds form between the hydrogen atoms of one molecule and the oxygen atoms of another.
- **F-H-F**: This sequence is less common, but in a hypothetical scenario where fluorine atoms are bonded with hydrogen in a chain-like structure, hydrogen bonds could form between hydrogen atoms and adjacent fluorine atoms.
- Additionally, here are some more examples of hydrogen-bonding sequences:

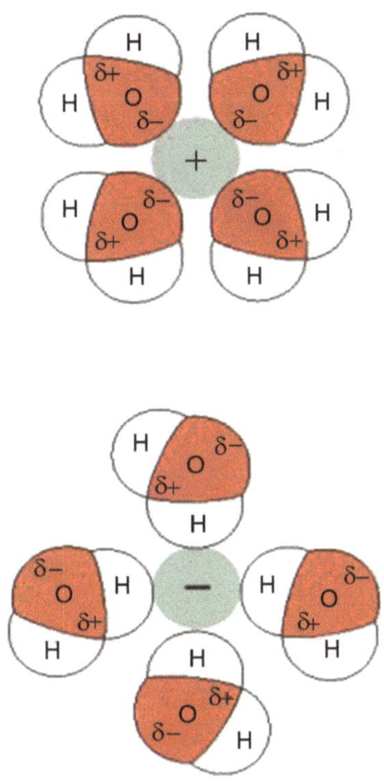

Figure 5.14. *How the dipoles on water molecules orient to reduce the affinity of oppositely charged ions or polar groups for each other.*

Source: https://www.ncbi.nlm.nih.gov/books/NBK26883/bin/ch2f14.jpg.

- **H-N-H-O-H**: An example of this would be an ammonia-water interaction.
- **H-O-H-N-H**: This could occur in a system where water molecules interact with ammonia molecules.
- **H-O-H-F**: This could occur in a system where water interacts with hydrogen fluoride.

- **Van Der Waals Attractions:** Any nonpolar atom will experience fluctuations in the electron cloud, which will result in a flickering dipole. Such dipoles will momentarily cause an atom nearby to exhibit an oppositely polarized flickering dipole. A very weak attraction is created between the atoms as a result of this interaction. However, when two surfaces fit closely, many atoms can come into contact at once, and the end result is frequently significant. Water does not lessen these so-called van der Waals attractions (see Table 5.2).

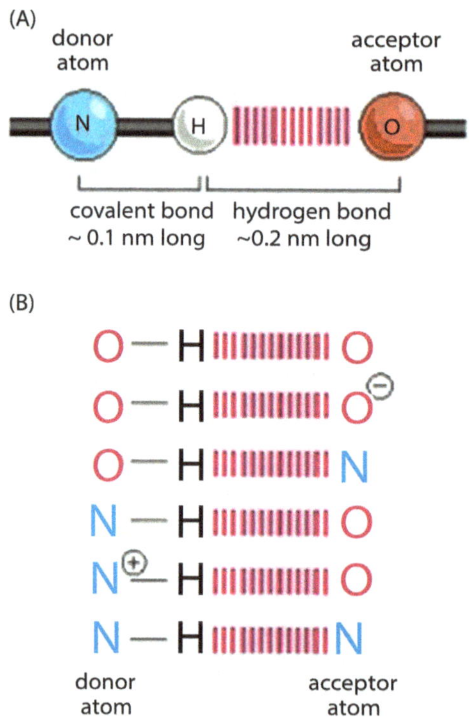

Figure 5.15. *Hydrogen bonds.*

Note: (A) Ball- and-stick model of a typical hydrogen bond. The distance between the hydrogen and the oxygen atom here is less than the sum of their van der Waals radii, indicating a partial sharing of electrons. (B) The most common hydrogen bonds in cells.

Source: https://www.ncbi.nlm.nih.gov/books/NBK26883/bin/ch2f15.jpg.

A hydrophobic force is the fourth effect that can be crucial in bringing molecules together in water. The pushing of nonpolar surfaces out of the hydrogen-bonded water network, where they would physically obstruct the extremely beneficial interactions between water molecules, is what generates this force. The force is rather nonspecific because it reduces the contact between two nonpolar surfaces when they are brought together.

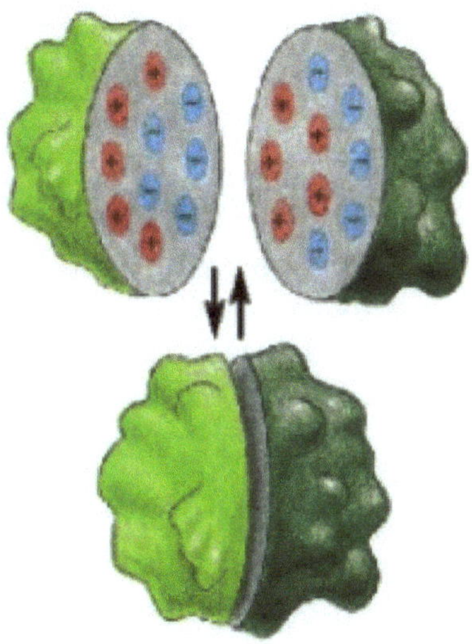

Figure 5.16. *How two macro-molecules with complementary surfaces can bind tightly to one another through noncovalent interactions. In this schematic illustration, plus and minus are used to mark chemical groups that can form attractive interactions when paired.*

Source: https://www.ncbi.nlm.nih.gov/books/NBK26883/bin/ch2f16.jpg.

5.3. THE CHEMISTRY OF CELLS

Cells are incredibly complex and diverse structures, capable not only of self-replication—the very essence of life—but also of performing a wide range of specialized tasks in multicellular organisms. Yet cells obey the same laws of chemistry and physics that determine the behavior of nonliving systems. Consequently, modern cell biology seeks to understand cellular processes in terms of chemical and physical reactions.

5.3.1. A Cell Is Formed From Carbon Compounds

After examining the formation of small molecules from atoms and their behavior in an aqueous medium, we will now look at the primary classes of small molecules that are present in cells and their biological functions. We will see that all of the remarkable diversity of form and behavior exhibited by living things is derived from a small number of fundamental classes of molecules. When water is excluded, almost every molecule in a cell is made of carbon. When it comes to the ability of any element to form large molecules, silicon is a poor second to carbon. A carbon atom can form four covalent bonds with other atoms due to its small size, four electrons, and four vacancies in its outermost shell. Most importantly, one carbon atom can join with other carbon atoms to form chains and rings through extremely stable covalent C-C bonds, creating large, complex molecules with no discernible maximum size. Organic molecules are the large and small carbon compounds that are produced by cells.

Certain combinations of atoms, such as the methyl ($-CH_3$), hydroxyl (-OH), carboxyl (-COOH), carbonyl (-C=O), phosphate ($-PO_3{}^{2-}$), and amino ($-NH_2$) groups, occur repeatedly in organic molecules. Each such chemical group has distinct chemical and physical properties that influence the behavior of the molecule in which the group occurs.

5.3.2. Cells Contain Four Major Families of Small Organic Molecules

With molecular weights between 100 and 1,000 and up to 30 carbon atoms or so, the small organic molecules that make up the cell are carbon-based substances. They typically have a variety of outcomes and are discovered free in solutions. A portion of them serve as monomer building blocks for the cell's enormous polymeric macromolecules, which include proteins, nucleic acids, and large polysaccharides. A complex network of intracellular metabolic pathways breaks down and transforms others into other small molecules, serving as sources of energy. A multitude of small molecules play multiple roles within the cell, such as serving as an energy source and a possible subunit for a macromolecule. Merely making up a 10th of the total mass of organic matter in a cell, small organic molecules are far less common than their macromolecule counterparts (Table 5.3). An approximate estimate of the variety of these tiny molecules in a typical cell would be one thousand.

Table 5.3. The Approximate Chemical Composition of a Bacterial Cell

	Percent of Total Cell Weight	Number of Types of Each Molecule
Water	70	1
Inorganic ions	1	20
Sugars and precursors	1	250
Amino acids and precursors	0.4	100
Nucleotides and precursors	0.4	100
Fatty acids and precursors	1	50
Other small molecules	0.2	~300
Macromolecules (proteins, nucleic acids, and polysaccharides)	26	~3000

Source: https://www.ncbi.nlm.nih.gov/books/NBK26883/table/A201/?report=objectonly.

The same group of simple compounds serve as the basis for the synthesis and breakdown of all organic molecules. Both their synthesis and disintegration are carried out by precise, regulated chemical reactions that follow certain rules.

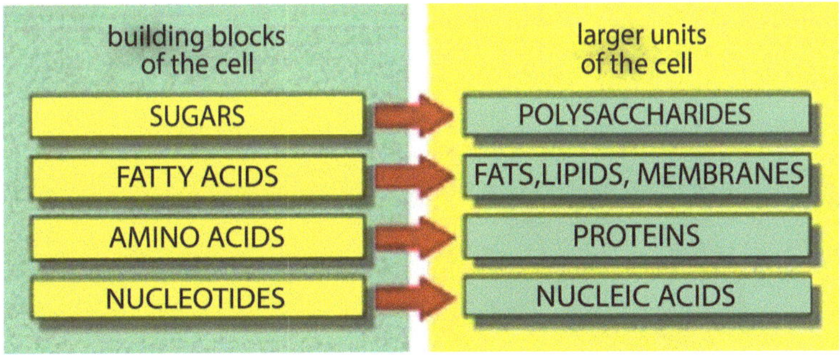

Figure 5.17. *The four main families of small organic molecules in cells. These small molecules form the monomeric building blocks, or subunits, for most of the macromolecules and other assemblies of the cell. Some, like the sugars and the fatty acids, are also energy sources.*

Source: https://www.ncbi.nlm.nih.gov/books/NBK26883/bin/ch2f17.jpg.

Because of this, the molecules that make up a cell are related chemically, with the majority falling into a limited number of unique families. In general, there are four main groups of small organic molecules found in cells: sugars, fatty acids, amino acids, and nucleotides (Figure 5.17). These four families of small organic molecules, along with the macromolecules created by joining them into long chains, make up a significant portion of the mass of cells, even though many substances found in cells do not fall into these categories (see Table 5.3).

5.3.3. Sugars Provide an Energy Source for Cells and Are the Subunits of Polysaccharides

The compounds with the general formula $(CH_2O)_n$, where n is typically 3, 4, 5, 6, 7, or 8, are the simplest sugars, or monosaccharides. Because of this straightforward formula, sugars and the molecules derived from them are also referred to as carbohydrates. For instance, the formula for glucose is $C_6H_{12}O_6$ (Figure 5.18). The formula, however, does not completely describe the molecule because different covalent bonding configurations can join the same set of carbons, hydrogens, and oxygens to form structures with distinct shapes. For instance, glucose can be changed into mannose or galactose by simply rotating some OH groups in relation to the rest of the molecule. Furthermore, each of these sugars can exist in one of two mirror-image forms known as the d-form or the l-form. Isomers are groups of molecules having the same chemical formula but distinct structures; optical isomers are a subset of these molecules that are mirror-image pairs. Isomers are present in a wide range of organic molecules and are primarily responsible for the vast diversity of sugars.

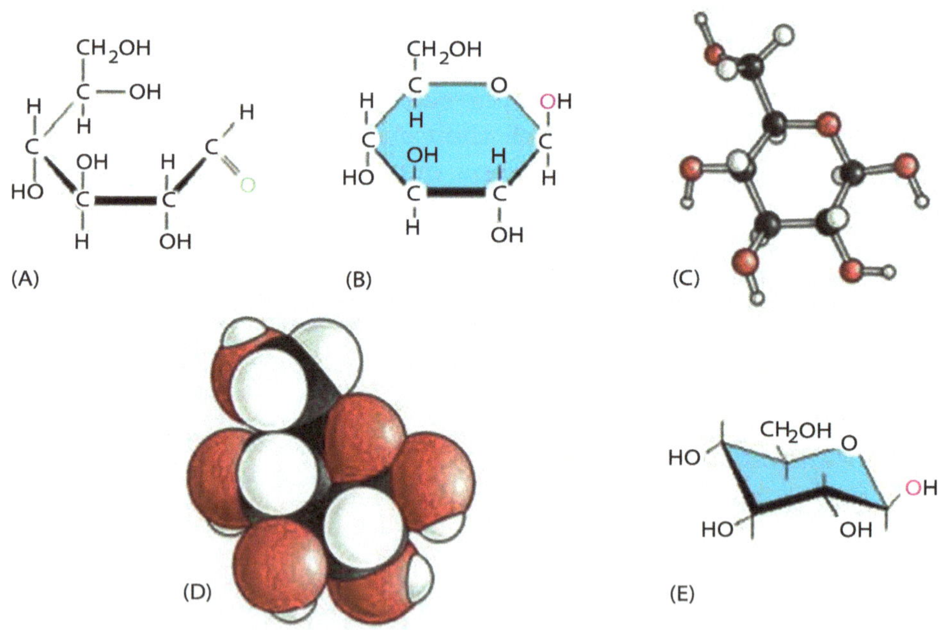

Figure 5.18. *The structure of glucose, a simple sugar.*

Source: https://www.ncbi.nlm.nih.gov/books/NBK26883/bin/ch2f18.jpg.

There are multiple ways to represent a molecule, as shown with water (see Figure 5.12). The atoms are depicted in the structural formulas in (A), (B), and (E) as chemical symbols connected by lines that stand in for covalent bonds. Here, the -H and -OH groups are not in the same plane as the sugar ring, as indicated by the thickened lines that represent the ring's plane. (A) This sugar's open-chain form, which is in balance with (B)'s more stable cyclic or ring form. (C) A ball-and-stick model that illustrates the atoms' three-dimensional spatial arrangement. (D) A space-filling model that uses the van der Waals radii to represent the molecule's surface contours in addition to showing the atoms' three-dimensional arrangement. (E) A different method of drawing the cyclic molecule that more closely captures the geometry than the structural formula in (B) is the chair form. The atoms in (C) and (D) are depicted using the standard atomic color scheme. As an illustration, consider the colors H, white; C, black; O, red; and N, blue (also see Figure 5.8).

Sugars can be found in two different forms: rings and open chains. Sugars have several hydroxyl groups and either one ketone (C=O) or one aldehyde (H>C=O) group when they are in their open-chain form. A special role is played by the aldehyde or ketone group. The molecule may first change into a ring form by reacting with a hydroxyl group within it; in this form, the carbon of the original aldehyde or ketone group is the only one bonded to two oxygen atoms. Second, after the ring is formed, the same carbon can bind to another carbon on a sugar molecule that has a hydroxyl group, forming a disaccharide. An example of this would be sucrose, which is made up of glucose and fructose units. Larger sugar polymers include giant polysaccharides, which can have thousands of monosaccharide units, and oligosaccharides, which include trisaccharides, tetrasaccharides, and so forth. Some common characteristics of biochemical bond formation are illustrated by the way sugars are bonded together to form polymers. A condensation reaction, in which a water molecule is released as the bond is formed, forms a bond between an -OH group on one sugar and an -OH group on another (Figure 5.19). Other biological polymers, like proteins and nucleic acids, also have subunits connected by water-ejecting condensation reactions. The opposite process of hydrolysis, in which a water molecule is consumed, can break the bonds formed by all of these condensation reactions (see Figure 5.19).

As a result of the loss of a water molecule, two molecules come together in a general class of reactions known as condensation reactions. We call the opposite reaction—hydrolysis—when water is added. As you can see in Figure 5.18, one of the two partners—the one on the left in this instance—is the carbon bonded to two oxygen atoms, which forms the sugar ring.

Glycosidic bonds, as shown, are a common kind of covalent bond between two sugar molecules (see also Figure 5.20). Sugar polymers can branch, and an enormous variety of polysaccharide structures are possible due to the numerous free hydroxyl groups that each monosaccharide possesses that can link it to another monosaccharide or to another compound. 11 distinct varieties of a basic disaccharide with two glucose residues can be found (Figure 5.20), and three distinct hexoses (C6H12O6) can combine to form several 1,000 trisaccharides. Because each unit of a polysaccharide is joined to the next in an identical manner, figuring out the arrangement of sugars in a polysaccharide is far more difficult than figuring out the nucleotide sequence of a DNA molecule.

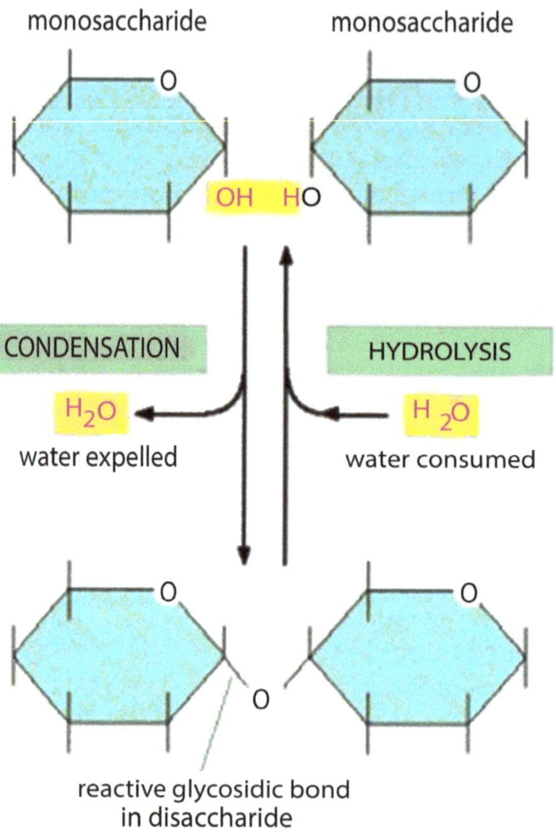

Figure 5.19. *The reaction of two monosaccharides to form a disaccharide.*

Source: https://www.ncbi.nlm.nih.gov/books/NBK26883/bin/ch2f19.jpg.

They are chemically different, even though the only difference between them is the kind of linkage that connects the two glucose units. The number of different types of oligosaccharides that can be used in cells is very large because the oligosaccharides associated with proteins and lipids can contain six or more different kinds of sugar joined in both linear and branched arrangements through glycosidic bonds like those illustrated here. One of the most important monosaccharides for cells to use as an energy source is glucose.

It is reduced to smaller molecules through a sequence of reactions, which releases energy that the cell can use to perform helpful tasks, as we will discuss later. Simple polysaccharides, which are solely made up of glucose units and are used by cells as long-term energy stores, are primarily glycogen in animals and starch in plants. Sugars are used for more than just making and storing energy. They can also be utilized to create mechanical supports, for instance. Consequently, cellulose, the most prevalent organic substance on Earth, is a polysaccharide of glucose and is found in plant cell walls. Chitin, which is found in the cell walls of fungi and the exoskeletons of insects, is another incredibly common organic material. It is a polysaccharide, specifically a linear polymer of the sugar derivative N-acetylglucosamine. The primary constituents of slime, mucus, and gristle are polysaccharides of different kinds.

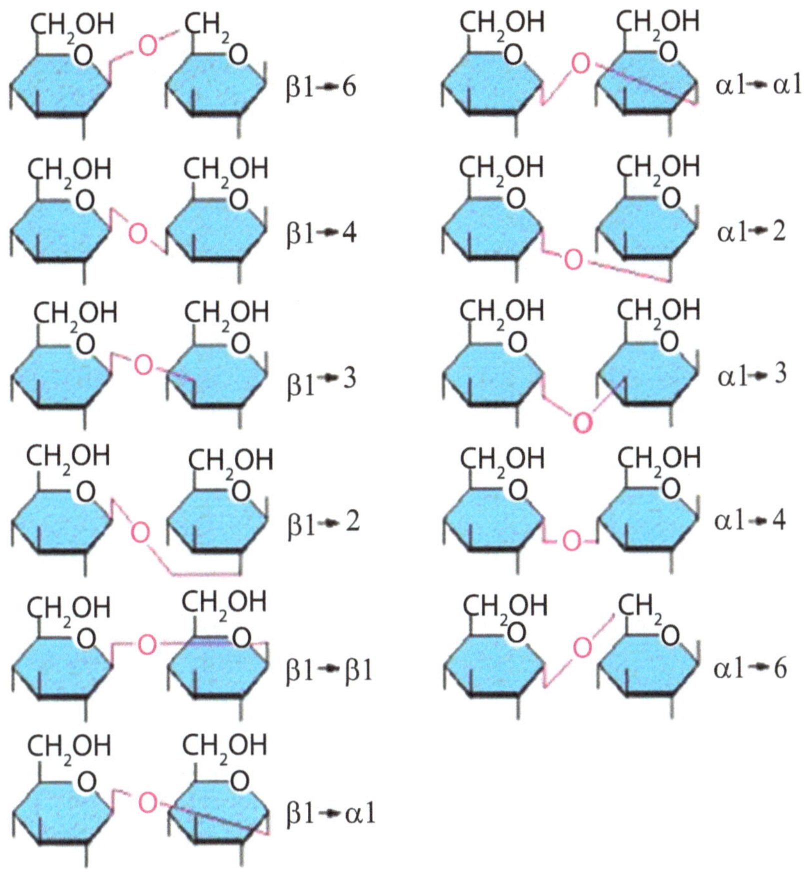

Figure 5.20. *Eleven disaccharides consisting of two D-glucose units.*

Source: https://www.ncbi.nlm.nih.gov/books/NBK26883/bin/ch2f20.jpg.

Glycolipids, which are present in cell membranes, and glycoproteins, which are proteins bound together covalently, are formed from smaller oligosaccharides. The majority of cells have surfaces covered in and ornamented with sugar polymers that are part of glycoproteins and glycolipids in the cell membrane, as explained in Chapter 10. Other cells frequently recognize these sugar side chains selectively. Additionally, variations in the specifics of each person's cell-surface sugars serve as the molecular foundation for the main blood groups found in humans.

5.3.4. Fatty Acids Are Components of Cell Membranes

A fatty acid molecule has two chemically separate regions, like in the case of palmitic acid (Figure 5.21). A lengthy hydrocarbon chain is one example; it is chemically non-reactive and hydrophobic. A carboxylic (-COOH) group is the other, and it functions such as an acid (carboxylic acid) because it is highly hydrophilic, chemically reactive, and ionized in solution (-COO-). The carboxylic acid group on almost all fatty acid molecules in a cell forms covalent bonds with other molecules.

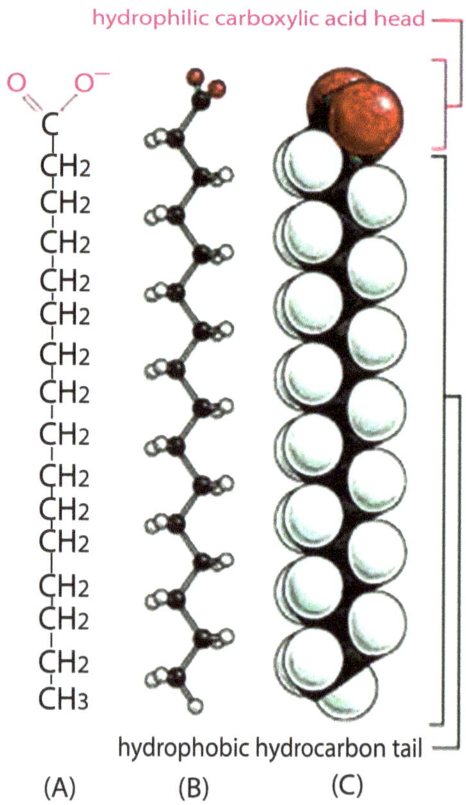

Figure 5.21. *A fatty acid is composed of a hydrophobic hydrocarbon chain to which is attached a hydrophilic carboxylic acid group. A fatty acid is composed of a hydrophobic hydrocarbon chain to which is attached a hydrophilic carboxylic acid group. Palmitic acid is shown here. Different fatty acids have different hydrocarbon tails.*

Note: (A) Structural formula. The carboxylic acid group is shown in its ionized form. (B) Ball-and-stick model. (C) Space-filling model.

Source: https://www.ncbi.nlm.nih.gov/books/NBK26883/bin/ch2f21.jpg.

With the greatest number of hydrogen atoms and no double bonds between carbon atoms, the hydrocarbon tail of palmitic acid is saturated. Another common fatty acid found in animal fat is stearic acid, which is likewise saturated. Other fatty acids, like oleic acid, have unsaturated tails that contain one or more double bonds. The molecules' ability to pack together in a solid mass is hindered by the kinks caused by the double bonds. This explains the distinction between soft (polyunsaturated) and hard (saturated) margarine. The length of their hydrocarbon chains and the quantity and location of their carbon-carbon double bonds are the only differences between the numerous types of fatty acids present in cells. Because they can be broken down to produce roughly six times as much usable energy, weight for weight, as glucose, fatty acids function as a concentrated food reserve in cells. These are the animal fats found in meat, butter, and cream, as well as the plant oils like corn oil and olive oil. They are stored in the cytoplasm of many cells as droplets of triacylglycerol molecules, which are made up of three fatty acid chains connected to a glycerol molecule. The fatty acid chains are freed from triacylglycerols and split into two-carbon units when energy is needed. These two-carbon units follow the same pathways leading to energy production as those that result from the breakdown of glucose.

Lipids include fatty acids and their derivatives, such as triacylglycerols. A loosely defined group of biological molecules known as lipids share the property of being soluble in fat and organic solvents like benzene but insoluble in water. They usually consist of several linked aromatic rings, as in the case of steroids, or long hydrocarbon chains, as in the case of fatty acids and isoprenes. The synthesis of cell membranes is the primary role of fatty acids in cells. All cells are surrounded by these thin sheets, which also enclose their internal organelles. They are mostly made up of phospholipids, which are tiny molecules primarily made of fatty acids and glycerol, similar to triacylglycerols. But unlike in triacylglycerols, where the glycerol is attached to three fatty acid chains, in phospholipids it is joined to two. A hydrophilic phosphate group connected to a small hydrophilic compound like choline is the "third" site on the glycerol. Because of this, each phospholipid molecule consists of a hydrophilic head that contains a phosphate and a hydrophobic tail made up of the two fatty acid chains. Their physical and chemical characteristics differ from those of triacylglycerols, which are primarily hydrophobic, as a result. Amphipathic molecules are those that have both hydrophobic and hydrophilic regions, such as phospholipids.

The membrane-forming property of phospholipids results from their amphipathic nature. Phospholipids will spread over the water surface to form a monolayer of phospholipid molecules, with the hydrophobic tails facing the air and the hydrophilic heads in contact with the water. Two such molecular layers can readily combine tail-to-tail in water to make a phospholipid sandwich, or *lipid bilayer*. This bilayer is the structural basis of all cell membranes (Figure 5.22).

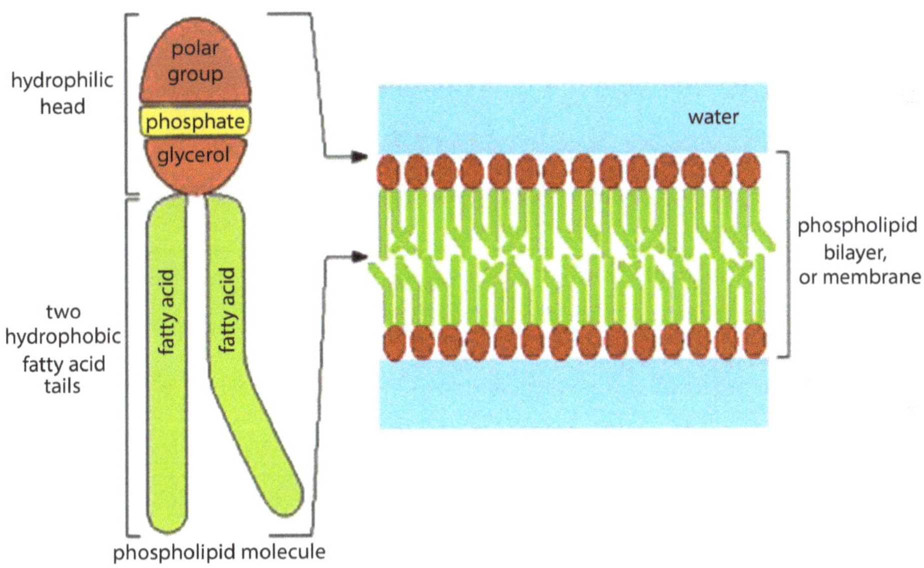

Figure 5.22. *Phospholipid structure and the orientation of phospholipids in membranes.*

Source: https://www.ncbi.nlm.nih.gov/books/NBK26883/bin/ch2f22.jpg.

In an aqueous environment, the hydrophobic tails of phospholipids pack together to exclude water. Here they have formed a bilayer with the hydrophilic head of each phospholipid facing the water. Lipid bilayers are the basis for cell membranes.

5.3.5. The Chemistry of Cells is Dominated by Macromolecules With Remarkable Properties

Of the carbon-containing molecules in a living cell, macromolecules are by far the most prevalent in terms of weight (Figure 5.23 and Table 5.4). They are the main constituents that give life its most distinguishing characteristics and serve as the foundation upon which a cell is built. The polymers that make up the macromolecules in cells are called monomers, or subunits, and they are created by covalently joining small organic molecules into lengthy chains (Figure 5.24). However, they possess a multitude of extraordinary qualities that were not foreseeable from their basic components.

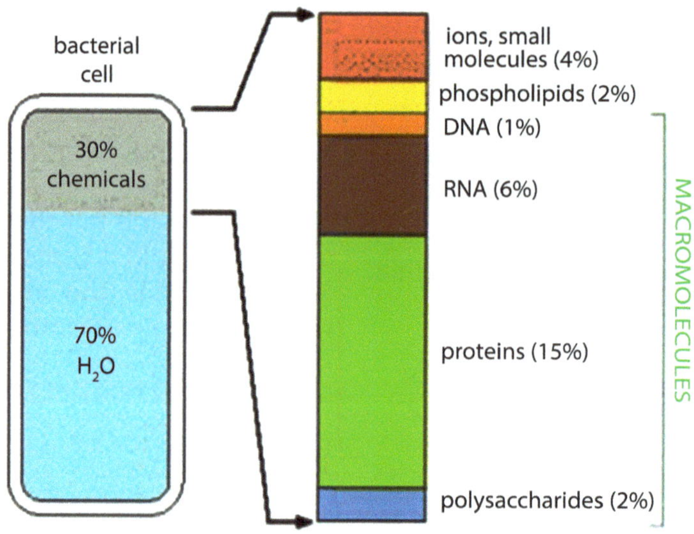

Figure 5.23. *Macromolecules are abundant in cells. The approximate composition of a bacterial cell is shown by weight. The composition of an animal cell is similar (see Table 5.4).*

Source: https://www.ncbi.nlm.nih.gov/books/NBK26883/bin/ch2f29.jpg.

Table 5.4. Approximate Chemical Compositions of a Typical Bacterium and a Typical Mammalian Cell

	Percent of Total Cell Weight	
Component	***E. Coli* Bacterium**	**Mammalian Cell**
H_2O	70	70
Inorganic ions (Na^+, K^+, Mg^{2+}, Ca^{2+}, Cl^-, etc.)	1	1
Miscellaneous small metabolites	3	3
Proteins	15	18
RNA	6	1.1
DNA	1	0.25
Phospholipids	2	3
Other lipids	–	2

Polysaccharides	2	2
Total cell volume	2×10^{-12} cm^3	4×10^{-9} cm^3
Relative cell volume	1	2000

Proteins, polysaccharides, DNA, and RNA are macromolecules. Lipids are not generally classed as macromolecules even though they share some of their features; for example, most are synthesized as linear polymers of a smaller molecule (the acetyl group on acetyl CoA), and they self-assemble into larger structures (membranes). Note that water and protein comprise most of the mass of both mammalian and bacterial cells.

Source: https://www.ncbi.nlm.nih.gov/books/NBK26883/table/A224/?report=objectonly.

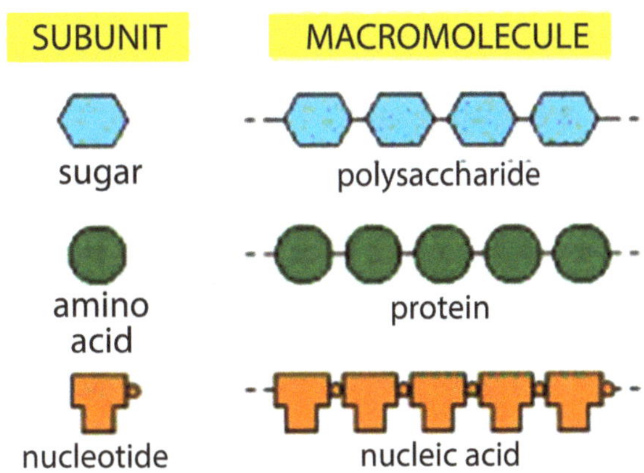

Figure 5.24. *Three families of macromolecules. Each is a polymer formed from small molecules (called monomers, or subunits) linked together by covalent bonds.*

Source: https://www.ncbi.nlm.nih.gov/books/NBK26883/bin/ch2f30.jpg.

Particularly plentiful and adaptable are proteins. In cells, they carry out thousands of different tasks. Numerous proteins function as enzymes, acting as catalysts to drive the numerous covalent bond formation and breaking reactions required by the cell. For example, proteins acting as enzymes catalyze all of the reactions by which cells extract energy from food molecules. The enzyme ribulose bisphosphate carboxylase, for example, converts CO_2 into sugars in photosynthetic organisms, generating the majority of the organic matter required for life on Earth. Other proteins, like tubulin, which self-assembles to form the long microtubules in cells, or histones, which are proteins that compact the DNA in chromosomes, are employed to construct structural elements. However, other proteins, like myosin in muscle, function as molecular motors to generate force and movement. In this book, we will explore the molecular basis of numerous additional functions that proteins can perform. We only touch on a few broad macromolecular chemistry concepts here that enable these kinds of operations.

While the specific chemical reactions for incorporating subunits into each polymer vary among proteins, nucleic acids, and polysaccharides, they all share significant characteristics. In a condensation reaction, where a water molecule is lost with each additional subunit added, a monomer is added to the end of a growing polymer chain to create a new polymer (see Figure 5.19). A large, complex molecule can be made easily by stepwise polymerization of monomers into a long chain, where the same set of enzymes perform the same reaction repeatedly to add subunits. The procedure is comparable to the monotonous running of a factory machine, save for one very important detail. With the exception of a few polysaccharides, the majority of macromolecules are composed of a collection of slightly distinct monomers, such as the 20 different amino acids that make up proteins. It is essential to life that the subunits are added to the polymer chain in a specific order, or sequence, rather than being assembled at random.

5.3.6. Noncovalent Bonds: Precise Shape of a Macromolecule and its Binding to Other Molecules

The majority of covalent bonds in macromolecules permit the atoms they join to rotate, greatly enhancing the flexibility of the polymer chain. This essentially enables a macromolecule to take on nearly infinitely many different forms, or conformations, as the polymer chain writhes and rotates in response to random thermal energy. Nonetheless, the numerous weak noncovalent bonds that form between the various components of the molecule severely restrict the shapes of the majority of biological macromolecules. The polymer chain can strongly prefer one specific conformation if enough of these noncovalent bonds are formed; this conformation is determined by the linear sequence of monomers in the chain. This leads to the tight folding of almost all protein molecules and a large number of small RNA molecules into a single, highly preferred conformation (Figure 5.25).

This chapter has already provided an explanation of the four categories of noncovalent interactions that are significant for biological molecules. Even though these interactions are very weak on their own, they can combine to fold biological macromolecules into unusual shapes. When two molecules fit together tightly, such as a hand in a glove, they can also strongly attract one another. Because the multipoint contacts necessary for strong binding allow a macromolecule to selectively bind to just one of the thousands of different types of molecules present inside a cell, this type of molecular interaction offers high specificity. Furthermore, interactions of nearly any affinity are possible because the strength of the binding is dependent on the quantity of noncovalent bonds that are formed, enabling quick dissociation when needed.

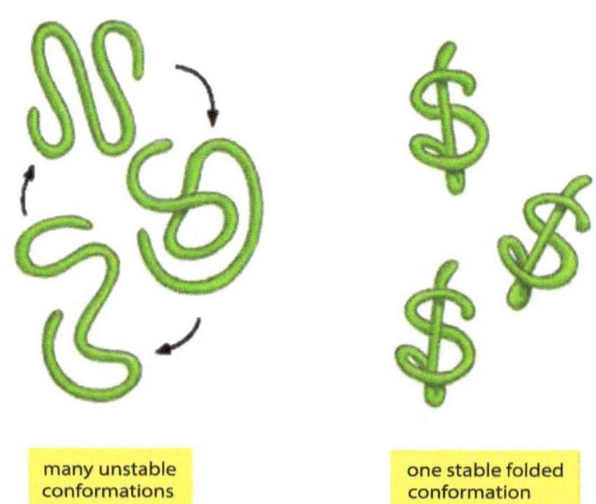

Figure 5.25. *Most proteins and many RNA molecules fold into only one stable conformation. If the noncovalent bonds maintaining this stable conformation are disrupted, the molecule becomes a flexible chain that usually has no biological value.*

Source: https://www.ncbi.nlm.nih.gov/books/NBK26883/bin/ch2f31.jpg.

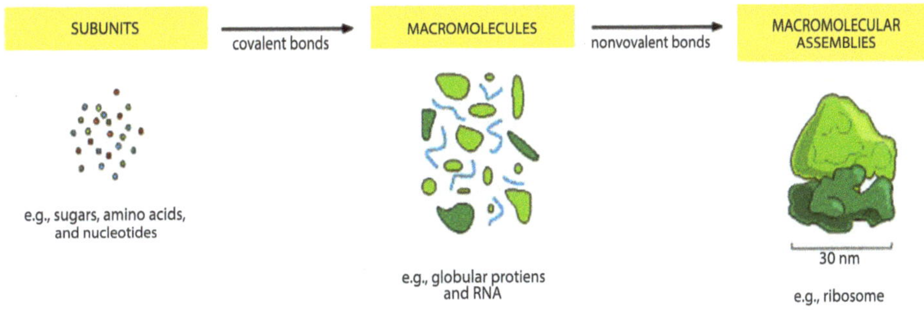

Figure 5.26. *Small molecules, proteins, and a ribosome drawn approximately to scale. Ribosomes are a central part of the machinery that the cell uses to make proteins: each ribosome is formed as a complex of about 90 macromolecules (protein and RNA molecules).*

Source: https://www.ncbi.nlm.nih.gov/books/NBK26883/bin/ch2f32.jpg.

All biological catalysis is dependent on this kind of binding, which enables proteins to act as enzymes. Furthermore, macromolecules can be utilized as building blocks to form larger structures thanks to noncovalent interactions. Large complexes of macromolecules are frequently formed in cells, enabling the creation of complex machines with numerous moving parts that carry out intricate processes like protein synthesis and DNA replication (Figure 5.26).

CASE STUDY

CHEMICALS THAT DISRUPT YEAST VACUOLAR TRAFFICKING ALSO TARGET PLANT VACUOLAR TRAFFICKING PATHWAY

The chemical screening using yeast as a system, because it is easy to grow and suitable for large-scale and automated screening. Although vacuoles are dispensable for yeast survival compared to plant cells, there are some homologous vesicle trafficking genes between the two systems. It turned out that Sortin1, one of the chemicals that disrupts yeast vacuolar CPY trafficking, also affects plant vacuolar trafficking and vacuole morphology. This provides clear evidence that chemicals derived from phenotypic screens are translatable between different systems.

Sortin1 Disrupts Vacuolar Trafficking of both Proteins and Flavonoids

Vacuoles are storage sites for different cellular materials, such as proteins and secondary metabolites. We showed that Sortin1 affects not only CPY protein but also flavonoid transport. Flavonoids are secondary metabolites that are not essential for plant survival but are involved in multiple cellular processes such as auxin transport, defense, modulating levels of reactive oxygen species, and providing different flower colors for attracting pollinators. The enzymes for flavonoid biosynthesis are localized in the cytoplasm, ER, and nucleus, the proposed sites of flavonoid biosynthesis. Flavonoids are located in different subcellular locations such as the vacuole, the cytosol, the ER, and the apoplast. Flavonoids are transported from sites of synthesis to storage via two proposed models. One model is called vesicle-trafficking-mediated flavonoid transport, wherein flavonoids are transported through the same pathway as protein trafficking and does not need specialized flavonoid carriers or tonoplast transporters. The other model suggests that flavonoid transport occurs through ABC-type transporters or the multidrug and toxic compound extrusion (MATE) antiporters, both of which need energy.

Sortin1 affects the vacuolar trafficking of both proteins and flavonoids, indicating that flavonoid transport to the vacuole could share the same membranemediated pathway as vacuolar proteins. But this does not rule out the possibility that one of the trafficking defects is direct, whereas the other one is indirect. The fact that Sortin1 mimics BSO to produce an oxidized cell environment to affect ABC transporter activity supports a model that a transporter-mediated flavonoid transport defect induces miss-trafficking of other vacuolar contents such as the proteins. On the other hand, mutants defective in flavonoid metabolism, such as tt4, show hypersensitivity to Sortin1 and have altered vacuolar morphology without treatment.This indicates that a properly regulated flavonoid pathway participates in the maintenance of plant vacuole integrity.

Mechanism of Sortin1 Action

Sortin1 decomposes in plant growth media where one of the products resembles the effects of BSO, a GSH biosynthesis inhibitor. This indicates that a series of observed Sortin1 effects might be due to an oxidized cellular environment that it creates. It could be that Sortin1 or an active decomposition product sequesters GSH or inhibits GSH production. This may induce the transcription of cytochrome P450-dependent monooxygenases and mixed function oxidases, as shown in the transcription profile after Sortin1 treatment. Reduced GSH level then affects flavonoid transport. The mechanism of flavonoid regulation of vacuolar biogenesis is not understood. However, the existence of vacuole biogenesis and flavonoid accumulation phenotypes in different mutants, such as mtv6, aha10, tt12, and tds4 indicates that there is a connection between the two pathways. Sortin1 might serve as a tool in solving this question, not all the s1h mutants have flavonoid accumulation/transport defects; this indicates that there are flavonoidindependent pathways that involve the regulation Sortin1-targeted vacuolar transport and vacuole biogenesis.

CLASS ACTIVITY

ENZYME CATALYSIS AND INHIBITION

- **Objective:** To understand enzyme catalysis and the effect of inhibitors on enzyme activity through a hands-on experiment.
- **Materials:** Fresh potatoes, Hydrogen peroxide (3% solution), Measuring cylinders, Test tubes, Stopwatches, pH buffer solutions, Inhibitors, Safety goggles and gloves, Lab coats
- **Learning Outcomes:** Understand the principles of enzyme catalysis and the factors affecting enzyme activity. Gain hands-on experience in conducting biochemical experiments. Analyze and interpret experimental data. Appreciate the relevance of enzyme inhibition in therapeutic contexts.

SUMMARY

- Chemical biology of cells is an interdisciplinary field that explores the interface between chemistry and biology, aiming to understand and manipulate cellular processes at the molecular level.
- The integration of chemical biology with advanced imaging techniques has further enhanced our understanding of cellular processes. Fluorescent probes and labels enable the visualization of molecular interactions and dynamics in real time within living cells.
- Chemical biology occupies a niche that is not adequately filled by traditional biological sciences. Biochemistry/enzymology are suited to understand proteins in isolation, or in lysates.
- Array technologies, microfluidics, and miniaturization are transformative fields that have significantly advanced research and applications in biology, chemistry, and medicine. These areas intersect and synergize to enable high-throughput analysis, precise control of biological and chemical processes, and the development of compact and efficient devices.
- Microfluidics is the science and technology of manipulating fluids at the microscale, typically in channels with dimensions of tens to hundreds of micrometers.
- Miniaturization refers to the trend of developing smaller and more efficient devices and systems. In the context of array technologies and microfluidics, miniaturization enhances portability, reduces material and reagent costs, and allows for the integration of multiple functionalities into compact platforms.
- Neutrons are uncharged subatomic particles of essentially the same mass as protons. They contribute to the structural stability of the nucleus—if there are too many or too few, the nucleus may disintegrate by radioactive decay—but they do not alter the chemical properties of the atom.
- The electron arrangement of an atom is most stable when all the electrons are in the most tightly bound states that are possible for them—that is, when they occupy the innermost shells.
- Ionic bonds are most likely to be formed by atoms that have just one or two electrons in addition to a filled outer shell or are just one or two electrons short of acquiring a filled outer shell.

REVIEW QUESTIONS

1. What is chemical biology, and how does it differ from traditional molecular biology?
2. Describe the role of small molecules in regulating cellular processes. Provide an example.
3. What are chemical probes, and how are they used in chemical biology?
4. How can chemical biology contribute to drug discovery and development?

5. Explain the concept of 'chemical genetics' and its applications.
6. What is the significance of post-translational modifications in chemical biology?
7. Discuss the role of chemical inhibitors in studying enzyme function.
8. How does chemical biology intersect with systems biology?
9. What are some common techniques used in chemical biology to study protein-ligand interactions?
10. Describe the importance of chemical biology in understanding cellular signaling pathways.

REFERENCES

1. Ansari, A. M., Ahmed, A. K., Matsangos, A. E., Lay, F., Born, L. J., Marti, G., & Ting, A. E. (2016). Cellular GFP toxicity and immunogenicity: Potential confounders in in vivo cell tracking experiments. *Stem Cell Reviews and Reports, 12*(4), 553–559. doi:10.1007/s12015-016-9670-8.
2. Ariotti, N., Hall, T. E., Rae, J., Ferguson, C., McMahon, K.-A., Martel, N., & Parton, R. G. (2015). Modular detection of GFP-labeled proteins for rapid screening by electron microscopy in cells and organisms. *Developmental Cell, 35*(5), 513–525. doi: 10.1016/j.devcel.2015.10.016.
3. Ariotti, N., Rae, J., Giles, N., Martel, N., Sierecki, E., Gambin, Y., & Parton, R. G. (2018). Ultrastructural localization of protein interactions using conditionally stable nanobodies. *PLoS Biology, 16*(3), e2005473. doi: 10.1371/journal.pbio.2005473.
4. Aye, Y., & Stubbe, J. (2011). Clofarabine 5′-di and -triphosphates inhibit human ribonucleotide reductase by altering the quaternary structure of its large subunit. *Proceedings of the National Academy of Sciences, 108*(24), 9815–9820. doi: 10.1073/pnas.1013274108.
5. Aye, Y., Brignole, E. J., Long, M. J. C., Chittuluru, J., Drennan, C. L., Asturias, F. J., & Stubbe, J. (2012). Clofarabine targets the large subunit (α) of human ribonucleotide reductase in live cells by assembly into persistent hexamers. *Chemical Biology, 19*(6), 799–805. doi: 10.1016/j.chembiol.2012.05.015.
6. Aye, Y., Li, M., & Long, M. J. C., Weiss, & R. S. (2015). Ribonucleotide reductase and cancer: Biological mechanisms and targeted therapies. *Oncogene, 34*(16), 2011–2021. doi:10.1038/onc.2014.155.
7. Banaszynski, L. A., Chen, L.-C., Maynard-Smith, L. A., Ooi, A. G., & Wandless, T. J. (2006). A rapid, reversible, and tunable method to regulate protein function in living cells using synthetic small molecules. *Cell, 126*(5), 995–1004. doi: 10.1016/j.cell.2006.07.025.
8. Bellucci, A., Fiorentini, C., Zaltieri, M., Missale, C., & Spano, P. (2014). The "in situ" proximity ligation assay to probe protein–protein interactions in intact tissues. In A. I. Ivanov (Ed.), *Exocytosis and Endocytosis* (pp. 397–405). New York, NY: Springer.
9. Belousov, V. V., Fradkov, A. F., Lukyanov, K. A., Staroverov, D. B., Shakhbazov, K. S., Terskikh, A. V., & Lukyanov, S. (2006). Genetically encoded fluorescent indicator

for intracellular hydrogen peroxide. *Nature Methods, 3*(4), 281–286. doi:10.1038/nmeth866.

10. Bertino, J. R., Göker, E., Gorlick, R., Li, W. W., & Banerjee, D. (1996). Resistance mechanisms to methotrexate in tumors. *The Oncologist, 1*(4), 223–226.
11. Bondeson, D. P., Mares, A., Smith, I. E. D., Ko, E., Campos, S., Miah, A. H., & Crews, C. M. (2015). Catalytic in vivo protein knockdown by small-molecule PROTACs. *Nature Chemical Biology, 11*(8), 611–617. doi:10.1038/nchembio.1858.
12. Branon, T. C., Bosch, J. A., Sanchez, A. D., Udeshi, N. D., Svinkina, T., Carr, S. A., & Ting, A. Y. (2018). Efficient proximity labeling in living cells and organisms with TurboID. *Nature Biotechnology, 36*(9), 880–887. doi:10.1038/nbt.4201.
13. Caras, I. W., & Martin Jr., D. W. (1988). Molecular cloning of the cDNA for a mutant mouse ribonucleotide reductase M1 that produces a dominant mutator phenotype in mammalian cells. *Molecular and Cellular Biology, 8*(7), 2698–2704. doi:10.1128/MCB.8.7.2698.
14. Hedstrom, L. (2017). The bare essentials of antibiotic target validation. *ACS Infectious Diseases, 3*(1), 2–4. doi:10.1021/acsinfecdis.6b00185.
15. Kawasumi, M., & Nghiem, P. (2007). Chemical genetics: Elucidating biological systems with small-molecule compounds. *Journal of Investigative Dermatology, 127*, 1577–1584.
16. Knowles, J. R., & Albery, W. J. (1977). Perfection in enzyme catalysis: The energetics of triosephosphate isomerase. *Accounts of Chemical Research, 10*(4), 105–111. doi:10.1021/ar50112a001.
17. Lipinski, C. A., Lombardo, F., Dominy, B. W., & Feeney, P. J. (2001). Experimental and computational approaches to estimate solubility and permeability in drug discovery and development settings. *Advanced Drug Delivery Reviews, 46*, 3–26.
18. Lokey, R. S. (2003). Forward chemical genetics: Progress and obstacles on the path to a new pharmacopoeia. *Current Opinion in Chemical Biology, 7*, 91–96.
19. Matthaei, J. H., Jones, O. W., Martin, R. G., & Nirenberg, M. W. (1962). Characteristics and composition of RNA coding units. *Proceedings of the National Academy of Sciences, 48*(4), 666–677. doi:10.1073/pnas.48.4.666.
20. Ogasawara, D., Ichu, T.-A., Vartabedian, V. F., Benthuysen, J., Jing, H., Reed, A., & Cravatt, B. F. (2018). Selective blockade of the lyso-PS lipase ABHD12 stimulates immune responses in vivo. *Nature Chemical Biology, 14*(12), 1099–1108. doi:10.1038/s41589-018-0155-8.
21. Osz-Papai, J., Radu, L., Abdulrahman, W., Kolb-Cheynel, I., Troffer-Charlier, N., Birck, C., & Klaholz, B. P. (2015). Insect cells–baculovirus system for the production of difficult to express proteins. In E. García-Fruitós (Ed.), *Insoluble Proteins: Methods and Protocols* (pp. 181–205). New York, NY: Springer.

CHAPTER

6 Diffraction and Microscopy

LEARNING OBJECTIVES

After studying this chapter, you will be able to:

- Discuss about diffraction and microscopy in chemical biology
- How to electron diffraction using transmission electron microscopy

KEY TERMS FROM THIS CHAPTER

Crystallography	Electron beam
Electron density	Magnetic materials
Metal oxides	Neutron scattering
Optical microscopy	Scanning electron microscopy (SEM)
Wavelength	X-ray diffraction

INTRODUCTION

Fundamental methods in imaging and structural analysis, diffraction and microscopy are important in many scientific fields, from biology to materials science. The term "diffraction" describes how waves—like sound, light, or electrons—bend and spread when they come into contact with a barrier or opening. Many microscopy techniques, especially those that depend on wave interactions to generate high-resolution images, operate on this phenomenon. Diffraction establishes a fundamental resolution limit in optical microscopy, or the diffraction limit, which is roughly half the wavelength of the illuminating light. Ernst Abbe was the first person to describe this limitation in the 1800s. The smallest distance that can be clearly resolved between two points is proportional to the wavelength of light divided by the lens system's numerical aperture, according to Abbe's criterion. Because of this, the resolution of traditional optical microscopes has traditionally been limited to 200 nanometers. Modern developments have overcome this limit, though, by using fluorescence and non-linear optical effects to create super-resolution microscopy methods like STED (Stimulated Emission Depletion) and PALM (Photoactivated Localization Microscopy), which achieve resolutions on the order of tens of nanometers.

Compared to optical microscopy, electron microscopy—which comprises transmission electron microscopy (TEM) and scanning electron microscopy (SEM)—achieves much higher resolutions through the use of electron diffraction. Because electrons have much shorter wavelengths than visible light, features at the atomic level can be resolved using electron microscopes. Through the use of electrons passing through a thin sample, TEM is able to obtain detailed information about the internal composition and arrangement of the sample through the formation of diffraction patterns that result from interactions with its atomic structure. SEM, on the other hand, creates high-resolution images of the sample's topography by scanning a concentrated electron beam across the sample surface. Secondary electrons are emitted from the surface as a result of this process.

Another diffraction-based method that uses X-rays to target a crystalline sample is called X-ray crystallography. The atomic structure of the crystal can be ascertained by analyzing the pattern formed by the diffracted rays. Understanding the structures of intricate biological molecules, like proteins and nucleic acids, is essential for comprehending biological processes and functions. This technique has played a key role in clarifying these structures. Science and technology have advanced significantly as a result of the revolutionary changes that diffraction and microscopy techniques have brought about in our understanding and visualization of the micro- and nanoworld. As they continue to advance, more scientific frontiers will be crossed and even deeper insights into the minute details of matter will be revealed.

6.1. DIFFRACTION AND MICROSCOPY IN CHEMICAL BIOLOGY

In chemical biology, diffraction and microscopy are essential tools that help researchers visualize and comprehend biological molecules' molecular and atomic structures and functions. By offering in-depth insights into the complex workings of cellular processes and molecular interactions, these techniques have completely changed the field.

Chemical biology has relied heavily on diffraction, especially X-ray crystallography, to determine the three-dimensional structures of biomolecules. Crystallized samples of proteins, nucleic acids, or complexes can be exposed to X-rays, which scientists can then use to map the atomic arrangement within the molecule by analyzing the diffraction patterns that result. Understanding the molecular underpinnings of biological processes like enzyme activity, signal transduction, and molecular recognition requires knowledge of these intricate structural details. Thanks to groundbreaking discoveries like the double helix structure of DNA and the complex architecture of ribosomes, X-ray crystallography has greatly advanced our understanding of genetics and the functioning of biological machinery.

Advancements in microscopy techniques, particularly those with super-resolution capabilities, have revolutionized the field of chemical biology. Traditional optical microscopy is limited by the diffraction limit of about 200 nanometers, but methods like STED and PALM have surpassed this barrier. By utilizing specific fluorescence markers and non-linear optical processes, researchers can now visualize molecular dynamics within living cells at resolutions as fine as tens of nanometers. This breakthrough has allowed for the observation of the spatial arrangement and interactions of proteins, lipids, and other biomolecules in real time, offering valuable insights into processes such as protein trafficking, membrane dynamics, and signal transduction pathways.

Without the need for crystallization, electron microscopy—including transmission electron microscopy (TEM) and cryo-electron microscopy (cryo-EM)—offers even greater resolution, down to the atomic level. The ability to obtain high-resolution images of biomolecules in their natural, hydrated states has made cryo-EM particularly well-known. Large, complex assemblies like viruses, ribosomes, and membrane proteins have proven difficult to study without this method, which has allowed researchers to gain previously unattainable insights into their structure and function. Furthermore, chemical biology has benefited from atomic force microscopy (AFM) since it makes it possible to image and work with individual molecules. AFM provides a distinct viewpoint on molecular dynamics and function by measuring the mechanical characteristics of molecules and tracking their interactions. Diffraction and sophisticated microscopy methods have worked together to provide a previously unheard-of comprehension of biological molecules and processes. They are still at the forefront of chemical biology innovation, contributing to the development of biomimetic materials, the discovery of new drugs, and the understanding of basic biological processes.

6.1.1. Key Principles of X-Ray Diffraction

A strong analytical method for figuring out a crystal's atomic and molecular structure is X-ray diffraction (XRD). The interactions between X-rays and the periodic atomic arrangements within a crystalline material form the foundation of X-ray diffraction theory.

Here are the main principles:

1. **Bragg's Law:** Bragg's Law is fundamental to understanding X-ray diffraction. It describes the condition for constructive interference of X-rays scattered by a crystal lattice. The law is given by the equation:

   $n\lambda = 2d\sin\theta$

Where;

- n is the order of the diffraction,
- λ is the wavelength of the incident X-rays,
- d is the distance between adjacent planes in the crystal lattice (interplanar spacing),
- θ is the angle of incidence at which diffraction occurs (Bragg angle).

2. **Crystalline Structure:** The regular, periodic arrangement of atoms within a crystal is essential to XRD. The incident X-rays scatter in distinct, predictable directions as a result of this regularity. The diffraction pattern, which is determined by the crystal structure, can be utilized to infer the arrangement of atoms within the crystal.
3. **Diffraction Pattern:** The electrons in the atoms scatter X-rays when they interact with a crystal. These dispersed waves combine to form a diffraction pattern, which usually takes the form of several spots or rings due to constructive interference. The crystal structure can be inferred from the location and intensity of these spots.
4. **Reciprocal Lattice and Ewald Sphere:** The concept of the reciprocal lattice is central to XRD. The reciprocal lattice is a mathematical construct used to visualize the diffraction conditions. The Ewald sphere is a geometric representation that shows the relationship between the incident and scattered X-rays. The intersection of the reciprocal lattice points with the Ewald sphere corresponds to the directions of diffracted beams.
5. **Fourier Transform:** Using a Fourier transform, the diffraction pattern and the electron density inside the crystal are connected. Through the examination of the diffraction spots' location and intensity, researchers can piece together the crystal's electron density map. The locations of the atoms inside the unit cell are shown on this map.
6. **Unit Cell and Symmetry:** The smallest repeating unit in the crystal lattice that defines the structure is called a unit cell. The dimensions and symmetry characteristics of the unit cell can

be ascertained using XRD data. The crystal can be categorized into one of the 230 space groups with the aid of the symmetry elements, such as rotation axes and mirror planes.

7. **Phase Problem:** The phase problem, which results from the diffraction pattern's inability to reveal the phase of the diffracted waves in addition to their amplitude, is a major obstacle in XRD. Several techniques, including molecular replacement, anomalous dispersion, and direct methods, are applied to resolve the phase problem and acquire the entire electron density map.
8. **Sample Preparation:** To obtain accurate XRD data, high-quality single crystals or polycrystalline samples must be prepared. While powder diffraction (used for polycrystalline samples) is helpful for identifying phases and analyzing mixtures, single crystals offer more detailed information.

6.1.2. Applications of X-Ray Diffraction

XRD is widely used in materials science, chemistry, and biology to:

- Determine crystal structures of small molecules and complex biomolecules.
- Identify unknown crystalline substances.
- Study phase transitions and the effects of temperature and pressure on materials.
- Analyze the quality and purity of materials.

By understanding and applying these principles, X-ray diffraction enables detailed insights into the atomic structure of materials, contributing significantly to advances in science and technology.

6.1.3. Structural Information From X-Ray Diffraction

A strong and adaptable method for obtaining comprehensive structural details about crystalline materials is X-ray diffraction (XRD). This method is based on the orderly array of atoms in a crystal scattering X-rays. Vital hints regarding the atomic structure, such as the arrangement of atoms, interatomic distances, and crystallographic symmetry, are revealed by the diffraction pattern created by the scattered X-rays.

- **Basic Principles:** When a crystal is exposed to X-rays, the X-rays are scattered by the electrons surrounding the atoms. The constructive interference of these scattered X-rays, as described by Bragg's Law, results in a diffraction pattern unique to the crystal structure. Bragg's Law is given by:

  $n\lambda=2d\sin\theta$

  Where λ is the wavelength of the X-rays, d is the distance between crystal planes, θ is the angle of incidence, and n is the order of the diffraction. By analyzing the angles and intensities of the diffracted beams, one can infer the positions of the atoms within the crystal.
- **Determining Crystal Structure:** The tiniest repeating unit that characterizes the crystal structure, the unit cell dimensions, can be precisely measured using XRD. The lengths of the cell edges and the angles between them are among the unit cell parameters, and they are essential for determining the

crystal system (e.g., orthorhombic, tetragonal, and cubic). The crystal's symmetry is also shown by the diffraction pattern, which aids in placing it in one of the 230 space groups.

- **Electron Density Maps:** The diffraction pattern is made up of reflections, with each one representing a distinct set of crystallographic planes. The brightness of these reflections is connected to the electron density in the crystal. By utilizing Fourier transform methods on the diffraction information, researchers can create an electron density map that shows a three-dimensional representation of the electron distribution within the unit cell. This map is essential for pinpointing the locations of individual atoms.
- **Phase Problem:** A significant challenge in XRD is the phase problem, which arises because the diffraction pattern only provides information about the amplitude of the scattered waves, not their phases. Several methods are employed to solve this problem:
    - o **Direct Methods**: Utilize statistical relationships between the observed diffraction intensities.
    - o **Molecular Replacement**: Uses a known structure of a similar molecule as a starting point.
    - o **Anomalous Dispersion**: Exploits differences in scattering at different wavelengths to obtain phase information.
- **Applications in Chemical Biology:** X-ray diffraction (XRD) has been a valuable tool in chemical biology for clarifying the structures of intricate biological macromolecules. For example, Rosalind Franklin's XRD data served as the basis for Watson and Crick's discovery of the double helix structure of DNA. Proteins are also extensively studied using X-ray diffraction (XRD), which reveals the three-dimensional structures of proteins and offers insights into their functional mechanisms. Designing medications that can alter biological activities and comprehending biochemical pathways depend on the structure-function relationship discovered in XRD data.

6.1.4. Neutron Diffraction

The use of neutron scattering to ascertain a material's atomic and/or magnetic structure is known as elastic neutron scattering or neutron diffraction. A sample to be studied is exposed to a thermal or cold neutron beam in order to produce a diffraction pattern that reveals details about the material's structure. This method is comparable to X-ray diffraction, but because neutrons and X-rays scatter differently, they provide complementary information: X-rays are best used for surface analysis, strong X-rays from synchrotron radiation are best used for thin specimens or shallow depths, and high penetration depth neutrons are best used for bulk samples.

6.1.4.1. Instrumental and Sample Requirements

A neutron source is necessary for the procedure. Usually, a nuclear reactor or spallation source produces neutrons. Other parts of a research reactor are required, such as filters to choose the desired neutron wavelength and a crystal monochromator (in the case of thermal neutrons). The setup might also include some movable components. Since crystals are not suitable for long-wavelength neutrons, gratings are employed as diffractive optical components instead. Since higher energy neutrons strike faster at a spallation source, the time-of-flight technique is used to sort the incident neutron energies. This eliminates the need for a monochromator and instead uses a series of synchronized aperture elements to filter neutron pulses with the desired wavelength.

The method that is most frequently used is powder diffraction, which just needs a powder that is polycrystalline. It is also feasible to do single-crystal work, but the crystals must be substantially larger than those needed for single-crystal X-ray crystallography. One mm3 or less crystals are frequently used. A device that can detect the neutrons after they have been scattered is also necessary for the technique. In summary, the primary drawback of neutron diffraction is its dependence on nuclear reactors. The method requires relatively large crystals for single crystal work, which are typically difficult to grow. The method has many benefits, including a penetration depth of several centimeters, sensitivity to light atoms, and the capacity to differentiate between isotopes without suffering radiation damage.

6.1.4.2. Nuclear Scattering

Neutrons, like all other quantum particles, are capable of exhibiting wave phenomena that are usually connected to sound or light. One of these phenomena is diffraction, which happens when waves come into contact with objects that are the same size as the wavelength. A particle's wavelength can be sufficiently short for atoms or their nuclei to act as diffraction barriers. The wavelength of a neutron beam coming from a reactor is close to one angstrom, or 0 point1 nanometer, which is the usual distance between atoms in a solid material when the beam is appropriately slowed and selected by its speed. One can then conduct an experiment on diffraction using such a beam. Applying pressure to a crystalline sample causes it to scatter within a finite range of precisely defined angles, as per the same Bragg's law that explains X-ray diffraction.

The way X-rays and neutrons interact with matter is different. Each atom's surrounding electron cloud is the main medium through which X-rays interact. Therefore, atoms with a higher atomic number (Z) will contribute more to the diffracted x-ray intensity. On the other hand, because neutrons interact directly with atoms' nuclei, each isotope contributes to the diffracted intensity differently; regular hydrogen and deuterium, for instance, contribute in different ways. Additionally, even in the presence of large Z atoms, it is frequently the case that light (low Z) atoms significantly contribute to the diffracted intensity. Instead of increasing linearly with the atomic number, the scattering length varies from isotope to isotope. Vanadium is a common material for containers because it is an element that strongly scatters X-rays but barely scatters neutrons in its nuclei. The locations of the atoms' nuclei directly affect non-magnetic neutron diffraction. Atoms have tiny nuclei, which are the source of neutron scattering. Additionally, the shape of an atom's electron cloud can be described without the use of an atomic form factor, and unlike X-rays, the scattering power of an atom does not decrease with scattering angle. Diffractograms, therefore, especially when conducted at low temperatures, can display distinct and strong diffraction peaks even at high angles. Liquid helium cooling systems found in many neutron sources enable data collection at as low as 4.2 K. The excellent high angle, or i.e., high resolution) data, which enables the precise determination of the atomic positions within the structure. On the other hand, series termination errors can cause Fourier maps (and to a lesser extent, difference Fourier maps) derived from neutron data to be meaningless at times.

6.1.4.3. Magnetic Scattering

Despite not being charged, neutrons interact with other magnetic moments, such as those that originate from the electron cloud surrounding an atom, because they possess a magnetic moment. Therefore, the microscopic magnetic structure of a material can be revealed by neutron diffraction. Since magnetic scattering is caused by the much larger electron cloud surrounding the tiny nucleus, it does require an atomic form factor. As a result, at higher angles, the magnetic contribution to the diffraction peaks will become less intense.

6.1.4.4. Uses

Gases, liquids, and amorphous solids can all have their static structure factors found using neutron diffraction. However, the majority of experiments focus on the crystalline solids' structure, which makes neutron diffraction a crucial crystallographic tool. X-ray powder diffraction and neutron diffraction are closely related concepts. As large single crystals are difficult or impossible to find for the majority of materials, the single crystal version of the technique is actually less frequently used because current neutron sources require relatively large samples. However, this image could very well change in the future. Rietveld refinement is typically used to process the data because it is typically a 1D powder diffractogram. Actually, the latter was developed for use in X-ray diffraction after having its roots in neutron diffraction at Petten, Netherlands.

The extremely accurate measurement of the lattice constant of metals and other crystalline materials is one useful application of elastic neutron scattering/diffraction. It is possible to derive a map of the lattice constant through the metal in conjunction with a precisely aligned micropositioner. This is readily translated to the stress field that the material is subjected to. To name just two applications, this has been used to analyze stresses in automotive and aerospace components. Because of the high penetration depth, residual stresses in large components like crankshafts, pistons, rails, and gears can be measured. Because of this method, specialized stress diffractometers have been created, like the ENGIN-X device at the ISIS neutron source.

Any material that diffracts can also be used with neutron diffraction to reveal information about its three-dimensional structure. Determining the solvation number of ion pairs in electrolyte solutions is another application. Since the development of the neutron diffraction technique, the magnetic scattering effect has been used to measure magnetic moments in materials and investigate the orientation and structure of magnetic dipoles. The study of magnetic dipole orientations in antiferromagnetic transition metal oxides, such as oxides of manganese, iron, nickel, and cobalt, was one of the first uses of neutron diffraction. Clifford Shull conducted the initial experiments that demonstrated the presence of magnetic dipoles arranged in an antiferromagnetic manner within a material structure. Neutron diffraction is still being used today to describe recently developed magnetic materials.

6.1.4.5. Hydrogen, Null-Scattering, and Contrast Variation

Low atomic number materials, such as proteins and surfactants, are much easier to determine their structures by neutron diffraction, which requires less flux than synchrotron radiation. This is due to the fact that certain materials with low atomic weights have a larger cross section for neutron interaction than materials with higher atomic numbers.

The fact that X-ray diffraction is less sensitive to the presence of hydrogen (H) in a structure than neutron diffraction is one of its main advantages over it. In contrast, the nuclei 1H and 2H (i.e., Neutrons are strongly scattered by deuterium (D). Neutron diffraction provides a more precise method of determining the position of hydrogen in a crystal and its thermal motions due to the greater scattering power of protons and deuterons. The structures of metal hydride complexes, e.g., Mg_2FeH_6 have been assessed by neutron diffraction.

The neutron scattering lengths b_H = −3.7406(11) fm and b_D = 6.671(4) fm, for H and D respectively, have opposite signs, which allows the technique to distinguish them. In fact, there is a particular isotope ratio for which the contribution of the element would cancel, this is called null scattering.

Working with a sample that has a comparatively high concentration of H is not recommended. Large inelastic components in the scattering intensity of H-nuclei result in a large continuous background that is essentially independent of the scattering angle. If the sample is crystalline, the elastic pattern usually consists of sharp Bragg reflections. In the inelastic background, they frequently drown. When applying this method to the study of liquid structure, the situation becomes even more problematic. Nonetheless, the scattering contrast can be sufficiently varied to bring attention to a single element within an otherwise complex structure by creating samples with varying isotope ratios. While it is usually rather expensive, it is possible to vary other elements. Because it is difficult to study structurally in other ways and plays a significant role in biochemical structures, hydrogen is particularly interesting and inexpensive.

6.1.5. Key Principles of Electron Microscopy

The limitations of light microscopes, which rely on the physics of light, led to the development of electron microscopes. Ernst Ruska assembled the electron microscope using the knowledge that electron wavelengths are much shorter than light wavelengths. Electron microscopes use signals produced by the interaction of a sample with an electron beam to determine the composition, morphology, and structure of the material. Because electrons are so tiny, they behave like waves, just like photons do in light. After going through the specimen, an electron beam is directed through several lenses to enlarge the image. The specimen's atoms scatter electrons to create the image. When heavy atoms are present, the contrast of the image will be enhanced because they scatter more effectively than low-atomic-number atoms.

Electrostatic and/or electromagnetic lenses—which are made of a wire coil wrapped around the outside of a tube and are often called solenoid—are used in all electromagnetic machines (EMs). Since an electron can't normally travel very far in air at atmospheric pressure, a conventional electron microscope requires the electron beam to be in a vacuum. Pumps are used to remove the electron microscope's column, and air locks are used to bring the specimens and any additional equipment into the vacuum. The distance between the specimen and the objective lens as well as the distance between the lenses stay constant thanks to the variable-focus lenses in the EM. The amount of current flowing through the intermediate and projector lens coils primarily determines the magnification. By adjusting the current flowing through the objective lens coil, the image can be focused. Instead of using an eyepiece to view the image or electron micrograph, a thermionic gun is employed as the electron source. Additionally, EMs employ low vacuum or variable pressure chambers, which maintain the pressure differential between the high vacuum levels necessary for the gun and column area and the low pressure required in the chamber, along with digital displays, computer interfaces, software for image analysis, and other tools.

The term "electron microscopy" (EM) refers to a group of sophisticated imaging methods that produce high-resolution images of objects by using an electron beam. The fundamental ideas of electron microscopy center on how electrons interact with matter to produce images of atomic or molecular structures.

Here are the main principles:

1. **Electron-Optical System:** Similar to how optical microscopes use glass lenses to control light, electron microscopes use electromagnetic lenses to focus and direct the electron beam. The electron source (or gun), condenser lenses, objective lenses, and projector lenses are the main parts. Together, these components create and amplify the specimen's image.
2. **Wave Nature of Electrons:** Electrons have both particle and wave properties, with their wavelength given by the de Broglie equation:

$$\lambda = \frac{h}{p}$$

   where λ is the wavelength, h is Planck's constant, and p is the momentum of the electron. At high velocities, electrons have much shorter wavelengths than visible light, allowing electron microscopes to achieve significantly higher resolution, down to the atomic scale.
3. **Interaction with Specimen:** Different signals are produced when the specimen interacts with the electron beam. These signals include transmitted electrons, backscattered electrons, secondary electrons, and distinctive X-rays. Different kinds of information regarding the specimen's composition, structure, and attributes are revealed by these interactions.
4. Types of Electron Microscopy:
   - **Transmission Electron Microscopy (TEM)**: In TEM, the electron beam is transmitted through a thin specimen. The electrons that pass through form an image that reveals the internal structure of the specimen. TEM can achieve atomic resolution and is used to study the fine details of cell structures, viruses, and materials at the nanoscale.
   - **Scanning Electron Microscopy (SEM)**: SEM scans a focused electron beam across the specimen's surface. The

emitted secondary electrons are collected to produce detailed images of the surface topography. SEM is widely used for examining the surface structure and composition of materials.

5. **Resolution and Magnification:** The electrons' wavelength and the electron lenses' quality control the resolution of an electron microscope. Compared to optical microscopes, which are constrained by light diffraction, electron microscopes can achieve much higher magnifications (up to several million times) and resolutions (down to fractions of a nanometer).
6. **Sample Preparation:** For electron microscopy to produce high-quality images, sample preparation must be done correctly. For TEM, samples need to be thin enough (usually less than 100 nanometers) so that electrons can flow through them. Biological specimens are frequently prepared using methods like cryo-fixation, ultramicrotomy, and staining. Samples may require coating with a conductive material for SEM in order to shield them from charging under the electron beam.
7. **Contrast Mechanisms:** In electron microscopy, contrast is produced by variations in electron scattering that result from changes in the specimen's thickness, density, and atomic number. Phase contrast, amplitude contrast, and energy-filtered imaging are a few techniques that improve the visibility of various features within the sample.
8. **Detection and Imaging Systems:** Several detectors are used by electron microscopes to record the signals that the specimen emits. Charge-coupled devices (CCDs) and phosphor screens are examples of detectors used in TEM that create images. SEM detectors that pick up secondary electrons, backscattered electrons, and X-rays give detailed surface and compositional information about the sample.
9. **Applications:** Electron microscopy is indispensable in many fields, including materials science, biology, nanotechnology, and electronics. It is used to:
   - Visualize cellular structures and macromolecular complexes in biology.
   - Analyze the microstructure and defects in materials science.
   - Investigate nanostructures and devices in nanotechnology.
   - Study the morphology and composition of surfaces and interfaces in various scientific disciplines.

The key principles of electron microscopy revolve around the manipulation of electron beams and their interactions with specimens to achieve high-resolution imaging and detailed structural analysis. This has established EM an essential tool for advancing scientific knowledge and technological innovation.

6.1.6. Key Principles of Scanning Probe Microscopy

The method known as scanning probe microscopy (SPM) is a flexible and effective way to measure the properties of surfaces at the nanoscale. It includes many techniques, such as Scanning Tunneling Microscopy

(STM) and Atomic Force Microscopy (AFM), which are now essential tools in biology, materials science, and nanotechnology. The fundamental ideas of SPM center on the interaction of a sharp probe with the sample's surface to reveal intricate details about its topography and other characteristics.

1. **Basic Principles:** The fundamental concept of SPM is to scan a very sharp probe, typically with a tip radius of just a few nanometers, across the surface of a sample. The probe interacts with the sample surface in different ways depending on the specific SPM technique used. These interactions are monitored to generate high-resolution images and gather quantitative data about the surface properties.
2. **Atomic Force Microscopy (AFM):** One of the most popular SPM methods is AFM. It measures the forces that exist between the surface of the sample and the sharp probe. The probe is supported by a flexible cantilever, and it encounters both repulsive and attractive forces while scanning the surface. A laser beam reflected from the cantilever's back onto a position-sensitive photodetector is used to measure the deflection caused by these forces. AFM can create precise topographic maps of the surface at atomic or molecular resolution by either oscillating the cantilever at its resonant frequency and measuring changes in amplitude or phase, or by applying a constant force between the probe and the surface (contact mode).
3. **Scanning Tunneling Microscopy (STM):** The quantum tunneling effect is essential to STM. It makes use of a conductive probe that is pressed up against the surface of a semiconductive or conductive material. A tunneling current is produced when electrons tunnel through the vacuum gap when a bias voltage is applied between the probe and the sample. The distance between the probe and the sample surface affects this current very much. Surface Topography Mapping (STM) allows atomic-level mapping of the surface topography by continuously scanning the probe across the surface with a constant tunneling current. STM works especially well for researching the electrical characteristics of individual atoms and surfaces.
4. **Resolution and Sensitivity:** The sharpness of the probe tip and the scanning mechanism's accuracy are the main factors that affect the resolution of SPM techniques. SPM has the ability to visualize individual atoms and molecules at atomic resolution. Small forces, currents, and displacements can be found thanks to the sensitivity of SPM techniques, which also provide a wealth of information about surface characteristics like mechanical stiffness, electrical conductivity, and roughness.
5. **Versatility and Applications:** SPM is highly versatile and can be adapted to study a wide range of samples under various environmental conditions, including ambient air, vacuum, and liquid environments. This versatility makes SPM suitable for applications in numerous fields:
   - **Materials Science**: Characterizing the surface

structure, mechanical properties, and electronic behavior of materials at the nanoscale.

- **Biology**: Imaging biological specimens, such as cells and biomolecules, in their native environments, and measuring their mechanical properties.
- **Nanotechnology**: Fabricating and characterizing nanostructures and nanodevices.
- **Surface Chemistry**: Investigating chemical reactions at surfaces and mapping chemical composition.

6. **Advanced Techniques:** Advanced methods like Kelvin probe force microscopy (KPFM), which measures surface potential, and magnetic force microscopy (MFM), which maps magnetic properties, are now included in SPM. These specialized methods increase SPM's capacity to offer thorough analysis of a range of surface phenomena. Utilizing the interactions between a sharp probe and a sample surface, scanning probe microscopy provides detailed surface characterization and high-resolution imaging. It is an essential tool for scientific research and technological advancement due to its versatility and capacity to function in various environments.

6.2. ELECTRON DIFFRACTION USING TRANSMISSION ELECTRON MICROSCOPY

Utilizing the transmission electron microscope for electron diffraction is a potent technique for analyzing the composition of materials, encompassing both flawless crystals and imperfect structures. The advantages of electron diffraction over other methods, e.g., the incredibly short wavelength (≈2 pm), the strong atomic scattering, and the capacity to study very small volumes of matter (≈10 nm^3) give rise to x-rays and neutrons. Through the use of electron diffraction alone or in conjunction with other diffraction techniques, the NIST Materials Science and Engineering Laboratory has a long history of using this technique to discover and characterize new structures.

B. pioneered the use of electron diffraction in the Soviet Union to solve crystallographic problems. K. Vainshtein and associates dating back to the 1940s. Numerous organic and inorganic substance structures were solved using the elektronograf, which focused 50 keV to 100 keV electrons to produce diffraction with scattering angles up to 3° to 5°. The modern transmission electron microscope (TEM), in which the scattered transmitted beams can also be recombined to form an image, and the electronograf are very similar devices. Modern TEMs are capable of a resolution of 1.65 Å for 300 kV (and below 1 Å for 1000 kV) electron energy-loss combined with chemical analysis (through x-ray energy and electron-loss energy spectroscopy) and a bright coherent field emission source of electrons due to numerous advancements in optics and microscope design.

Optical radiation diagrams, like the one in Figure 6.1, can be used to understand the fundamental concepts of electron microscopy. Six Points. After the objective lens focuses the wave fragments scattered by the atomic potential, diffraction spots form on the back focal plane. On the image plane, the diffracted waves are recombined to create an image. Diffracted electrons can be focused into a regular pattern of diffraction spots with the aid of electromagnetic lenses. This pattern is then projected and recorded as the electron diffraction pattern. An enlarged image of the sample can be seen if there is interference on the image plane between the diffracted and transmitted beams. Real space is the area at the image plane or at a specimen, whereas reciprocal space is the area where the diffraction pattern forms. The Fourier transform provides the mathematical formula for the transformation from real space to reciprocal space.

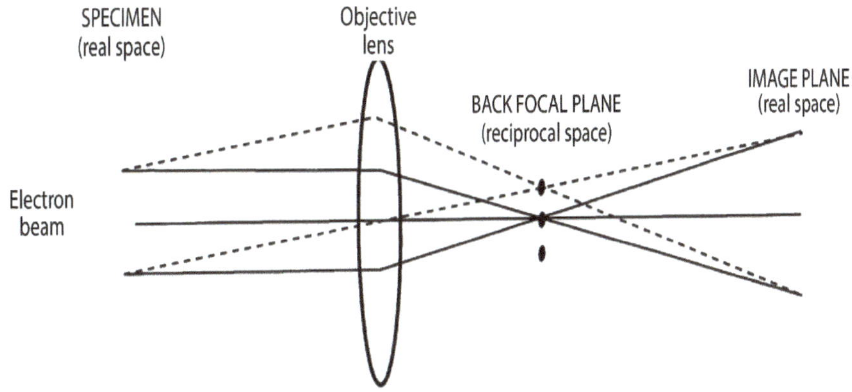

Figure 6.1. *Optical ray diagram with an optical objective lens showing the principle of the imaging process in a transmission electron microscope.*

Source: https://www.ncbi.nlm.nih.gov/pmc/articles/PMC4865294/pdf/j66ben.pdf.

The ability to view both electron microscope images (information in real space) and diffraction patterns (information in reciprocal space) for the same region by adjusting the electron lenses is a great advantage of the transmission electron microscope. We can obtain a diffraction pattern from a designated area as small as 100 nm in diameter by inserting a chosen area aperture and applying parallel incident beam illumination. One can now obtain a diffraction pattern from an area as small as a few nm in diameter using the newly developed microdiffraction techniques, which involve converging incident electrons on a specimen. Utilizing a conical beam ($\alpha > 10–3$ rad), convergent beam electron diffraction (CBED) creates diffraction disks whose intensity distribution makes it possible to uniquely determine all point groups and the majority of space groups. Reciprocal lattices (crystal structures) and mutual crystal orientation relationships can be readily obtained because nearly every grain in a polycrystalline material can record a selected area diffraction pattern. Thus, for many materials for which single crystals of the sizes appropriate for x-ray or neutron diffraction are not available, single crystal structural information can be obtained. These materials can be unstable or metastable phases, the results of phase transitions at low temperatures, fine precipitates, nanoparticles, etc.

Obtaining the electron diffraction pattern is the first step towards examining an electron microscope image. The image with increased contrast can then be viewed by passing either the transmitted beam or one of the diffracted beams through a small objective aperture (placed in the back focal plane) and switching lenses to the imaging mode. The bright-field method (correspondingly, a bright-field image) is the observation mode when only the transmitted beam is utilized, as shown in Figure 6.2a. Once a single diffracted beam has been chosen (Figure 6.2b), along with a dark field image, is referred to as the dark field method. The amplitude of the transmitted or diffracted beam changes as a result of absorption and dynamic scattering in the specimens, which is responsible for the contrast in these images. As a result, the contrast in the image is known as the amplitude contrast or absorption-diffraction contrast. The study of mesoscopic microstructures can benefit from the use of amplitude-contrast images. g. precipitates, interfaces, domains, and lattice defects. To determine the specific crystallographic details of these heterogeneities, kinematic and dynamic scattering theories are developed.

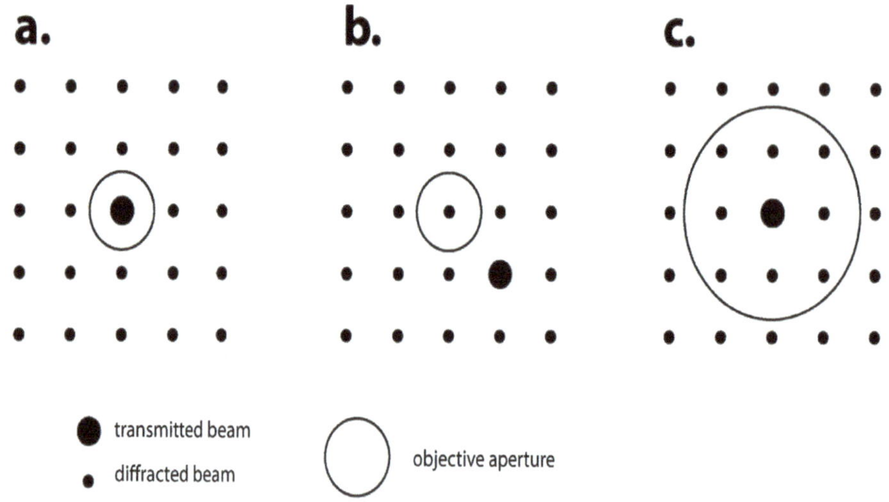

Figure 6.2. *Three observation modes in electron microscope using an objective aperture. The center of the objective aperture is on the optical axis.*

Note: (a) Bright-field method; (b) dark-field method; (c) high-resolution electron microscopy (axial illumination).

Source: https://www.ncbi.nlm.nih.gov/pmc/articles/PMC4865294/bin/j66benf2.jpg.

Electron microscope images can be formed by selecting multiple beams on the back focal plane using a large objective aperture, as shown in Figure 6.2c. This method, known as high-resolution electron microscopy (HREM), produces images through multiple beam interference which results in a phase contrast image. In optimal conditions, the phase contrast closely corresponds to the projected potential of a structure for a thin specimen with aberration compensation in the microscope. For thicker specimens and less favorable conditions, the phase contrast must be compared with calculated images. The theory of dynamic scattering and phase contrast formation has been well developed for methods such as multislice and Bloch waves. HREM is particularly useful for determining disordered or defected structures that may be challenging to detect or analyze using other methods. Additional refinement of the structural model can be done using higher resolution powder X-ray or neutron diffraction techniques.

Some key advantages of using electron scattering for crystallographic studies are the significantly larger scattering cross section compared to x-rays and neutrons (10^3 to 10^4 times larger), the much smaller typical wavelengths (approximately 2 pm), and the ability to focus the electron beam to extremely fine probe sizes (approximately 1 nm). These features allow for the study of much smaller objects as single crystals compared to other radiation sources. Additionally, electron scattering provides a high sensitivity to small deviations from an average structure caused by ordering, structural distortions, short-range ordering, or the presence of defects. These changes often result in very weak superstructure reflections or diffuse intensity, which can be challenging to detect using X-ray or neutron diffraction.

Moreover, a variety of supplementary features known as analytical electron microscopy are offered by contemporary transmission electron microscopes. In order to obtain information on chemical compositions and local atomic environments, various detectors analyze inelastically scattered electrons (Electron Energy-Loss Spectroscopy, or EELS), excited electromagnetic waves (Energy Dispersion Spectroscopy, or EDS), and Z-contrast. When coupled with elastic electron diffraction, this data is crucial for establishing structural models, particularly in cases where a material has different phases.

The subsequent sections showcase the diverse contributions made by NBS/NIST researchers in the materials research domain, where TEM plays a pivotal role in their investigations. The research's crystallographic components are highlighted. The Materials Science and Engineering Laboratory is the primary source of the contributions that are being presented.

6.2.1. Discovery of New Structures Using Selected Area and Convergent Beam Electron Diffraction

During the early 1980s, the Metallurgy Division of NBS conducted extensive research on the rapid solidification of melts. This process involves the rapid cooling of materials, such as metallic alloys, at rates exceeding 10^4 °C/s. The high cooling rates lead to the formation of new and often unstable crystalline or glassy structures. Additionally, the rapid cooling results in the formation of small-grain polycrystalline microstructures due to a high nucleation rate in the liquid. The unique combination of unknown structures and small grain sizes makes these materials challenging to study using X-ray diffraction but highly suitable for transmission electron microscopy (TEM). One of the most significant findings in modern crystallography came from Dan Shechtman's study of quickly solidifying Al-Mn alloys: a quasiperiodic structure with icosahedral symmetry, which includes 3-fold, 5-fold, and 2-fold rotation axes of symmetry. At that time, the entire field of crystallography was incompatible with such symmetry. By carefully building a reciprocal lattice using a series of selected area electron diffraction, the icosahedral symmetry of the phase was revealed. For the first time, it was demonstrated that a well-ordered homogeneous structure—not twinned—with symmetry elements incompatible with translational periodicity existed. J. W. In a separate piece in this issue, Cahn and D. Shechtman talk about the history of this amazing discovery and its crystallographic aspects.

The icosahedral phase's discovery led to a flurry of activity in the emerging field of quasicrystal research. Early on, TEM was instrumental in many aspects of this activity, to which many NIST researchers made active contributions. Not too long after Shechtman et al. publication, L. A distinct kind of quasiperiodic structure was identified by Bendersky: the decagonal phase with a 10-fold rotation axis and an apparent 10/mm point group (Figure 6.3).

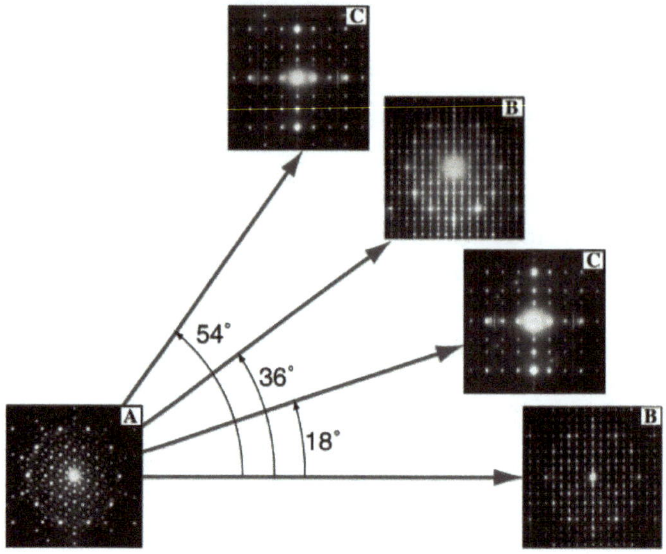

Figure 6.3. *A series of SAD electron diffraction patterns obtained from the $Al_{78}Mn_{22}$ rapidly solidified alloy by tilting a single grain. Based on these patterns, a unique non-crystallographic 10-fold axis and a one-dimensional periodicity of the decagonal phase were established.*

Source: https://www.ncbi.nlm.nih.gov/pmc/articles/PMC4865294/bin/j66benf3.jpg.

Electron diffraction analysis of this $Al_{80}Mn_{20}$ rapidly quenched alloy revealed a structure of two-dimensional quasiperiodic layers with a lattice parameter of c = 1.24 nm, stacked periodically along the 10-fold axis in the decagonal phase. Soon after, the Al-Pd system was discovered to have a decagonal phase that was comparable but had a different periodicity (c = 1.65 nm). Not only was a novel structure found, but the general principles of quasiperiodicity were also demonstrated, which added significance to the discovery. Many advances have been made in theory and experimentall procedures since the first quasiperiodic structures in Al-Mn alloys were found in 1984. More than a 100 distinct metallic systems have been found to have quasicrystalline phases, and it has been demonstrated that some of these phases are thermodynamically more stable than periodic crystals.

Among other significant discoveries at NIST associated with the new field of "quasi-crystallography" were:

- The first conclusive determination of the m^{3-} 5-point group for the icosahedral phase (for Al-38% Mn-5% Si (mass fraction) rapidly solidified alloy). Here the methods of convergent beam electron diffraction (CBED) were applied for the first time to a quasiperiodic structure. Figure 6.4 shows an example of such CBED patterns from which the whole (3-dimensional) pattern symmetries of fivefold [1τ0], threefold [111] and twofold [001] orientations were derived (τ-irrational "golden mean" number).

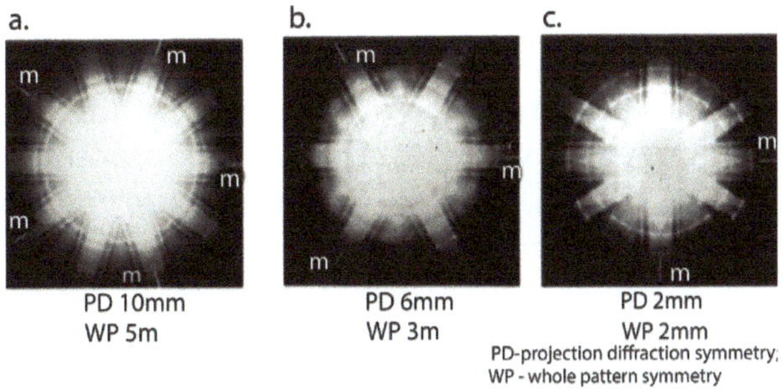

Figure 6.4. *CBED patterns taken along.*

Note: (a) fivefold [1τ0], (b) threefold [111] and (c) twofold [001] orientations. The lines indicate the mirror planes (m).

Source: https://www.ncbi.nlm.nih.gov/pmc/articles/PMC4865294/bin/j66benf4.jpg.

- Polycrystalline aggregates of a cubic phase [α-$Al_9(Mn,Fe)_2Si_2$] with an overall icosahedral symmetry were found in rapidly solidified $Al_{75}Mn_{15-x}Fe_xSi_{10}$ (x = 5 and 10) alloys. Through a twinning operation, the cubic axes undergo five-fold rotation about irrational <1,τ,0> axes; however only five orientations occur among hundreds of crystals (Figure 6.5). This is a special orientation relationship without any coincidence (or twin) lattice, and it is dictated by the non-crystallographic symmetry of a motif (in the case of the α phase—the 54-atom Mackay-icosahedron motif). The motifs are parallel throughout the entire polycrystalline aggregate, and the crystal axes change across grain boundaries. Based on this finding, the entire concept of twinning and special grain boundaries was re-examined. A new definition of special orientations, including hypertwins, based on reduction of the number of arithmetically independent lattice vectors was proposed. This new classification of special orientations within crystalline structures includes both old and new special orientations and can be easily interpreted in terms of quasilattices.

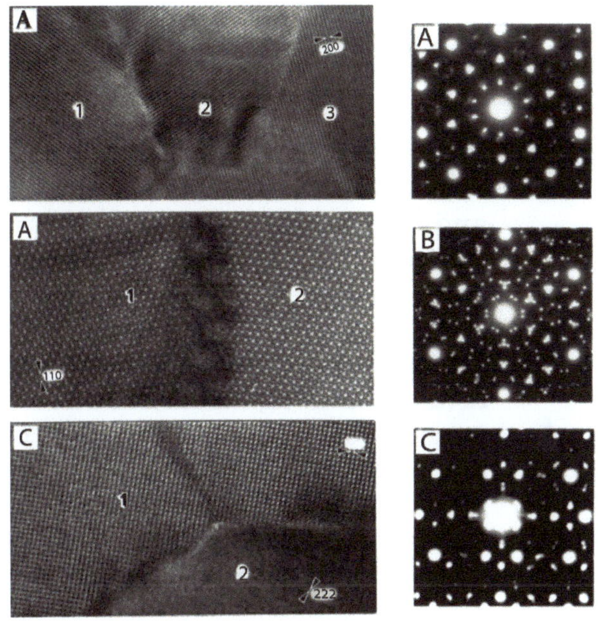

Figure 6.5. *Polycrystalline aggregates of a cubic phase (α-$Al_9(Mn,Fe)_2Si_2$) with their overall icosahedral symmetry were found in rapidly solidified $Al_{75}Mn_{15-x}Fe_x Si_{10}$. A, B, and C are high-resolution images of five orientational variants and corresponding SAD patterns in five-, three-, and two-fold orientation, respectively.*

Source: https://www.ncbi.nlm.nih.gov/pmc/articles/PMC4865294/bin/j66benf5.jpg.

- The nucleation and growth properties of icosahedral and decagonal phases were examined. The study showed that dendritic Al-Mn icosahedral crystals grow along three-fold axes, and the decagonal phase nucleates epitaxially on the icosahedral phase, with coinciding 5- and 10-fold axes. F. Gayle gave beautiful images of the faceted icosahedral crystals in his study of an Al-Li-Cu alloy, which showed growth of icosahedral phase dendrites along 5-fold axes.

- TEM was used to study nucleation behavior of the icosahedral phase in submicron size droplets of Al-14% Mn (atom fraction) produced by electrohydrodynamic atomization. The icosahedral phase was found to nucleate very easily; this phenomenon was explained by the possible topological similarities of atomic packing in liquids and icosahedral quasicrystals. Observation of a metallic glass-like structure for the highest cooling rates was explained by the microquasicrystalline structure rather than by a conventional structure of frozen melt. Modeling of a X-ray diffraction spectra (obtained from a similar but vapor deposited structure of Al-Mn) and analysis of heat evolution during DSC heating of as-deposited samples proved the concept of microquasicrystalline state.

Utilizing the strong electron scattering, it has been demonstrated that incommensurate modulations exist in a number of compounds. For example, one-dimensional modulations of atomic distortions were found in Zr_3Rh_4 (which has a basic rhombohedral structure). The modulation vector in this structure is incommensurate in both direction and k-value, and it is normal to the c-axis (Figure 6.6). Additional investigation into the $Zr_3(Rh_{1-x}Pdx)_4$ alloys has shown a fascinating relationship between the magnetic and superconducting characteristics and the occurrence of the incommensurate modulation.

Incommensurate modulations were also found in a series of the layered $Sr_n(Nb,Ti)_nO_{3n+2}$ compounds. These structures are composed of pseudo-two-dimensional slabs of a distorted perovskite structure. The slabs are n-octahedra thick and extend parallel to the {110} perovskite plane. The compounds with n = 4, 5, 6, and 7 were observed to undergo a commensurate-incommensurate phase transition in the temperature range 150°C to 250°C. The wave-vector of the incommensurate modulation is parallel to [100] direction of the basic orthorhombic lattice. The incommensurate phase transition observed in several $Sr_n(Nb,Ti)_nO_{3n+2}$ compounds were attributed to the structural distortion approximated by the alternating tilting of the $(Ti,Nb)O_6$ octahedra, with amplitudes varying in the direction of the modulation.

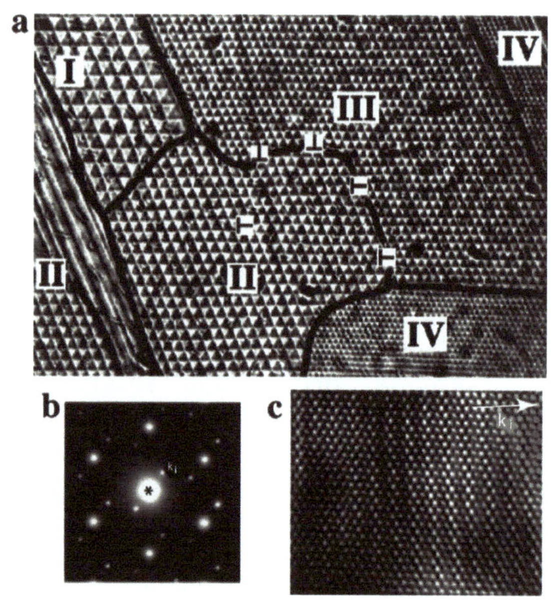

Figure 6.6. *(a) Electron diffraction and (b) corresponding high-resolution image of the Zr_3Rh_4 compound having a basic rhombohedral structure and one-dimensional incommensurate modulation k_i. (c) Optical image showing a regular array of triangular domains forming a macrolattice with defects. The domains are related to crystallographic variants (different directions of a k-vector) of the incommensurate phase.*

Source: https://www.ncbi.nlm.nih.gov/pmc/articles/PMC4865294/bin/j66benf6.jpg.

Very different types of incommensurate modulation were recently found in Ruddlesden-Popper ($n = 2$) $La_{2-2x}Ca_{1+2x}Mn_2O_7$ compounds. For the composition range $0.6 \leq x \leq 0.8$, two sets of near-orthogonal incommensurate k-vectors normal to the tetragonal c-axis were identified by electron diffraction. High-resolution imaging confirmed the presence of the incommensurate two-dimensional lattice. The lattice can be approximated with $a_{2D} = 5a_p$ (a_p—perovskite unit cell). A model based on the previously unobserved type of 4:1 charge/orbital ordering between the Mn^{4+} and Mn^{3+} cations was suggested. Investigation of the magnetic and electric properties of the compounds also indicates the presence of the charge ordering.

6.2.2. Structural High-Resolution Microscopy

Early in the 1970s, methods for imaging the extended defects and unit cell structures of inorganic crystals were developed through high-resolution electron microscopy. The primary goal of the approach in these early studies, which employed electron microscopes with relatively low resolution, was to identify one-dimensional structural sequences. The potential of TEM structural imaging to investigate the structures of complex oxides, and in particular "solid solutions," which really resulted from the intergrowth of layers of closely related structures/phases, was first recognized by Robert S. Roth at NBS. He actively worked in tandem with J., one of the pioneers of multi-beam structural imaging, in the 1970s. Australia's Allpress and Japan's S. Iijima (who was affiliated with Arizona State University at the time). In the Fifth Materials Research Symposium, which R. S. and NBS organized. Roth as well as S. J. Schneider, Jr. a large number of the publications focused on using innovative multi-beam imaging to solve structural problems.

Collaboration between S. Iijima and R. S. Roth resulted in the publication of one of the first high-resolution two-dimensional images where the intergrowth of 4 × 4, 4 × 3, and 3 × 3 blocks in $TiO_2{\cdot}7Nb_2O_5$ was imaged, thus demonstrating the possibility of solving complex structures directly (Fig. 6.7). This and similar efforts were essential in establishing the important structural principles of crystallographic shear and Wadsley defects (coherent intergrowth of one member of the structural family with another).

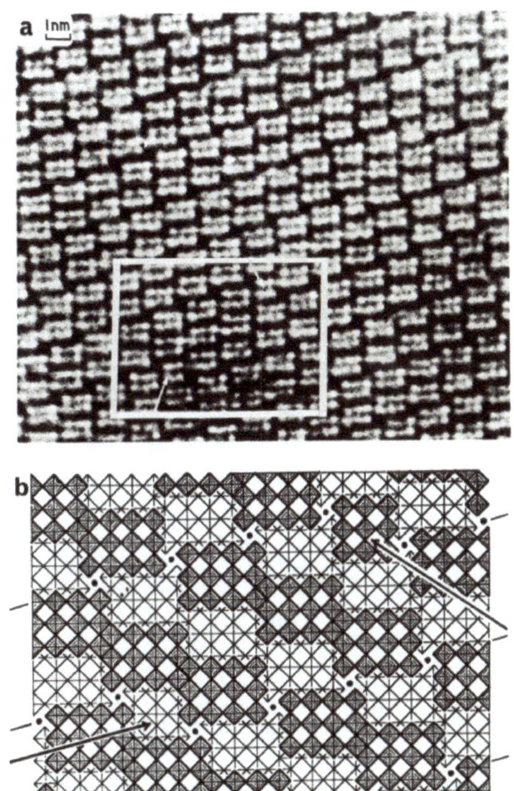

Figure 6.7. *(a) High-resolution two-dimensional lattice image from a fragment of TiO_2:$7Nb_2O_5$, showing several displacements associated with the presence of isolated 3 × 3 blocks (arrowed) in the matrix of 4 × 3 blocks. (b) Idealized model*

of the area outlined in (a). The crystallographic shear (CS) planes (containing tetrahedral metal atoms) are marked. They suffer a displacement at the points where 3 × 3 blocks (arrowed) occur.

Source: https://www.ncbi.nlm.nih.gov/pmc/articles/PMC4865294/bin/j66benf7.jpg.

Within the systems studied by Roth were polymorphs of ZrO_2:$12Nb_2O_5$, off-stoichiometric $xNb_2O_5 \cdot yWO_3$ compounds and mixed-block structures of Rb_2O-Ta_2O_5, Rb_2O-Nb_2O_5, and K_2O-Ta_2O_5.

High-quality microwave dielectrics with a dielectric constant (>30), minimal losses, and a temperature coefficient of dielectric constant that is almost zero were more in demand in the 1980s. Barium titanates were the most widely accepted materials at the time. R. Dot S. Roth along with his postdoc P. K. With a point-to-point resolution of 0.23 nm, Davies examined structural variations of these compounds using the recently installed Philips 430 microscope1 at NIST. They determined the processes of intergrowth defect formation in two significant compounds, $Ba_2Ti_9O_{20}$ and $BaTi_5O_{11}$, using structural imaging. They discovered that the most common flaw was the emergence of a novel triclinic polytype, whose ionic arrangement was quite similar to the structure that was eventually accepted. Nonstoichiometric defect formation was negligible, while stoichiometric defects with significant microtwinning and the formation of face-sharing octahedra were also observed. In the 1990s, TEM structural studies of microwave dielectrics were carried out in tandem with phase diagram analyzes of these materials conducted by T in the NIST Ceramics Division. R. Roth and Vanderah.

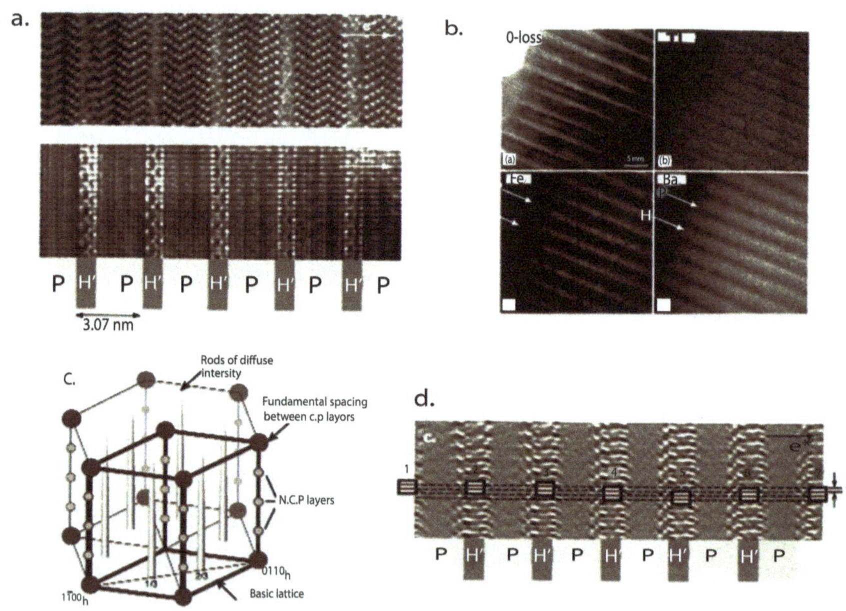

Figure 6.8. *(a) HRTEM images of the M phase taken in orientations corresponding to [112¯0] (upper micrograph) and [11¯00] (lower micrograph) zone axes of the hexagonal Ba(Ti,Fe)O_3 layer. On the background the yellow and pink stripes correspond to P (perovskite) and H-type (magnetite) slabs, respectively. (b) Composional maps of the M phase (in [112¯0] zone axis orientaion) obtained with Ti $L_{2,3}$, Ba $M_{4,5}$, and Fe $L_{2,3}$ edges. The maps show the Ti composition is relatively uniform throughout the crystals whereas the Fe elemental distribution reveals an enhancement corresponding*

to the H slabs. Similarly, the Ba distribution is decreased in the H slabs relative to the P slabs. (c) A schematic drawing of a reciprocal lattice of the compounds showing the typical rods of continuous intensity in the c-direction. (d) The image obtained by inverse FFT shows only the H slabs. Rectangles correspond to a characteristic structural element of the H slab and emphasize random shift of the H slabs.

Source: https://www.ncbi.nlm.nih.gov/pmc/articles/PMC4865294/bin/j66benf8.jpg.

The Materials Science and Engineering Laboratory (MSEL) of NIST acquired the ultra-high resolution JEOL3010 microscope in the 1990s, which made high-resolution imaging routine for examining defects and structures. When studying recently discovered structures, the typical methodology involved creating an approximate model using electron diffraction and high-resolution experimental images. Next, image simulations that were compatible with the experimental images were created, and lastly, the model was refined using higher resolution x-ray or neutron powder diffraction.

Recently, high-resolution imaging has allowed for the discovery of a very interesting and novel structural architecture for a set of six new ferrimagnetic phases in the $BaO:TiO_2:Fe_2O_3$ system. It was established that the compounds are members of a class of intergrowth structures with order. Two kinds of alternating structural two-dimensional slabs were found, despite the fact that the structures generally show a close-packed framework of [O + (Ba/O)] layers (with stacking sequences differing for different phases). The first has the hexagonal h-$BaTiO_3$ structure, while the second has a spinel-like, Fe-rich structure. Fig 6.8.

An illustration of such structure's high-resolution imaging is provided by Figure 6.8a (observed for the M phase). Using energy-filtered mapping and EELS, the compositional distribution of Fe, Ti, and Ba was determined (Figure 6.8b). All six phases could be thought of as self-assembling magnetic multilayer structures with interesting physical properties due to the strongly heterogeneous distribution of Fe.

For most of the phases, one-dimensional structural disorder was seen (Figures 6.8c) and determined to be caused by the h-$BaTiO_3$ blocks' poor spatial correlation (translation within a closed-packed plane) (Figure 6.8). Even when utilizing a single crystal, the disorder was the cause of the difficulty in determining the structures through x-ray diffraction.

A structural study of the microwave dielectric calcium tantalate, $Ca_2Ta_2O_7$, showed diffraction patterns displaying strong subcell characteristics of the fluorite structure. The complex distributions of superlattice reflections depend on the nature and concentration of the dopant and on the reaction temperature. Polytypes with n = 3, 4, 5, 6, and 7 were prepared as single crystals by the flux method. High-resolution electron microscope images were used to determine a characteristic structural feature of the polytypes (Fig. 6.9).

The major feature was a fluorite-type cation array, which is periodically twinned on $(111)_{fluorite}$ to give mixed cubic-hexagonal stacking of the cation layers. The basic repeating units are slabs of different thicknesses comprising a Ca_3Ta layer and a Ta_3Ca layer. Single crystal x-ray and powder x-ray and neutron data were used to determine an exact structure of the polytypes. For n = 3, the structure is the 3-block trigonal weberite $P3_121$ (3T). In the 7M polytype the *h*-stacked Ta_3Ca layers divide the structure into two blocks of *c*-stacked layers which contain 3 and 4 repeating units and can be thus designated as 3M and 4M blocks. The stacking sequence of metal atom layers in the 7M polytype is *hccccchcccccch*... For the 5M polytype, the *h*-stacked layers divide the structure into two 2M and 3M blocks. The metal atom layer stacking sequence is then *hccchcccch*.

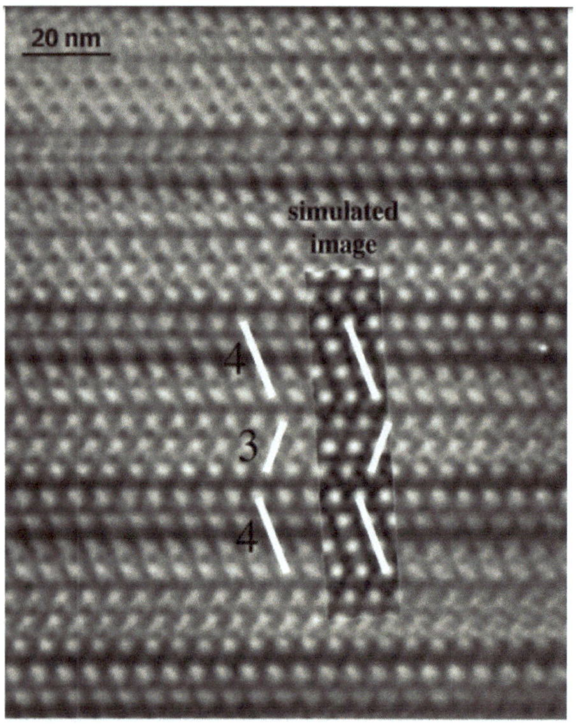

Figure 6.9. *High-resolution TEM image for 7M-$Ca_2Ta_2O_7$ viewed along [110] and the corresponding simulated phase-contrast image (inset). The simulated image was obtained for 35 nm thickness, –50 nm defocus value and the 10 nm^{-1} aperture. The 4- and 3-layers structural blocks are emphasized.*

Source: https://www.ncbi.nlm.nih.gov/pmc/articles/PMC4865294/bin/j66benf9.jpg.

6.2.2.1. Study of Phase Transitions

The shift in symmetry is the fundamental characteristic of a structural phase transition. Since it has less symmetry than the high temperature phase(s), the low temperature (room temperature) phase is typically the one under study. Its space group frequently corresponds to the high temperature phase's subgroup. Consequently, a single high temperature phase crystal (grain) at low temperature is split into several structural variants, or domains related to symmetry. The variations are distinguished from one another by particular interfaces and are either translational (anti-phase domains, or APDs) or rotational (twins). Predicting the structure of high temperature phases and potential intermediate phases is frequently possible with knowledge of the low temperature phase's structure and the crystallographic nature of the interfaces. The best method for this approach is to use TEM direct-space imaging (amplitude contrast) in conjunction with SAD and CB electron diffraction. High-resolution imaging and the Fast-Fourier Transform work best for very small domain structures.

Intermetallic compounds, specifically Ti-Al-Nb alloys, were the subject of research by scientists from the Metallurgy Division of NIST in the early 1990s due to interest in a new class of high-temperature materials for aerospace applications. Understanding the intricacy of the phase changes taking place in these materials was made possible thanks in large part to the use of TEM. (1) A martensitic-like transition from the high-temperature BCC phase, with formation of a Widmanstatten-type lamellar structure (an HCP-to-FCC transition plus ordering), and omega-type displacements are among the transformations observed in the Ti-Al-Nb system.

Study of the Ti_2AlNb and Ti_4AlNb_3 alloys showed the presence of two phases: the Ti/Nb-rich b.c.c. phase and the ternary orthorhombic Ti_2AlNb phase. Depending on the alloy composition, one of the phases was observed as a precipitate with plate morphology in the matrix of the other phase. For the Ti_4Al_3Nb alloy, formation of a sequence of metastable ω-related phases from the B2 (ordered BCC) phase was established and summarized in the form of maximal group-subgroup relationship (Fig. 6.10). An equilibrium low temperature phase has the $B8_2$ structure (and triple-cell $D8_2$ structure for the more Nb-rich compositions). The metastable phase and both $B8_2$ and $D8_2$ are structurally related to the B2 phase by displacement (collapse) of every two (out of three) (111) planes. The ω-type structures were verified by means of transmission electron microscopy and by single crystal x-ray diffraction.

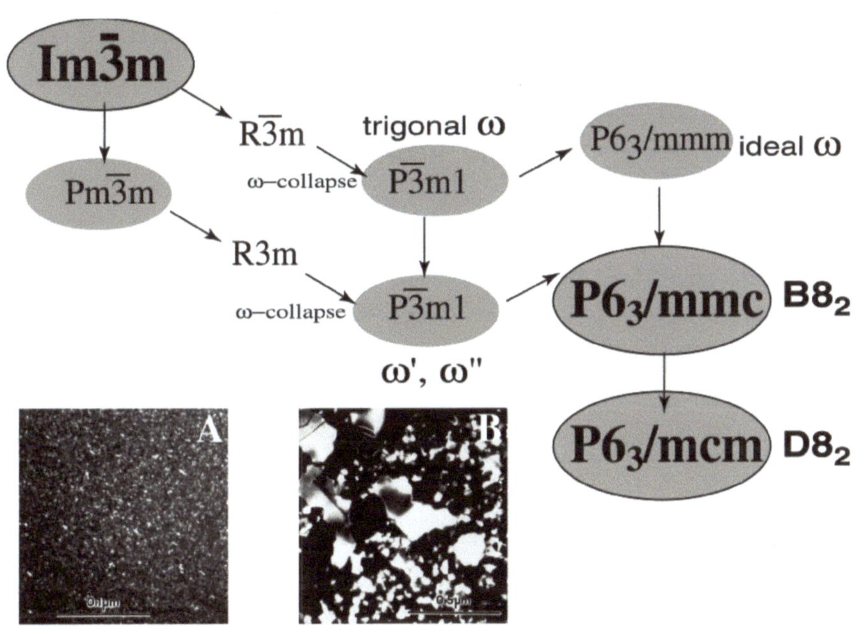

Figure 6.10. *A sequence of ω-related phases formed from the high temperature B2 (ordered BCC) phase in the Ti_4Al_3Nb alloy is summarized in the form of maximal group-subgroup relationship. Dark field images A and B show microstructures of (A) metastable ω′ phase and (B) a mixture of ω″ and $B8_2$ phases.*

Source: https://www.ncbi.nlm.nih.gov/pmc/articles/PMC4865294/bin/j66benf10.jpg.

For Ti-25% Al-12.5% Nb (atom fraction), the transformation path from high temperatures includes the formation of the intermediate hexagonal DO_{19} phase, and subsequent formation of the low-temperature orthorhombic O (Ti_2AlNb) phase. For alloys close to Ti-25% Al-25% Nb (atom fraction), the path involves an intermediate B19 structure and subsequent formation of a translational domain structure of the O phase. The experimental results were summarized as a sequence of crystallographic structural relationships developed from subgroup symmetry relations (Figure 6.11). Symmetry elements lost in each step of the sequence determine the possibilities for variants of the low symmetry phase and domains with characteristic interfaces that can be present in the microstructure.

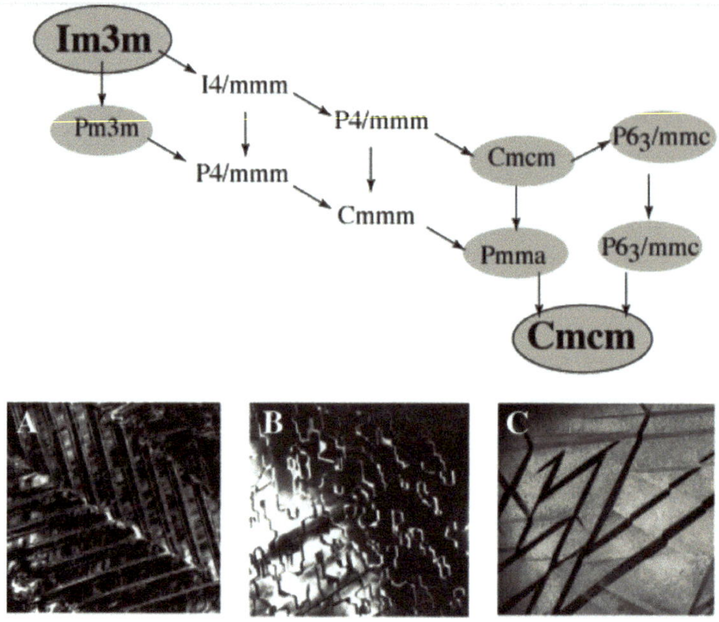

Figure 6.11. *A sequence of phases formed from the high temperature B2 (ordered BCC) Ti-Al-Nb phase is summarized in the form of maximal group-subgroup relationship.*

Note: (A) Ti_2AINb alloy, rotational domains of the O phase accommodated in the form of hierarchical twin structure; (B) Ti_2AINb alloy, interfaces between translational domain formed in the Pm3m(B2)-to-Pmma(B19) transition; (C) two-phase mixture of DO_{19} (matrix) and O-phase (precipitates).

Source: https://www.ncbi.nlm.nih.gov/pmc/articles/PMC4865294/bin/j66benf11.jpg.

Figure 6.11 gives three examples of such domain: (A) Ti_2AlNb alloy, rotational domains of the O phase accommodated in the form of a hierarchical twin structure; (B) Ti_2AlNb alloy, interfaces between translational domains formed in the course of the $Pm\bar{3}m$(B2)-to-Pmma(B19) transition; (C) a two-phase mixture of DO_{19} (a matrix) and three-variants of O-phase (precipitates). The orientation of inter-domain interfaces is determined by the requirement for a strain-free interface (as in Figure 6.11A and C). A structure of the O-phase (an important ternary phase, currently a base for new high-temperature composites) was refined using the preliminary model derived from the TEM studies and powder neutron diffraction data.

The approach of the maximal group-subgroup relationship was also applied to phase transformations in other systems, and in particular to transformations between ordered polymorphs of the microwave dielectric $Ca(Ca_{1/3}Nb_{2/3})O_3$ and $Ca(Ca_{1/3}Nb_{2/3})O_3$-$CaTiO_3$. Four $Ca_4Nb_2O_9$ polymorphs with perovskite-related structures and different arrangements of B-site cations were identified. Ordering of the B-cations (Ca^{2+} and Nb^{5+}) in this system is combined with displacement of oxygen ions by octahedral tilting. Three of the polymorphs have 1:1 (monoclinic $P2_1/c$), 2:1 (monoclinic $P2_1/c$) and 3:1 (triclinic $P1$) ordering.

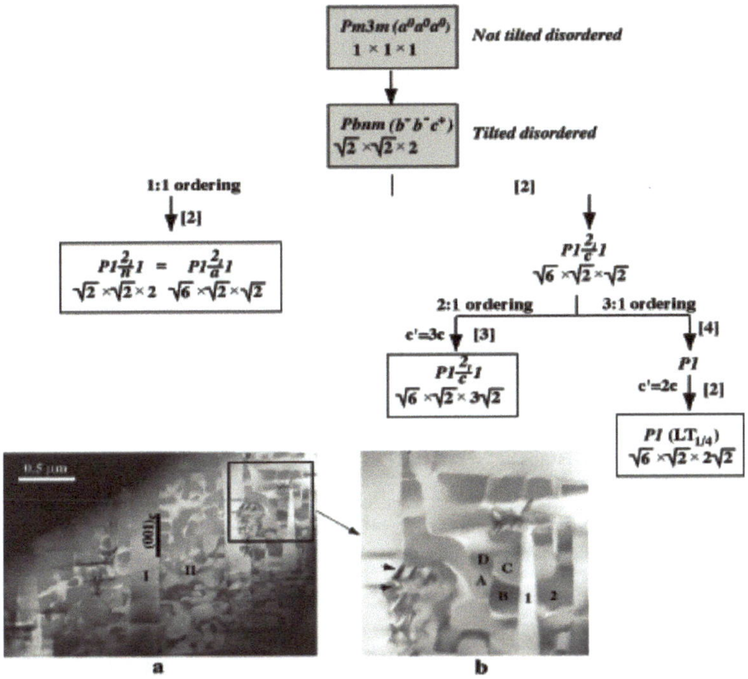

Figure 6.12. *Top. Symmetry tree describing group/subgroup relations between the three $Ca_4Nb_2O_9$ polymorphs with distinct (1:1, 1:2, and 1:3) arrangements of the B-cations; the numbers in brackets indicate the number of crystallographic variants generated by the corresponding minimal symmetry reductions. Since all three phases feature similar $b^-b^-c^+$ octahedral tilting, their space groups are subgroups of Pbnm, which describes a disordered perovskite structure with the same tilting. Bottom: a-Dark-field image of the single grain containing the metastable 1:3 ordered $Ca_4Nb_2O_9$ polymorph. b-Magnified view of the area outlined by rectangular in (a). Twin-type domains I and II in the area labeled a are related to the Pm3m→Pbnm octahedral tilting transition. The domains exhibit a substructure consisting of both rotational (1 and 2 in the area labeled b) and translational (A, B, C, and D) domains. The domains 1 and 2 can be described by the Pbnm→P1 symmetry reduction, while the domains A, B, C, and D can be accounted for by the loss of the c-glide plane ($P2_1/c$→P1) and doubling of the c lattice parameter. Additionally precipitates of the 1:2 ordered polymorph nucleated in the 1:3 ordered matrix are seen in b (indicated by arrows).*

Source: https://www.ncbi.nlm.nih.gov/pmc/articles/PMC4865294/bin/j66benf12.jpg.

The 1:1 ordered structure features alternating $\{111\}_c$ planes occupied exclusively by Nb, and the occupied by a disordered mixture of Ca and the remaining Nb cations, while the 1:2 ordered polymorph exhibits {CaNbNb...} sequence of the $\{111\}_c$ B-cation planes. The metastable 3:1 structure is derived from the 1:1 ordered structure by a partial ordering of Ca and Nb cations on those B-sites which were occupied by a random Ca/Nb mixture in the 1:1 ordered array; the resulting ordered arrangement can be viewed as 1:3 type. The models were subsequently validated by Rietveld refinements using combined x-ray and neutron powder diffraction data. The relationship between the polymorphs in the form of maximal subgroup relationship was derived from the complex hierarchy of interfaces observed by TEM (Figure 6.12). The space groups of the ordered $Ca_4Nb_2O_9$ polymorphs (deduced from electron diffraction data) do not exhibit group/subgroup relations, and, therefore, phase transitions between them must be first-order transitions.

Conventional TEM was used to solve a variety of microstructural formation-related issues. For instance, the effect of Cr_2O_3 diffusion into polycrystalline Al_2O_3 down grain boundaries and the concurrent generation of local stresses due to the change in lattice parameter were studied in the study of Diffusion-Induced Grain Boundary Migration (DIGM) phenomena. A strong relationship was discovered between the dislocation spacing and the chromium concentration in the alloyed regions when these dislocations were examined using bright and dark field transmission electron microscopy.

TEM was used to examine the areas surrounding a microhardness indentation to investigate another issue pertaining to the plastic deformation of $A1_2O_3$ by slip and twinning. The results establish: (1) the occurrence of pyramidal slip on $\{11\bar{2}3\}$ $\langle 1\bar{1}00\rangle$, and (2) the nature of basal twins in this material. The observations on basal twins, in particular, have led to a completely different description of the twinning process. In the study of important Ti-Al aerospace alloys, the morphology of discontinuous coarsening in the Ti_3Al (α_2)/TiAl (γ) fully lamellar structure was examined.

Three different shapes were observed in discontinuously coarsened lamellar structures (secondary lamellae). Type-(I) lamellae have a low energy habit plane as their interfaces and have the same direction as the original primary lamellae. Type-(II) lamellae have the same crystallographic orientation of α_2 plates as the original primary lamellae, but a different direction and irregular faceted interfaces. Type-(III) lamellae have a different direction and crystallographic orientation of the α_2 plates compared to the original primary lamellae, but with low energy habit plane interfaces. The growth kinetics of these lamellae types were analyzed using the Livingston and Cahn treatment to determine the relationship between misorientation of primary lamellae, secondary lamellae, and colony boundaries.

Precipitation of second phases in aluminum alloys presents another challenging application for electron diffraction due to the small dimensions and multiple variants of the precipitates, which are most often metastable phases in commercial alloys and tempers. In the late 1980s the Metallurgy Division had a collaborative effort with Martin Marietta Laboratory to determine the origins of the ultra-high strength in Weldalite aluminum alloys, later used to build the Space Shuttle External Tank to reduce launch weight.

This novel aluminum alloy was primarily strengthened with 1.2% Li and 4.0% Cu (mass fraction). Silver and magnesium were added, 0.4% each, to aid precipitation of second phases, and 0.1% Zr was added to inhibit recrystallization. TEM imaging and diffraction was required to sort out the complex microstructure responsible for the high strength of the alloy. Our microstructural studies established that the Weldalite alloys are strengthened primarily by the equilibrium T_1-Al_2CuLi phase, in contrast to other commercial aluminum alloys, which are strengthened by metastable precipitate phases. In addition to T_1, additional strengthening arises from 3 other lath or plate-like precipitates (S'-Al_2CuMg, η'-Al_2Cu and an unknown phase), all lying on distinct habit planes in the matrix (Fig. 6.13). When α', or $Al_3(Zr,X)$, which inhibits grain boundary motion, is included, it appears that this alloy in the highest strength condition may represent a metastable equilibrium between six phases.

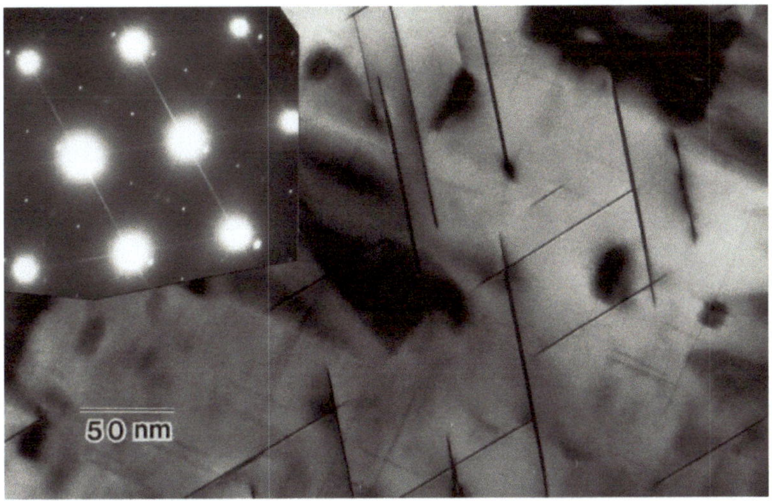

Figure 6.13. *This ultra-high strength aluminum alloy is strengthened by a mix of precipitates, including the prominent T1-Al_2CuLi platelets lying on matrix {111} planes, and platelike θ′ (on {100} planes) and lath S′ precipitates (on {210} planes), identified by the diffraction pattern.*

Source: https://www.ncbi.nlm.nih.gov/pmc/articles/PMC4865294/bin/j66benf13.jpg.

CASE STUDY

UTILIZING ELECTRON MICROSCOPY TO UNRAVEL THE STRUCTURE OF THE ZIKA VIRUS

- Background

The Zika virus, a mosquito-borne pathogen, garnered global attention due to its links to severe birth defects. Understanding its structure at a molecular level became crucial for developing treatments and vaccines.

- Objective

The primary objective was to determine the Zika virus's structural details using electron microscopy to aid in the development of therapeutic interventions.

- Methodology

1. **Sample Preparation**: Viral particles were isolated and purified from infected cells. Samples were then prepared for electron microscopy, which involved staining, fixing, and embedding the particles to preserve their structure.
2. **Transmission Electron Microscopy (TEM)**: TEM was employed to visualize the virus's internal structures. Ultra-thin sections of the virus were examined at high resolution to reveal fine structural details.
3. **Cryo-Electron Microscopy (Cryo-EM)**: This advanced technique was used to image the virus at near-atomic resolution. The virus was rapidly frozen to preserve its native state, and multiple images were taken from different angles to reconstruct a 3D model.

- Findings
  - **Structural Insights**: TEM and Cryo-EM provided detailed images of the Zika virus, revealing its icosahedral symmetry and the arrangement of its envelope proteins. The high-resolution 3D model showed the surface glycoproteins' spatial organization, crucial for understanding how the virus interacts with host cells.
  - **Protein Analysis**: Cryo-EM identified key epitopes on the viral surface proteins, which are potential targets for neutralizing antibodies. This information is vital for vaccine design, enabling the development of vaccines that elicit a strong immune response.
- Impact
  - **Vaccine Development**: The structural data obtained through electron microscopy directly informed the design of several candidate vaccines. By

targeting specific epitopes identified in the study, researchers could create more effective vaccines.

- **Drug Design**: Understanding the virus's structure facilitated the identification of potential drug targets, leading to the development of antiviral compounds that could inhibit the virus's ability to replicate.

- Conclusion

The case study of the Zika virus demonstrates the power of electron microscopy in virology. By providing detailed structural insights, electron microscopy not only advanced the scientific understanding of the virus but also played a critical role in the development of effective medical interventions, underscoring its value in addressing global health challenges.

CLASS ACTIVITY

For a class activity on microscopy, students can explore various microscopy techniques by rotating through stations set up with different microscopes: light, fluorescence, and electron microscopes. Each station should include prepared slides or samples relevant to the microscope type. Students can observe and sketch what they see, noting differences in resolution and detail. Additionally, brief explanations and videos at each station can enhance understanding. After the rotations, students can compare their observations and discuss the advantages and limitations of each microscopy technique, fostering a deeper appreciation of microscopy's role in scientific research.

SUMMARY

- Diffraction and microscopy are fundamental techniques in the field of imaging and structural analysis, playing crucial roles in various scientific disciplines, from materials science to biology.
- Electron microscopy, which includes transmission electron microscopy (TEM) and scanning electron microscopy (SEM), utilizes electron diffraction to achieve much higher resolutions than optical microscopy.
- X-ray diffraction (XRD) is a powerful analytical technique used to determine the atomic and molecular structure of a crystal.
- Neutron diffraction or elastic neutron scattering is the application of neutron scattering to the determination of the atomic and/or magnetic structure of a material.
- Electron microscopy (EM) encompasses several advanced imaging techniques that use a beam of electrons to obtain high-resolution images of specimens.
- Scanning Probe Microscopy (SPM) is a versatile and powerful technique used to image surfaces and measure their properties at the nanometer scale.
- High-resolution electron microscopy techniques were developed in the early 1970s for imaging structures of inorganic crystals at the unit cell level and their extended defects.
- The essential feature of a structural phase transition is the change of symmetry. Most often it is a low temperature (room temperature) phase that is studied, which has a lower symmetry than the high temperature phase(s).

REVIEW QUESTIONS

1. What is diffraction, and how does it relate to the wave nature of light? Provide an example of a diffraction phenomenon in everyday life.
2. Describe the diffraction pattern produced by a single slit and explain how the width of the slit influences this pattern. Use diagrams to illustrate your explanation.
3. Explain the setup and results of the double-slit experiment performed by Thomas Young. How does this experiment demonstrate the wave nature of light?
4. Explain how a diffraction grating works and how it is used to measure the wavelength of light. What is the effect of increasing the number of lines per millimeter on the diffraction pattern?
5. State Bragg's Law and explain how it is used in X-ray diffraction to determine crystal structures. Provide an example calculation to determine the interplanar spacing in a crystal.
6. Define resolution in the context of microscopy. How does the wavelength of the light or electrons used in a microscope affect the resolving power of the instrument?

7. Compare and contrast scanning electron microscopy (SEM) and transmission electron microscopy (TEM) in terms of their operating principles, image formation, and applications.
8. Why is sample preparation crucial in electron microscopy, and what are some common techniques used for preparing samples for TEM analysis?
9. What is confocal microscopy, and how does it improve upon traditional light microscopy in terms of resolution and imaging capabilities? Describe an application where confocal microscopy is particularly useful.
10. Discuss the application of X-ray diffraction (XRD) and electron microscopy in materials science. How do these techniques complement each other in the analysis of material properties and structures?

REFERENCES

1. Bendersky, L. A., Levin, I., Roth, R. S., & Shapiro, A. J. (2001). $Ca_4Nb_2O_9$–$CaTiO_3$: Phase Equilibria and Microstructures. *J Solid State Chem, 160*, 257.
2. Bennett, L. H., Waterstrat, R. M., Swartzendruber, L. J., Bendersky, L. A., Brown, H. J., & Watson, R. E. (2000). Magnetism and incommensurate waves in $Zr_3(Rh_{1-x}Pd_x)_4$. *J Appl Phys, 87*, 6016.
3. Fawcett, I. D., Kim, E., Greenblatt, M., Croft, M., & Bendersky, L. A. (2000). *Phys Rev B, 62*(6), 6485.
4. Grey, I. E., Roth, R. S., Mumme, W. G., Planes, J., Bendersky, L. A., Li, C., & Chenavas, J. (2001). to be published in *J Solid State Chem*.
5. Levin, I., Bendersky, L. A., Cline, J. P., Roth, R. S., & Vanderah, T. A. (2000). Octahedral Tilting and Cation Ordering in Perovskite-Like $Ca_4Nb_2O_9$=3•$Ca(Ca_{1/3}Nb_{2/3})O_3$ **Polymorphs.** *J Solid State Chem, 150*, 43.
6. Levin, I., Chan, J. Y., Geyer, R. G., Maslar, J. E., & Vanderah, T. A. (2001). *J Solid State Chem, 156*, 122.
7. Ojeda-May, P., Terrones, M., Terrones, H., Hoffman, D., et al. (2007). Determination of chiralities of single-walled carbon nanotubes by neutron powder diffraction technique. *Diamond and Related Materials, 16*(3), 473–476.
8. Page, K., Proffen, T., Niederberger, M., & Seshadri, R. (2010). Probing local dipoles and ligand structure in $BaTiO_3$ nanoparticles. *Chemistry of Materials, 22*(15), 4386–4391.
9. Piccoli, P. M. B., Koetzle, T. F., & Schultz, A. J. (2007). Single crystal neutron diffraction for the inorganic chemist—A practical guide. *Comments on Inorganic Chemistry, 28*, 3–38.

CHAPTER

7

Molecular Recognition and Binding

LEARNING OBJECTIVES

After studying this chapter, you will be able to:

- Understand the roles of molecular recognition and binding
- Define theoretical models of binding
- Discuss biological molecular recognition studies

KEY TERMS FROM THIS CHAPTER

Affinity
Binding site
Electrostatic interactions
Hydrogen bonding
Induced fit
Molecular docking
Receptor
Van der Waals forces
Allosteric regulation
Conformational change
Enzyme-substrate
Hydrophobic effect
Ligand
Non-covalent interactions
Specificity

INTRODUCTION

Molecular recognition and binding are fundamental concepts in chemical biology, playing a crucial role in various biological processes and applications. Molecular recognition refers to the specific and selective interaction between two or more molecules through non-covalent interactions such as hydrogen bonds, van der Waals forces, electrostatic interactions, and hydrophobic effects. This process is essential for the proper functioning of biological systems, as it underlies the specificity of enzyme-substrate interactions, antibody-antigen binding, and receptor-ligand associations. The ability of molecules to recognize and bind to specific targets with high affinity and selectivity is crucial for maintaining the integrity and regulation of biochemical pathways. Binding events are central to the mechanisms of molecular recognition, where the strength and specificity of these interactions determine the biological outcome. In chemical biology, understanding the principles of binding enables the design and development of drugs, diagnostics, and therapeutic agents. For instance, the binding affinity between a drug and its target protein can influence the drug's efficacy and specificity, minimizing off-target effects and adverse reactions. Techniques such as X-ray crystallography, nuclear magnetic resonance (NMR) spectroscopy, and surface plasmon resonance (SPR) are employed to study these interactions at the molecular level, providing insights into the structural and dynamic aspects of binding.

The exploration of molecular recognition and binding extends to the design of synthetic molecules that mimic natural biological functions. This includes the development of molecular sensors, artificial enzymes, and novel therapeutic agents that can intervene in disease pathways with high precision. By harnessing the principles of molecular recognition, researchers can create molecules that specifically target cancer cells, pathogens, or malfunctioning proteins, offering new avenues for treatment and diagnosis. Overall, the study of molecular recognition and binding not only enhances our understanding of biological systems but also drives innovation in the field of chemical biology, leading to the development of advanced biotechnological applications.

7.1. ROLES OF MOLECULAR RECOGNITION AND BINDING

Molecular recognition and binding events found in biology are ubiquitous, diverse and pivotal, but there are common themes and principles. In order to impart a flavor of this, we will take a brief look at a number of interlocking but diverse examples of biological molecular recognition and binding events. These examples come from fields as diverse as neurotransmission, bio-catalysis, immunity (antibody recognition), autoimmunity, inflammation and chromatin condensation all the way through to the control of gene expression (Baron & McCammon, 2013).

7.1.1. Acetyl Choline, Receptor, and Esterase

Brain tissue is comprised substantially of neuronal cells and glial cells. The former are the active component of brain tissue responsible for the transmission of sensory, emotional and motor information between brain regions and into the periphery (body) in the form of electrical pulses. The latter are support cells that ensure the metabolic stability and structural integrity of neuronal cells. Neuronal cells communicate with each other through synaptic junctions that comprise a narrow intercellular gap (synaptic cleft) between membrane regions of two adjacent neuronal cells known as the presynaptic membrane and the postsynaptic membrane (Figure 7.1). Small molecule messengers known as neurotransmitters provide communication across the gap between pre- and postsynaptic membranes. In short, electrical pulses traveling down the presynaptic neuron reach the presynaptic membrane, where they trigger Ca^{2+} influx across the membrane, leading to the migration of synaptic vesicles containing neuronal transmitters towards the presynaptic membrane. When they reach the membrane, synaptic vesicles fuse with the presynaptic membrane resulting in the release of substantial neurotransmitter into the synaptic cleft. Neurotransmitter must diffuse across the synaptic junction in order to be recognized by and become bound to binding sites found in neurotransmitter receptor proteins. The molecular recognition and binding of the neurotransmitter often facilitates the conversion of the receptor into an ion-channel protein able to re-initiate the propagation of electrical pulses but this time through the postsynaptic neuron. Such a process is critical to the function of all neuronal cells.

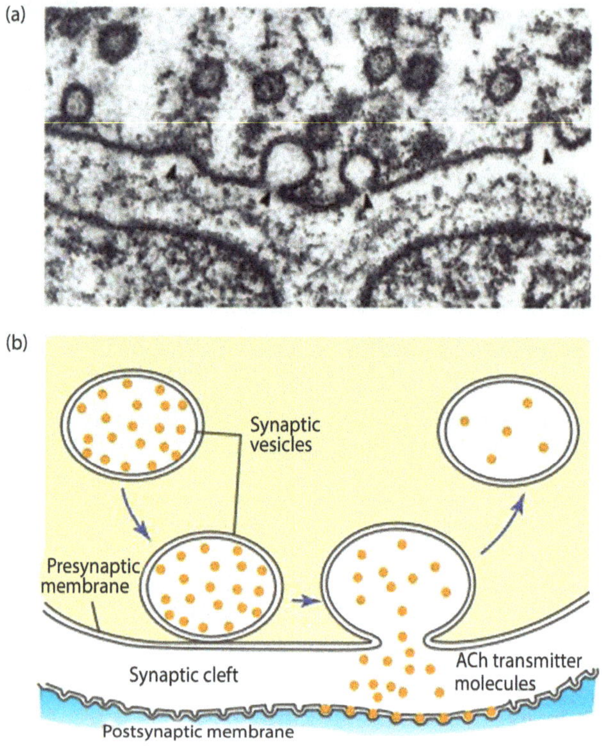

Figure 7.1. *Acetyl Choline (ACh) neuronal synapses.*

Note: (a) Electron microscope image. (b) Schematic illustration of synaptic junction drawn from electron microscope image illustrating the deposition of ACh transmitter in the synaptic cleft from synaptic vesicles.

Source: Andrew Miller, Julian Tanner, Essentials of Chemical Biology Structure and Dynamics of Biological Macromolecules; ISBN 978-0-470-84530-1.

One of the main neurotransmitters in the brain is known as acetylcholine (ACh), which is recognized and bound by the postsynaptic acetylcholine receptor (AChR) (Figure 7.2). The molecular recognition and binding of ACh by AChR is a seminal event in neurochemistry and as such is archetypical of all binding events between the many known neurotransmitters and their cognate receptor proteins. Neurotransmitters and their interactions with their cognate receptors are central to the chemistry of the brain, and deficiencies (or excesses) of neurotransmitter substances are frequently associated with neurological disease. Indeed, the famous neurodegenerative Alzheimer disease has long been associated with deficiencies in neurological levels of ACh, as have various types of clinical depression and memory loss.

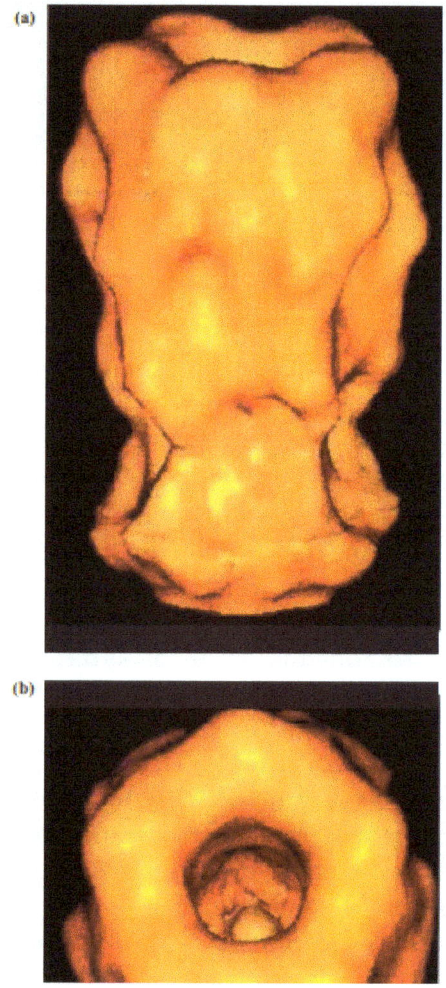

Figure 7.2. *Acetyl choline receptor (AChR) electron microscopy images of receptor R with (a) side view-bottom band is the membrane region; (b) top-view from the synapse side.*

Source: Andrew Miller, Julian Tanner, Essentials of Chemical Biology Structure and Dynamics of Biological Macromolecules; ISBN 978-0-470-84530-1.

Whilst the exact details and consequences of the molecular recognition and binding of ACh by AChR have yet to be properly elucidated (in spite of the clear importance), the molecular recognition and binding of ACh by enzyme biocatalyst acetylcholine esterase (AChE) is much better known, owing to excellent high resolution X-ray crystallographic data (Figure 7.3), and in many ways is just as significant. Therefore, all that we need to note here is that AChE has a crucial housekeeping function to metabolize excess ACh remaining in the synaptic cleft post-AChR activation, thereby minimizing the possibility of neurotransmitter-mediated excitotoxic damage to neighboring neuronal cells.

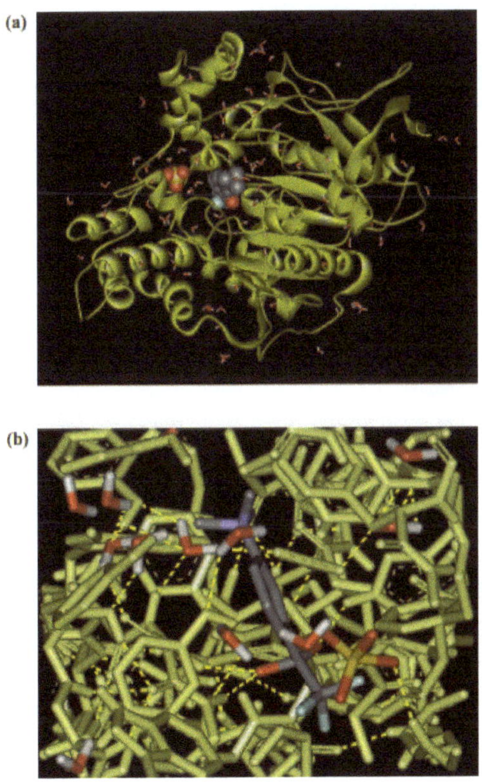

Figure 7.3. *AChE X-ray crystallography structures (pdb: 1amn).*

Note: (a) Ribbon display structure depiction with X-ray crystallographic water molecules illustrated and bound substrate depicted in the CPK display mode. Color coding is as follows: Hydrogen: White; Oxygen; Red; Nitrogen: Blue; Carbon: Grey. In this structure, sulfate (Sulfur: Yellow) is also included. (b) Close-up of active site to show dense hydrogen bonding network (bright yellow). Bound substrates and water molecules are shown in the stick display mode. The protein polypeptide is also shown in CA display mode.

Source: Andrew Miller, Julian Tanner, Essentials of Chemical Biology Structure and Dynamics of Biological Macromolecules; ISBN 978-0-470-84530-1.

7.1.2. Adaptive Immunity, Antibodies, and Myasthenia Gravis

The immune system can seem impossibly complex and indeed the study of the immune system (immunology) is without doubt one of the most complex arenas of biology, comprising a veritable feast of molecular recognition and binding events. In spite of this, there are some useful simplifications once again, in that key components of the immune system such as antibodies are amazingly similar in structure. The schematic diagram in Figure 7.4 shows the main architectural features of all antibodies. Four polypeptides, two heavy chains and two light chains, are linked together covalently by disulfide bonds. Critically, both heavy and both light chains have constant regions that differ little between antibodies, irrespective of subclass or type. In addition, both heavy and light chains have also variable regions that do vary substantially in amino-acid residue sequence between different types of antibody. Each variable region of a given type of antibody can act as a variable affinity binding site (antigen binding site) for the molecular recognition and binding of a complementary protein or peptide known as an antigen. Antigens are immune system stimulatory (immunogenic) compounds that mobilize the adaptive immune system in response to the formation of antibody–antigen complexes. Immune system diversity is so high in effect that every possible antigen can expect to be matched by a corresponding type of antibody able to recognize and bind to this specific antigen to the exclusion of all others, resulting in a completely specific immune response to this antigen. Such selectivity requires that the antibody type diversity of the immune system must exceed 10^{12}. The chemical biology reader can only marvel that so much diversity in molecular recognition and binding behavior is possible as a consequence of multiple subtle variations in the amino-acid residue compositions of different antibody variable regions! (Mobley & Dill, 2009).

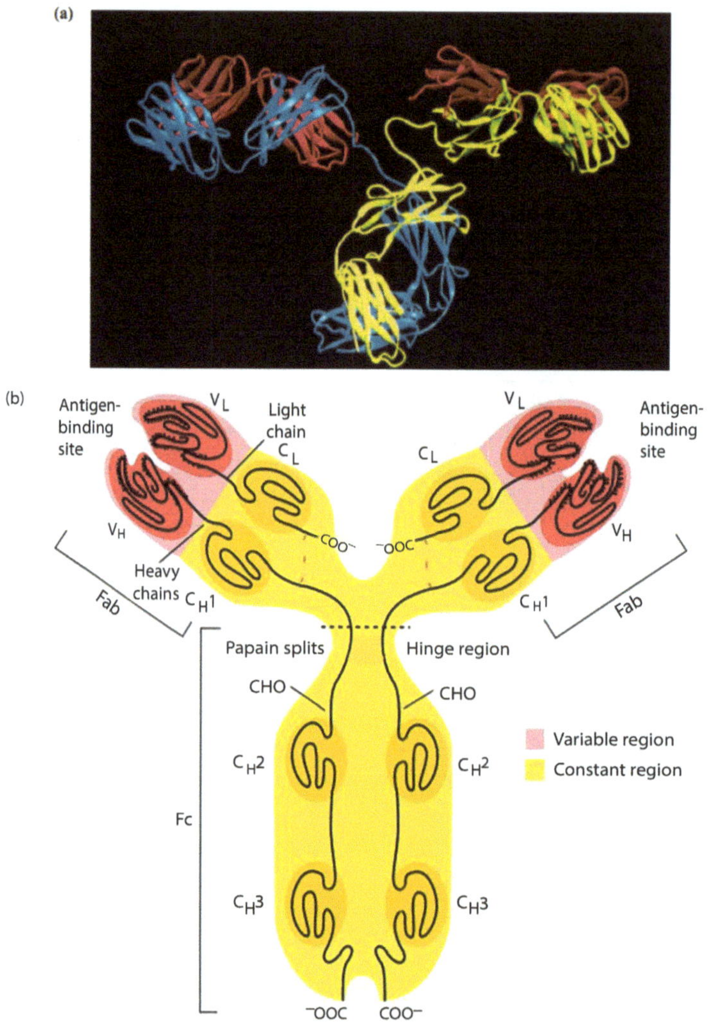

Figure 7.4. *Antibody Representations.*

Note: (a) X-ray crystal structure. (b) Schematic diagram drawn from X-ray crystal structure illustrating the conserved structural features of all antibodies.

Source: Andrew Miller, Julian Tanner, Essentials of Chemical Biology Structure and Dynamics of Biological Macromolecules; ISBN 978–0-470–84530–1.

Remarkably in the case of the normal functioning of the immune system, self-proteins and peptides do not normally become immune stimulatory antigens, but in autoimmune diseases they do. One such example is myasthenia gravis (MG), in which antibodies are generated that target peptide regions (epitopes) on the AChR protein, destroying the protein and loss of neuronal cell function (Figure 7.5). If only the process of molecular recognition and binding could be adequately understood in this case, then the ravaging consequences of this neurological disorder could be avoided by the design of appropriate inhibitors. An alternative approach might be to study the multi-center interaction of cytokine interleukin 1β (IL-1β) with its cognate receptor IL-1R.

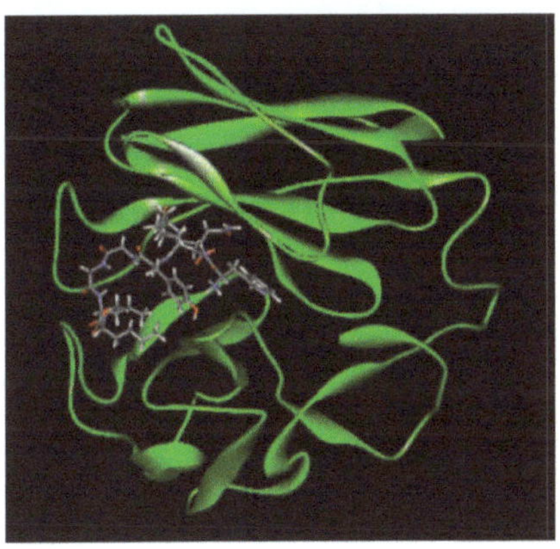

Figure 7.5. *X-ray structure of F_v (V_H + V_L; see Figure 7.4) binding AChR-derived peptide (pdb: 1f3r). This molecular recognition & binding event explains a cascade of events that leads to autoimmunity and self-destruction of the AChR.*

Source: Andrew Miller, Julian Tanner, Essentials of Chemical Biology Structure and Dynamics of Biological Macromolecules; ISBN 978–0-470–84530–1.

The molecular recognition and binding of IL-1β to IL-1R is an essential signal for inflammation by stimulating the proliferation of immune system cells as well as other agents of inflammation.

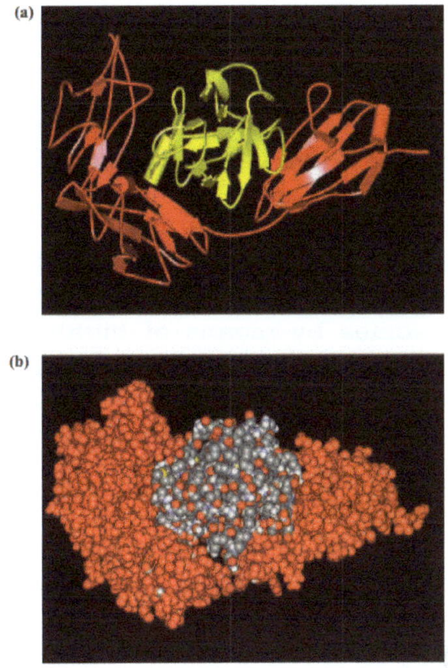

Figure 7.6. *Cytokine interleukin-1β (IL-1β) interacting with cognate IL-1 receptor (IL-1R) (pdb; 1itb).*

Note: (a) Side view of IL-1β (Schematic display; Yellow) interacting with IL-1R (Schematic display; Red). (b) Side view of the same but both IL-1β and IL-1R are rendered in CPK display structure mode to show density of molecular packing.

Source: Andrew Miller, Julian Tanner, Essentials of Chemical Biology Structure and Dynamics of Biological Macromolecules; ISBN 978–0-470–84530–1.

As with the immune system, the inflammatory system is equally diverse and complex but underpinned by sequences of molecular recognition and binding events. By studying the interaction between IL-1β and IL-1R (Figure 7.6), there is the promise of being able to identify inhibitors to a molecular interaction that plays a very significant role in generating potentially harmful inflammatory effects during the progress of such neurodegenerative diseases as AD and MG.

7.1.3. DNA Packaging and Expression Control

One of the great triumphs of molecular recognition and binding in biology is the controlled packaging for storage and unpackaging for function of DNA in eukaryotic cell nuclei. The secret to this is the phenomenal ordering of DNA into chromosomes by means of highly cationic proteins. Chromosomes comprise chromatin filaments that themselves comprise myriad nucleosome core particles (Figure 7.7). Each nucleosome particle contains four histone proteins (H2A, H2B, H3 and H4) that are tightly packed and able to distort DNA double helix to wrap around each set of four proteins twice. Each core particle is then clipped together by a rod of H1 histone protein. Four nucleosome core particles in a row form a zigzag arrangement that closes up to form a solenoid with six core particles per turn. This solenoid is the basic repeat structure of each chromatin filament. The phenomenal charge–charge recognition by H2A, H2B, H3 and H4 is largely responsible for the extraordinary capacity of these proteins to distort the DNA helical axis from linear to bent with a high degree of curvature. However, little remains known about the molecular processes of assembly and disassembly of chromatin filaments after and before gene expression events (Crick & Orgel, 1964).

In DNA terms, the other great triumph of molecular recognition and binding is the control of gene expression through DNA-sequence selective protein binding. Proteins that bind DNA and control expression are known as transcription factors. One such is the GCN4 protein of yeast that is perhaps the first to be studied in molecular detail (Figure 7.8). The extraordinary fact is that DNA has impressively few genes in the total length, especially where eukaryotes are concerned. In total there are 33 000 human genes as against approximately 30 000 nematode genes and a few 1000 in most bacteria.

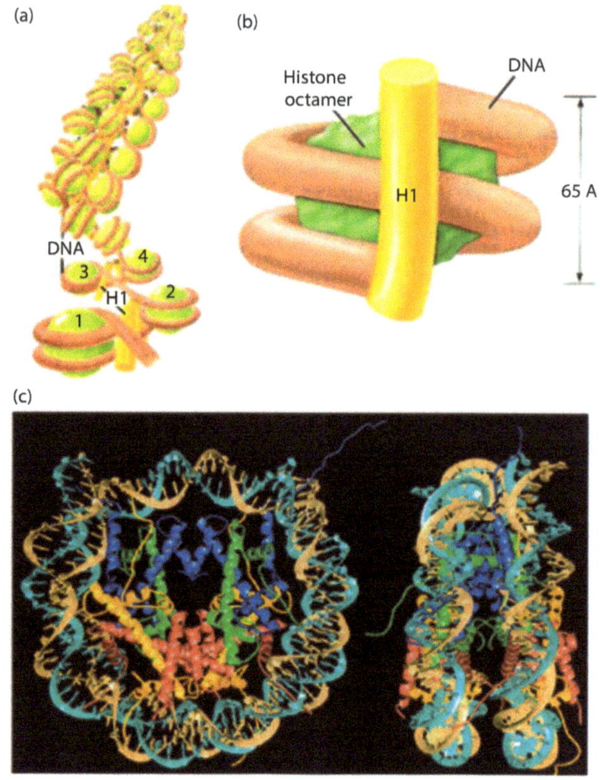

Figure 7.7. *Chromatin filament to chromosome.*

Note: (a) Proposed model of the 300 Å chromatin filament. The zigzag pattern of nucleosomes (1,2,3,4) closes up to form a solenoid with ~6 nucleosomes per turn. (b) Model of histone H1 binding to the DNA of the nucleosome. (c) X-ray crystal structure of the nucleosome core particle. Two views are illustrated (top-left and side-right) showing histone octamer in ribbon structure form (H2A yellow; H2B red; H3 blue; H4 green).

Source: Andrew Miller, Julian Tanner, Essentials of Chemical Biology Structure and Dynamics of Biological Macromolecules; ISBN 978-0-470-84530-1.

Given the fact that there is relatively little difference between the human genome and the nematode genome in terms of overall gene number, then the old adage must apply, that is it is not how many you have but how you use them. In other words, the high level of expression control (both temporal and cellular) during human development as compared to the situation in nematode development must be the presiding reason that humans are many times more complex as organisms than nematodes.

(a)

(b)

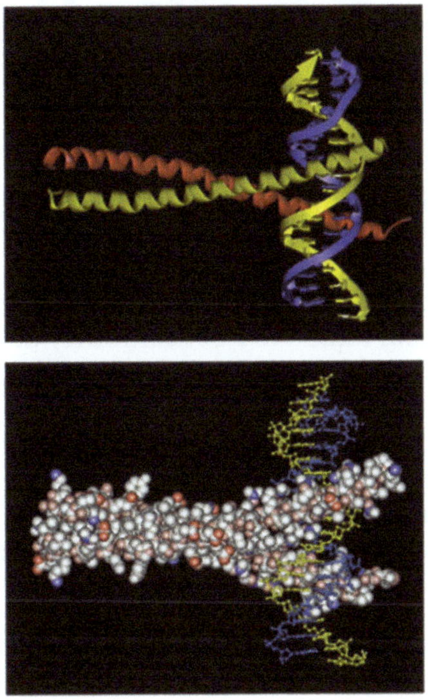

Figure 7.8. *X-ray structure of GCN4 eukaryotic transcription factor and interactions with double stranded DNA (pdb: 1ysa).*

Note: (a) Two helices of protein are shown (red and yellow) in ribbon display structure: Two strands of DNA are shown in rings display. (b) Alternative depiction in which protein is shown in CPK display mode (Carbon: Grey; Oxygen: Red: Nitrogen: Blue). Two strands of DNA (blue and yellow) are shown in ball and stick display structures.

Source: Andrew Miller, Julian Tanner, Essentials of Chemical Biology Structure and Dynamics of Biological Macromolecules; ISBN 978–0-470–84530–1.

Transcription factors are central to this process of expression control although not the complete reason. The molecular recognition and binding of DNA by GCN4 rests with the remarkable scissors grip, where cationic amino-acid residues in the N-terminal regions of each extended α-helix charge complement anionic phosphates of the phosphodiester backbone and these residues together with other amino-acid residues make intimate contact with base-pair-specific deoxynucleotide sequences within DNA involving interactions along the DNA major groove. These sequences are non-coding sequences that do not constitute part of any gene but represent control sequences involved in regulating and modulating gene expression through specific interaction with proteins such as transcription factors.

7.2. THEORETICAL MODELS OF BINDING

Having considered something of the theories of biological macromolecular motion, collisions and encounters in an aqueous buffer medium, we now require a framework with which to be able to study and measure the effectiveness of final binding events themselves. Fundamentally, 'everything can be said to bind to everything else' to a greater or lesser extent. Molecules in a biological milieu are continuously associating with and dissociating from neighboring molecules. The vast majority of these binding events are too weak and transient to measure effectively. These are known as non-specific binding events (or non-specific interactions). Studies of molecular recognition are primarily concerned with the tiny minority of strong, stable interactions otherwise known as specific binding events (or specific interactions). These stand 'head and shoulders' above the non-specific background, a bit like spectral peaks projecting above background noise. Specific binding events are the cornerstone of chemical biology (Crick & Orgel, 1964).

7.2.1. Single-Site, Single-Affinity Binding

In order to study binding, theoretical models, analytical equations and appropriate constants are needed to characterize the strength of a given binding event. Single-site, single-affinity binding is the easiest form of binding to model. In this case the binding equilibrium is represented by the binding scheme in Scheme 7.1, where R is receptor (enzyme, protein etc.) and L is ligand. For this simple binding equilibrium, we can define two equilibrium binding constants that give a measure of the equilibrium position, either to the right-hand side of the equation or to the left-hand side. These two constants are the association constant, K_a, given by:

$$K_a = \frac{[RL]}{[R][L]}(M^{-1}) \quad (1)$$

or the dissociation constant, K_d, given by

$$K_d = \frac{[R][L]}{[RL]}(M) \quad (2)$$

Either constant will do but they are reciprocals of each other. The principal units of K_a can be defined as per molar (M^{-1}) and the units of K_d as molar (M) (NB M = mol l^{-1} or mol dm^{-3}). These constants define the strength and integrity of molecular recognition at the single binding site under the equilibrium conditions being investigated. Generally speaking, Ka values of 10^5–10^6 M^{-1} or higher correspond to tight, specific binding equilibria. In other words, K_d values of 10^{-5}–10^{-6} M (or 10^{-1} μM, where μM = μmol dm^{-3}). Much of our understanding and appreciation of molecular recognition events derives from our being able to measure and interpret the magnitude of these constants. In order to measure any constant, we need an equation that relates this constant to variables that may be determined by experiment. For a single-site, single-affinity binding model, there is a relatively simple way of deriving such an equation. If we define total concentration of receptor $[R]_0$ according to:

$$[R]_0 = [RL] + [R] \quad (3)$$

So that total receptor concentration is a sum of the concentrations of bonded ligand and ligand-free states. If we now substitute for [R] from Eqn. (1) then we achieve the following:

$$[R]_0 = [RL] + \frac{[RL]}{K_a[L]} \quad (4)$$

Which rearranges to:

$$[RL] = [R]_0 \left(\frac{K_a[L]}{1 + K_a[L]}\right) \quad (5)$$

Finally, let us introduce B, which represents the number of moles of ligand L bound per mole of receptor R (B may also be defined as the combined mole fraction of ligand L bound to receptor). This is given mathematically by the simple ratio represented by:

$$B = \frac{[RL]}{[R]_0} \quad (6)$$

Substituting for [RL] from Eqn. (5) then gives us the following expression that happily relates K_a to experimental variables:

$$B = \frac{K_a[L]}{1 + K_a[L]} \quad (7)$$

Eqn. (7) is a classical rectangular hyperbola as shown by a typical plot of B versus [L] (Figure 7.9). This plot is sometimes known as a binding isotherm. A similar rectangular hyperbola is also seen in biocatalyst studies, reflecting the fact that if sufficient ligand is present then all the active sites in a receptor should be effectively occupied or saturated with ligand on a continuous basis with time. The binding isotherm and related functions is absolutely fundamental to biological molecular recognition and binding. Very commonly, Eqn. (7) is transposed into a linear form:

$$B = 1 - \frac{1}{K_a}\frac{B}{[L]} \quad (8)$$

which is known as a Scatchard equation. This linear form is also illustrated (Figure 7.10). It is impressive how often this very simple equation surfaces in the analysis of equilibrium binding constants.

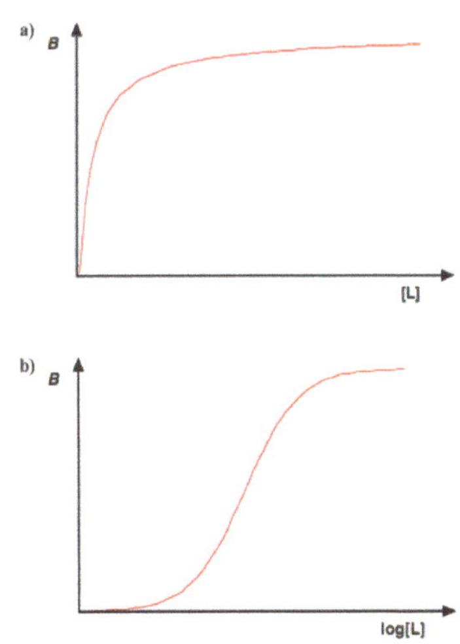

Figure 7.9. *Ideal binding isotherms.*

Note: (a) Classical hyperbolic binding isotherm obtained by plotting values of B against [L].

(b) Classical semi-log plot obtained by plotting values of B against log[L]. Appearance of sigmoidal shape implies that binding interactions between ligand and receptor are >70% saturated and binding data are therefore appropriate to derive accurate association constant K_a or dissociation constant K_d values.

Source: Andrew Miller, Julian Tanner, Essentials of Chemical Biology Structure and Dynamics of Biological Macromolecules; ISBN 978-0-470-84530-1.

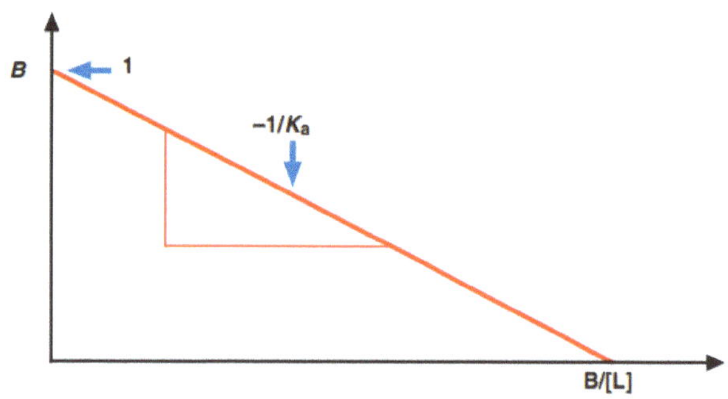

Figure 7.10. *Ideal single-site, single-affinity binding Scatchard plot.*

Source: Andrew Miller, Julian Tanner, Essentials of Chemical Biology Structure and Dynamics of Biological Macromolecules; ISBN 978-0-470-84530-1.

7.2.2. Independent Multiple-Site, Equal Affinity Binding

The Scatchard equation can be very simply adapted for a multiple-site, single-affinity binding model as well. In this case, the binding equilibrium equation can be represented by the binding scheme in Scheme 7.2, where n corresponds to the total number of independent binding sites located on the receptor R. Equal site affinity is critical. If the affinity of each binding site were different for each ligand L, then we would be forced to use a multiple site, variable affinity binding model which is a great deal more complicated! However, where independent binding sites are involved of approximately equal affinity then the combined mole fraction for binding can be simply expressed as:

$$B = \frac{\sum_{i=1}^{i=n} [\mathrm{RL}]_i}{[\mathrm{R}]_0} \tag{9}$$

By analogy to the derivation of Eqn. (8), we can then substitute for the sum of [RL]i terms with a sum of rectangular hyperbolic expressions as follows:

$$B = \sum_{1}^{n} \frac{K_a[\mathrm{L}]}{1 + K_a[L]} \tag{10}$$

where K_a is now the site association constant for each individual i-th binding site. K_a is identical for each individual binding site, hence the simple sum expression can be converted into the simple product expression:

$$B = n\frac{K_a[\mathrm{L}]}{1 + K_a[\mathrm{L}]} \tag{11}$$

Eqn. (11) can then be transposed into the linear form

$$B = n - \frac{1}{K_a}\frac{B}{[\mathrm{L}]} \tag{12}$$

This linear form, Eqn. (12), is also illustrated (Figure 7.11). Note that B becomes equal to n at complete saturation binding when [L] is in large excess and all receptor binding sites are occupied.

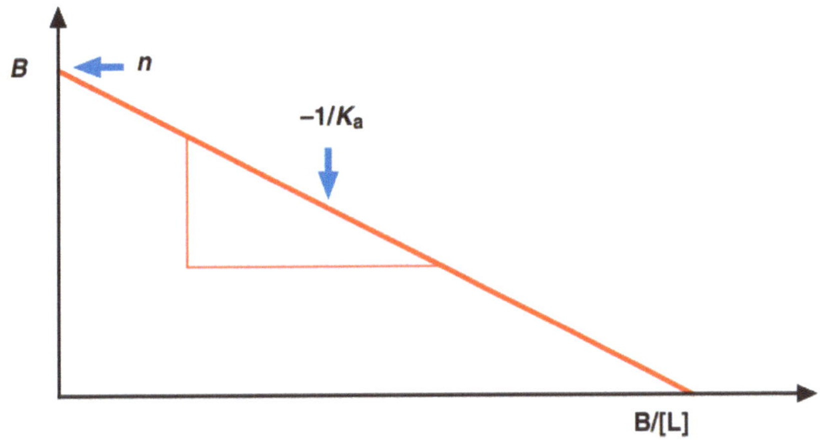

Figure 7.11. *Ideal independent multiple-site, equal affinity binding Scatchard plot.*

Source: Andrew Miller, Julian Tanner, Essentials of Chemical Biology Structure and Dynamics of Biological Macromolecules; ISBN 978–0-470–84530–1.

7.2.3. Independent Multiple-Site, Variable Affinity Binding

Multiple-site, variable affinity binding models are necessarily much more complicated and demand some quite fearsome mathematics to model correctly. For this reason, we would try and apply simpler binding models wherever possible. In the case of multiple-site, variable affinity binding models, binding behavior cannot be reduced to a single equilibrium binding event but is comprised of a series of binding equilibria that each fit the general scheme (Scheme 7.3), where the total number of independent binding sites of variable affinity located on receptor R is n and the final saturating binding equilibrium corresponds with the situation when i equals n. Each binding equilibrium for a given site has independent stoichiometric equilibrium binding constants. By analogy to Eqn. (1), the equation for each stoichiometric association constant, $K_{a,i}$, is as follows:

$$K_{a,i} = \frac{[RL_i]}{[RL_{i-1}][L]} \tag{13}$$

The relationship between B and the complete set of $K_{a,i}$ constants does not involve a simple sum converting through into a simple product expression, but is instead a complex polynomial expression of degree n:

$$B = \frac{K_{a,1}[L] + 2K_{a,1}K_{a,2}[L]^2 + \cdots i(K_{a,1}K_{a,2}\ldots K_{a,i})[L]^i + \cdots n(K_{a,1}K_{a,2}\cdots K_{a,n})[L]^n}{1 + K_{a,1}[L] + K_{a,1}K_{a,2}[L]^2 + \cdots (K_{a,1}K_{a,2}\ldots K_{a,i})[L]^i + \cdots (K_{a,1}K_{a,2}\cdots K_{a,n})[L]^n} \tag{14}$$

undeniably a complex expression! There are n terms in this equation corresponding to the n roots of the polynomial representing the number of independent binding sites in the receptor R. This is known as a real sites equation, with good reason as we shall see. In order to solve such an equation, the polynomial needs to be equated with another algebraic expression. This new expression is known as a ghost sites equation that has all the appearance of Eqn. (10), except comprised of stoichiometric ghost site association constants that are all different from each other:

$$B=\frac{K_\alpha[L]}{1+K_\alpha[L]}+\frac{K_\beta[L]}{1+K_\beta[L]}+\cdots\frac{K_v[L]}{1+K_v[L]}=\sum_{\alpha}^{v}\frac{K_\omega[L]}{1+K_\omega[L]} \tag{15}$$

Do not confuse K_α–K_ω with the stoichiometric real site association constants in Eqn. (14). The ghost site constants are not real numbers but are in fact imaginary numbers, pure mathematical conveniences that allow us to write Eqn. (16) in a form that is more amenable to solution and then equate this to the polynomial expression (15)!

A simple example should serve to illustrate how Eqns. of the form of (15) and (16) can be used in combination to solve for stoichiometric real site association constants. Imagine we have a receptor R with two independent binding sites of two different affinities. In a real scenario, four equilibria and two stoichiometric association constants account for the interaction between two separate ligands L and this receptor (Figure 7.12). Two independent equilibria then make up a parallel imaginary scenario (Figure 7.12). A real sites equation (polynomial of degree 2) may be written to account for the real scenario and then directly equated with a ghost sites equation containing two summed terms harboring two different imaginary numbers K_α and K_β as shown:

$$\frac{K_{a,1}[L]+2K_{a,1}K_{a,2}[L]^2}{1+K_{a,1}[L]+K_{a,1}K_{a,2}[L]^2}=B=\frac{K_\alpha[L]}{1+K_\alpha[L]}+\frac{K_\beta[L]}{1+K_\beta[L]} \tag{17}$$

Experimental values of B and [L] may then be used to solve for K_α and K_β in the right-hand equality, which are in turn used to extract values of $K_{a,1}$ and $K_{a,2}$ from the left-hand equality. Hopefully, it is clear how this approach can be expanded in general to include receptors with more than two independent binding sites of different affinities using additional roots of the polynomial on the left and additional imaginary terms on the right.

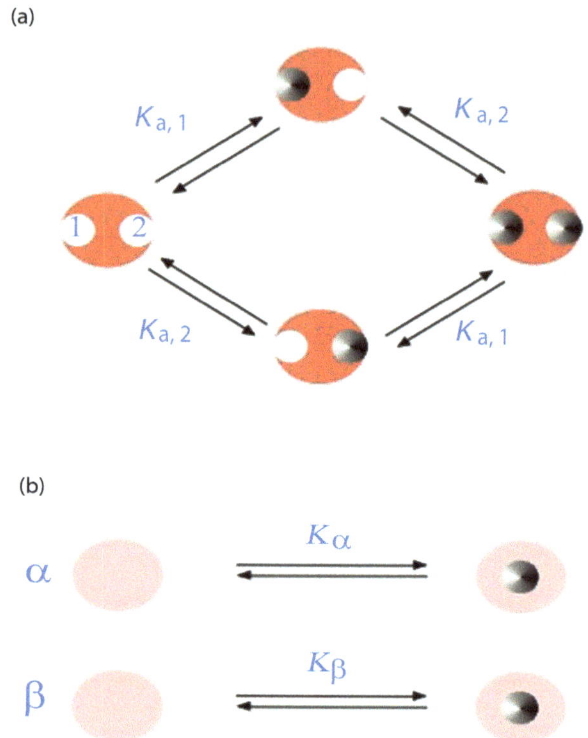

Figure 7.12. *Independent multiple-site variable affinity binding equilibria.*

Note: (a) Diagram illustrates the emergence of four different equilibrium constants to describe the binding interactions between ligands (gray) and two independent binding sites of significantly different affinities found in a biological macromolecular receptor (red). (b) Imaginary "parallel" single-site, single-affinity ghost sites equilibria used to help derive a unique solution for equilibrium constants applicable to situation shown in (a).

Source: Andrew Miller, Julian Tanner, Essentials of Chemical Biology Structure and Dynamics of Biological Macromolecules; ISBN 978–0-470–84530–1.

7.2.4. Dependent Multiple-Site Cooperative Binding and Hill Equation

The moment that we lift the constraint of independent binding sites, the number of possible interlinked equilibria, each with its own stoichiometric equilibrium binding constants, might be expected to jump alarmingly together with the complexity of the analysis. However, perhaps surprisingly, when there is a tendency for ligand binding at any one receptor binding site to enhance persistently ligand binding at neighboring binding sites (positive cooperativity), then binding behavior can be accounted for in terms of a considerably reduced form of Eqn. (14) as shown:

$$B = \frac{n(K_{a,1}K_{a,2}\cdots K_{a,n})[\mathrm{L}]^n}{1+(K_{a,1}K_{a,2}\cdots K_{a,n})[\mathrm{L}]^n} \qquad (18)$$

Such a simplification is possible because positive cooperativity ensures that only end point saturation binding equilibria can dominate ligand binding behavior and hence only the highest power terms are relevant in an expression for B. Eqn. (18) can be simplified even further to give the following equation:

$$\frac{B}{n} = \frac{[\mathrm{L}]^n}{K' + [\mathrm{L}]^n} \qquad (19)$$

Which is in the form of a sigmoid (Figure 7.13). The linearized form of this equation is:

$$\log\left(\frac{B}{n-B}\right) = n\log[\mathrm{L}] - \log K' \qquad (20)$$

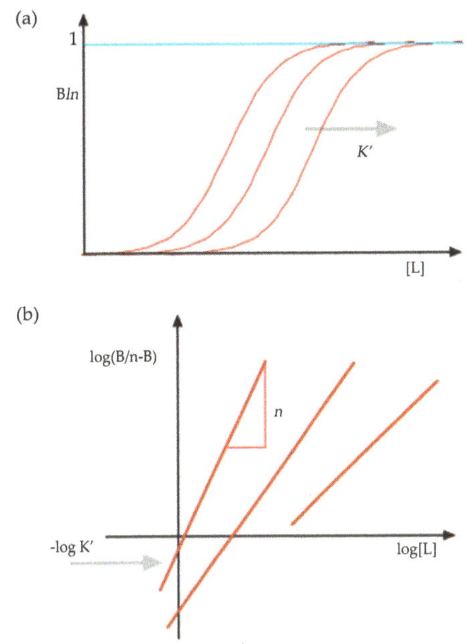

Figure 7.13. *Dependent multiple-site variable affinity cooperative binding equilibria.*

Note: (a) Classical sigmoidal binding isotherms indicative of ligand-receptor interactions that involve strong positive cooperativity. Curves move to right as composite equilibrium constant K increases. (b) Linear Hill plots derived from sigmoidal data illustrated in (a). Gradients and intercepts define values of n and K respectively

Source: Andrew Miller, Julian Tanner, Essentials of Chemical Biology Structure and Dynamics of Biological Macromolecules; ISBN 978-0-470-84530-1.

This is the Hill equation for cooperative binding of ligands by a receptor. The term n represents the maximum value of B at saturation as well as the effective number of binding sites involved in cooperative binding activity (Figure 7.13). For the sake of completeness, when there is a tendency for ligand binding at any one receptor binding site to reduce persistently ligand binding at neighboring binding sites (negative cooperativity) then binding behavior can also be accounted for in terms of a considerably reduced form of Eqn. (14), that is equivalent to Eqn. (7).

7.3. BIOLOGICAL MOLECULAR RECOGNITION STUDIES

7.3.1. LysU Enzyme Substrate Recognition

Many enzymes (protein bio-catalysts) are well characterized as biocatalysts but quite often key molecular recognition and binding events that precede biocatalysis are overlooked. This cannot be said to be true of LysU. LysU is a homodimeric lysyl tRNA synthetase enzyme, and one of the primary enzymes involved in Ap_4A synthesis in the bacterium *E. coli*. The enzymic surface mechanism of LysU-catalyzed Ap_4A synthesis is well known and is illustrated in Figure 7.14.

The first step (Step 1) is very specific and apparently only ATP or deoxy-ATP analogs are able to act as first nucleotide substrates to form a lysyl-adenylate intermediate. The second step (Step 2) is unusual in requiring the use of the Zn^{2+} ion. The X-ray crystal structures of LysU demonstrate a host of hydrogen bonding interactions responsible for determining the strength and specificity of binding between substrates and enzyme active site in Step 1 (Figure 7.15). Also, cationic amino-acid residues are present in the vicinity of the polyphosphate chain for electrostatic (long and short range) interactions; similarly, anionic amino-acid residues are present in the vicinity of l-lysine for equivalent electrostatic interactions. Complementarities between enzyme active sites and substrates are usually extensive at the level of charge and hydrogen bonding in order to ensure optimal specificity of molecular recognition and binding.

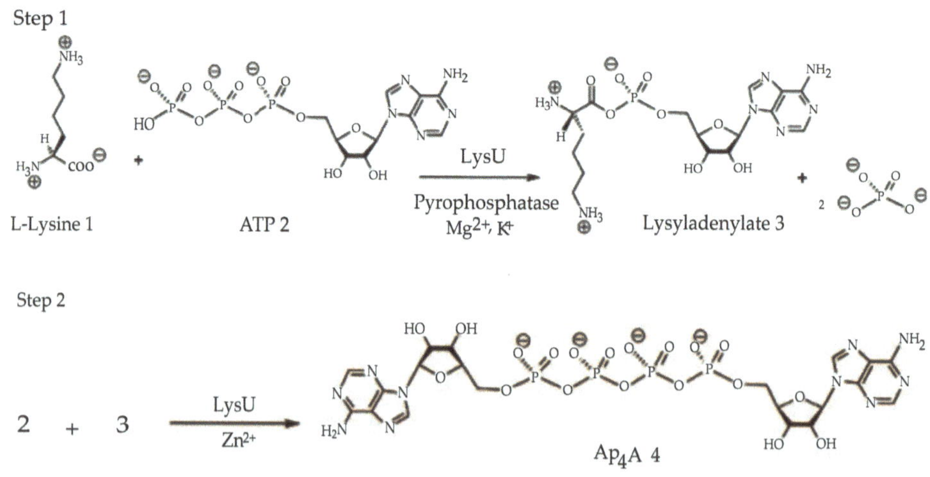

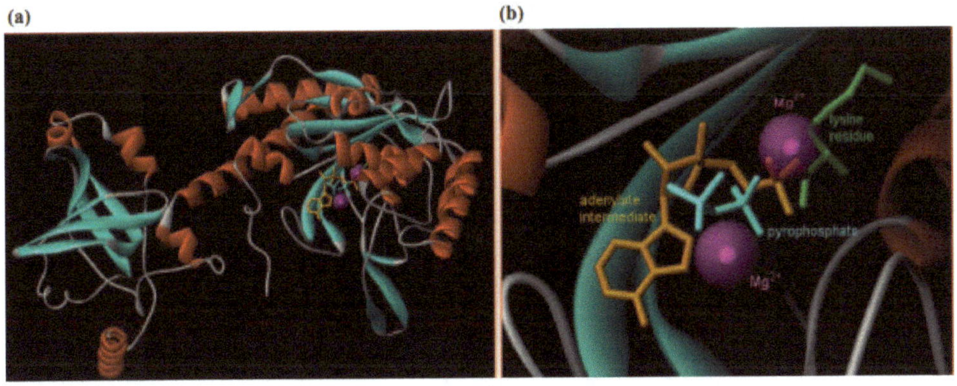

Figure 7.14. *Illustration of step 1 and step 2 of enzymic surface mechanism of LysU catalyzed diadenosine-5,' 5"'-P^1,P^4-tetraphosphate (Ap_4A) 4 formation.*

Source: Andrew Miller, Julian Tanner, Essentials of Chemical Biology Structure and Dynamics of Biological Macromolecules; ISBN 978–0-470–84530–1.

Catalysis requires a precise order of substrate binding that was demonstrated by a combination of fluorescence binding titration experiments and ITC binding titration experiments. Initially, the presence of Mg^{2+} ions was found to be essential for l-lysine binding to either active site of the LysU dimer. Thereafter, the binding of l-lysine at both sites was found to be responsible for a substantial structural rearrangement of the C-terminal active site domain of each monomer, as observed by CD spectroscopy (Figure 7.16). The change in LysU conformation elicited by l-lysine binding was also sufficient to induce a significant change in intrinsic tryptophan fluorescence upon l-lysine binding.

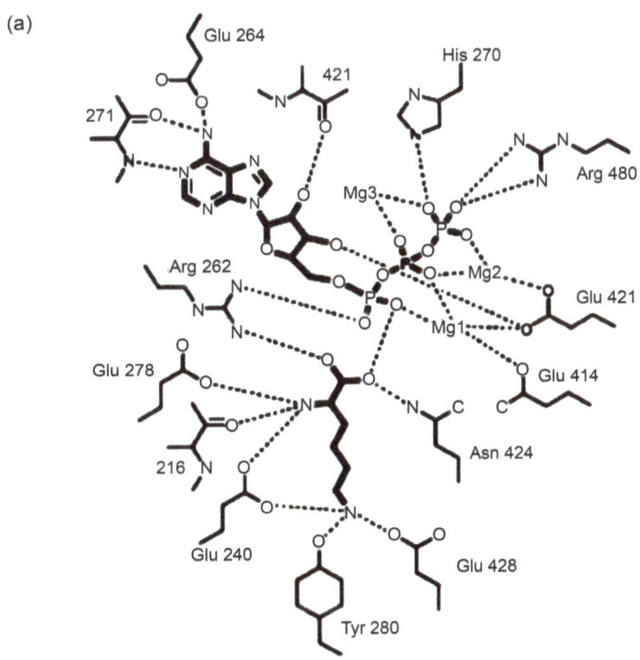

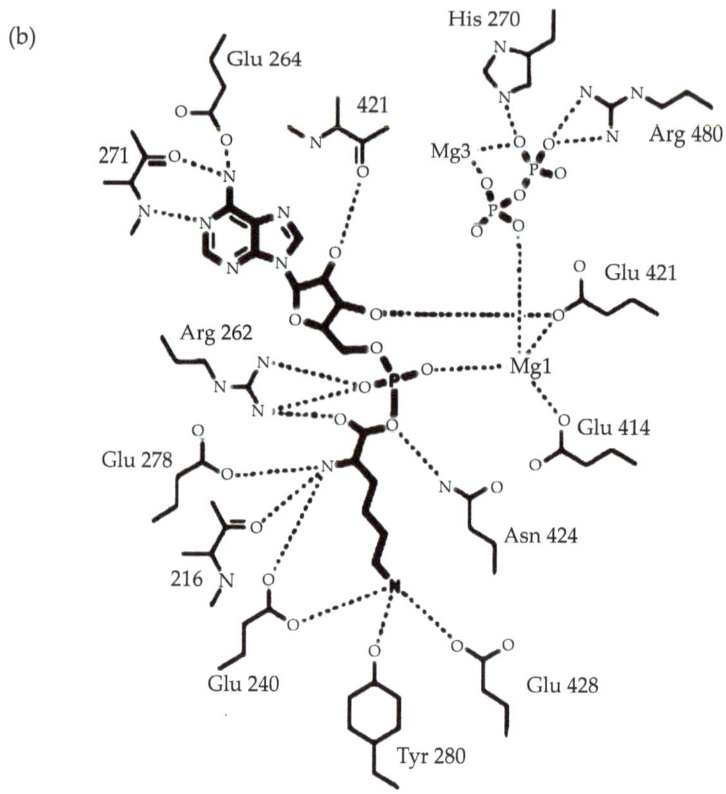

Figure 7.15. *Two dimensional schematic representation of the active site of LysU. H-bond interactions are shown for the ternary complex between L-lysine 1, ATP 2 and LysU (a), and for the complex between the lysyladenylate intermediate 3 and LysU (b). The motif 2 residue Arg262 is critical for the molecular recognition and binding of both L-lysine 1 and ATP 2 α-phosphate to LysU. A number of conserved residues in the motif 2 loop (residues 264–271) assume an ordered conformation only upon ATP binding. Positions of the Mg^{2+} binding sites are also shown. There are 3 Mg^{2+} ions Mg1, Mg2 and Mg3 in (a) and 2 Mg^{2+} ions Mg1 and Mg3 in (b).*

Source: Andrew Miller, Julian Tanner, Essentials of Chemical Biology Structure and Dynamics of Biological Macromolecules; ISBN 978-0-470-84530-1.

Therefore, intrinsic tryptophan fluorescence binding titration experiments were used successfully to determine the site dissociation constant K_d for l-lysine and the binding stoichiometry (K_d 7.96 ± 1.40 µmol L^{-1}; 0.95 ± 0.1 l-lysine/monomer) (Figure 7.16). ITC binding experiments revealed a similar value for K_d and binding stoichiometry (K_d 4.95 ± 2.54 µmol L^{-1}; 0.87 ± 0.14 l-lysine/monomer) (Figure 7.17). Critically, ATP was unable to bind to LysU until after the association of l-lysine and Mg^{2+} ions. Note how binding stoichiometry is determined either by curve fitting with the appropriate binding equation or else directly from plots (Figure 7.16).

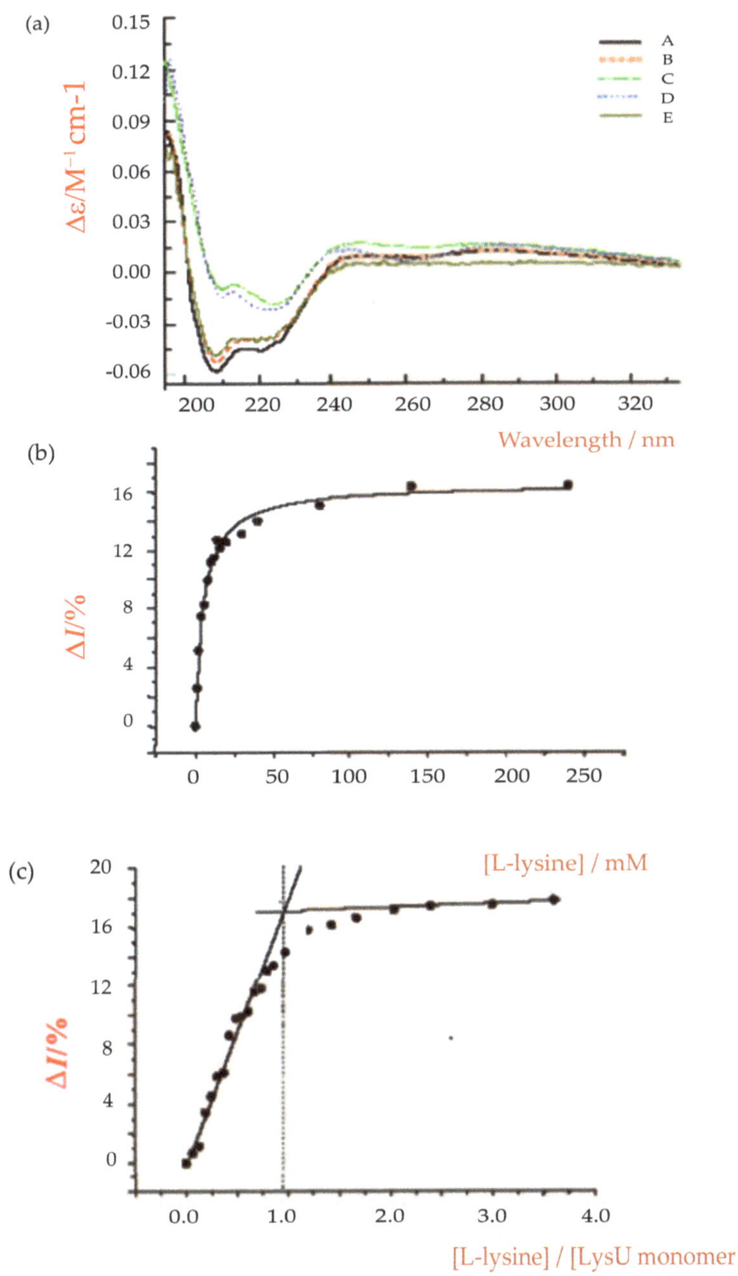

Figure 7.16. *Effects of L-lysine binding to LysU in the presence of Mg^{2+} and K^{+} ions.*

Source: Andrew Miller, Julian Tanner, Essentials of Chemical Biology Structure and Dynamics of Biological Macromolecules; ISBN 978–0-470–84530–1.

ATP binding was studied by an ITC binding experiment in the absence of any other spectroscopic signature that could be titrated to saturation. ITC studies were performed using hydrolysis resistant β,γ -methylene-ATP (AMPCP) in place of ATP in order to avoid complications in ITC measurements from ATP hydrolysis (Figure 7.17). In this case the value for K_d was found to be 1.22±0.15μM and the stoichiometry 0.51±0.01 AMPPCP/

monomer. Only one ATP molecule was found to bind per LysU dimer, suggesting that ATP binding was accompanied by strong negative cooperativity leading to half-of-sites binding. Consequently, the ordered process of LysU molecular recognition and binding of substrates requires that half-of-sites catalysis takes place subsequently.

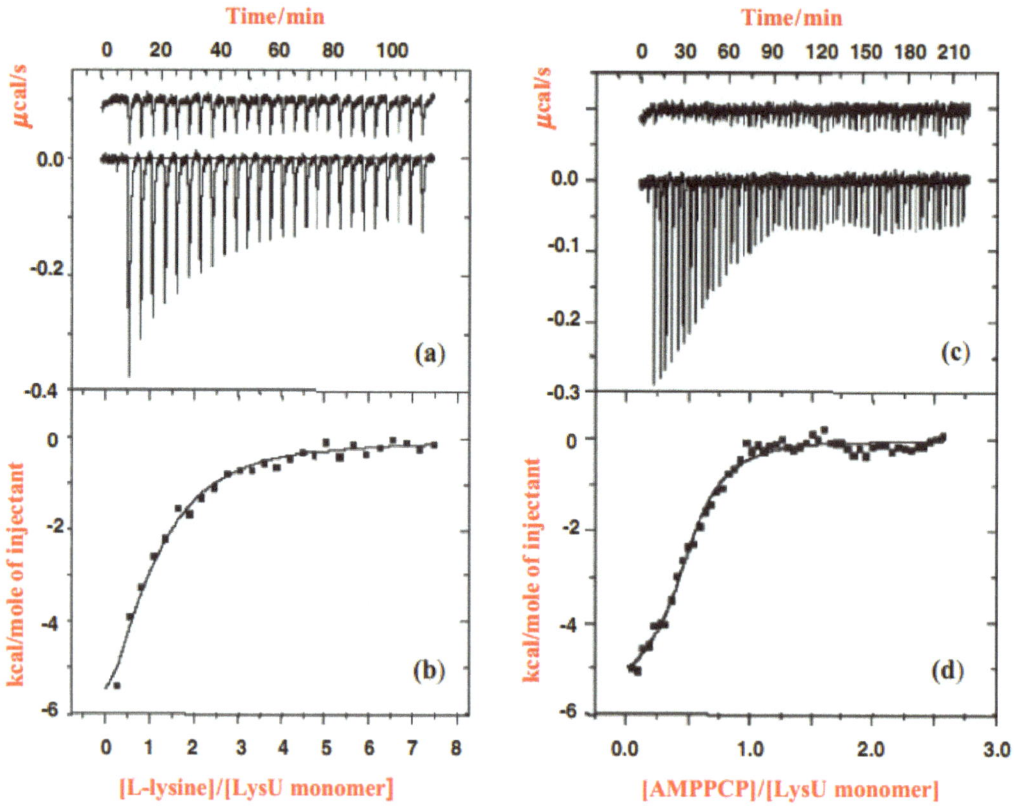

Figure 7.17. *ITC binding experiments showing the effects of L-lysine and AMPPCP binding to LysU in the presence of* Mg^{2+} *and* K^{+} *ions.*

Source: Andrew Miller, Julian Tanner, Essentials of Chemical Biology Structure and Dynamics of Biological Macromolecules; ISBN 978-0-470-84530-1.

In order to understand the basis of half-of-sites binding, theoretical molecular modeling studies were performed using the X-ray crystal structures of LysU as a starting point. Molecular modeling is largely beyond our scope here, but it is worth noting that molecular mechanics Poisson–Boltzmann surface area (MMPBSA) modeling experiments were able to demonstrate that the two LysU active sites are not structurally and functionally equivalent to each other even before the above experiments were performed. In the event, ATP was calculated to bind to one site (Site 2) about 25 kcal mol^{-1} more tightly than to the other (Site 1). This difference is more than adequate to account for the negative cooperativity and half-of-sites binding observed experimentally. Moreover, molecular dynamics simulations were able to indicate that binding of ATP to one active site (Site 2) was capable of inducing Arginine 269 (Arg 269) closure of the same site, thereby causing an opening up of the other active site (Site 1) via a trans-conformational relay system operating through the dimer interface by means of physical contacts between the C2–C3 loops of each monomer and corresponding Motif 2 loops in the corresponding monomer, proximal to each active site (Figure 7.18). Such a conformational opening of Site 1 would be sufficient explanation for impaired ATP binding affinity and half-of-sites reactivity!

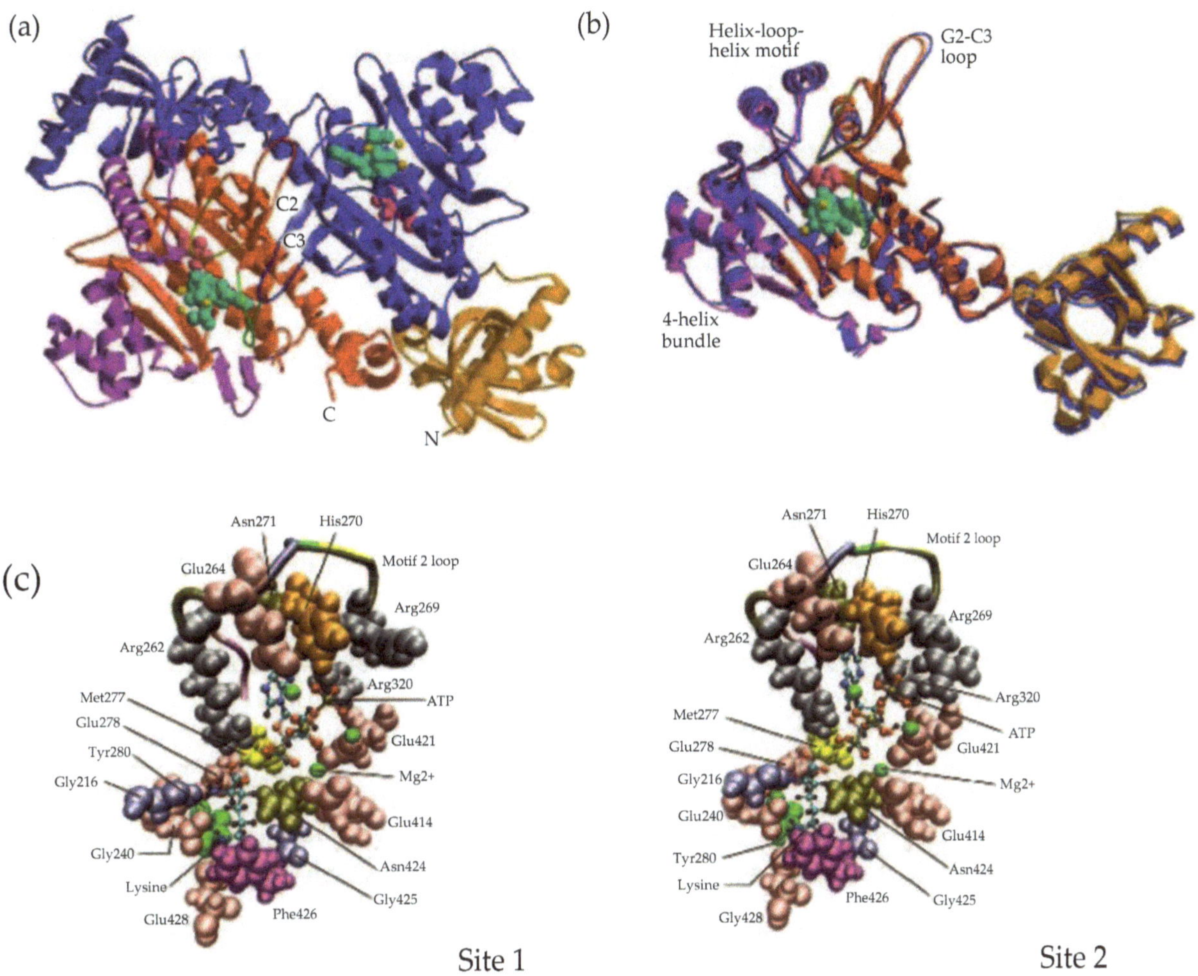

Figure 7.18. *LysU half-of-sites binding of ATP and half-of-sites reactivity.*

Source: Andrew Miller, Julian Tanner, Essentials of Chemical Biology Structure and Dynamics of Biological Macromolecules; ISBN 978–0-470–84530–1.

Clearly, the results of modeling studies should always be correlated where possible with experimental observations. However, the combination of experimental techniques with molecular modeling represents an important chemical biology case study for the investigation of the molecular recognition and binding events that precede catalysis in an enzyme biocatalyst mechanism. Not only has a clear substrate order of addition been established for LysU, but also a half-of-sites binding mechanism. Obviously, the precise combination of experimental techniques and modeling studies will vary according to the biocatalyst under investigation, but the outcomes should be illustrative of the need to combine as many techniques as possible to understand. Now, not only does conformational opening of LysU Site 1 appear to be an important reason that Site 1 has a lower affinity for ATP than Site 2 and does not experimentally bind ATP, but also lysyl-adenylate formation and pyrophosphate release in Site 2 appears likely to trigger a subsequent reverse set of trans-conformational changes across the dimer interface allowing Site 1 to bind ATP, close up and initiate Step 1 catalysis of lysyladenylate formation in turn. We appear to be making significant progress in understanding the molecular recognition and binding behavior of LysU!

7.3.2. Stress Protein Molecular Chaperones

Stress protein molecular chaperones have already been introduced. Such proteins have a rich and varied molecular recognition and binding behavior. Important examples include GroEL and the procollagen/collagen molecular chaperone protein known as heat shock protein 47 (Hsp47).

7.3.2.1. GroEL

GroEL is a paradox in molecular recognition and binding. A wide range of different unfolded protein substrates binds to GroEL with K_d values in the micromolar–nanomolar range. Such binding strength should be clearly characterized as specific, and yet GroEL has a promiscuous capacity with GroES to assist the folding/refolding of a diverse range of protein substrates. How is this possible? How is it possible to investigate this? In order to simplify the investigation but at the same time retain valuable information, extrinsic fluorescence binding titration experiments were performed using a series of peptide surrogates for unfolded protein substrates. Peptide binding interactions with GroEL were reported in each case by means of a dansyl (Dns) group covalently attached to each peptide. Dns groups show a marked shift in I_{max} and an increase in fluorescence emission intensity with increase in quantum yield φ_F upon transfer from a free, aqueous hydrophilic environment to a more hydrophobic ligand-bound environment.

Initially, GroEL was found to have a preference for interaction with peptides comprising sequences or arrays of amino-acid residues that are amphiphilic, comprising hydrophobic and polar neutral/cationic residues. Sequences or arrays exclusively polar in character or else harboring an excess of anionic residues were not recognized or bound at all (Figure 7.19). Following this, the role of secondary structure in molecular recognition and binding was investigated by the systematic preparation of a series of six peptides divisible into two groups known as the AMPH series, designed with a graded ability to form a cationic amphiphilic α-helical structure in solution, and the NON-AMPH series, designed with the graded ability to form a non-amphiphilic α-helical structure in solution (Figure 7.19).

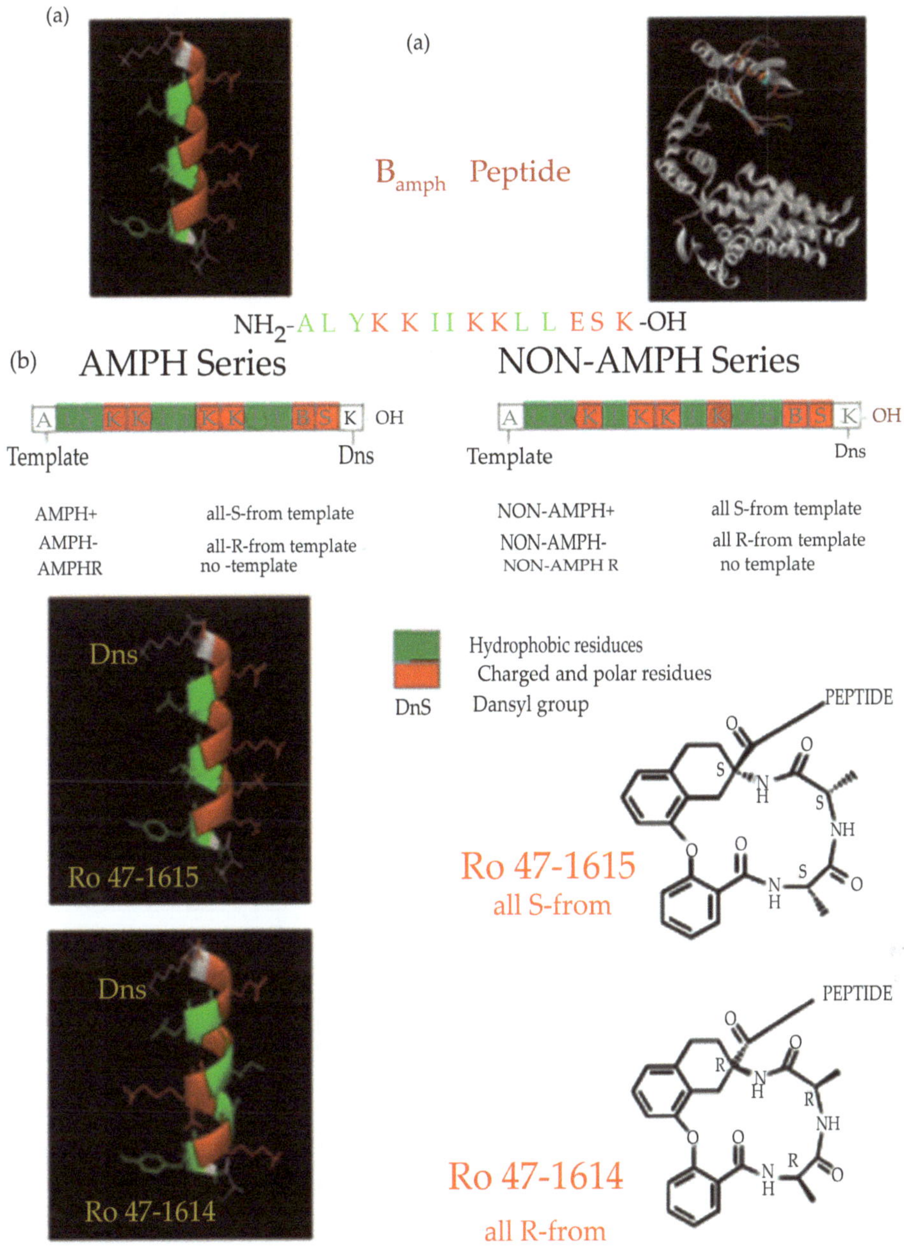

Figure 7.19. *GroEL peptide substrates.*

Source: Andrew Miller, Julian Tanner, Essentials of Chemical Biology Structure and Dynamics of Biological Macromolecules; ISBN 978–0-470–84530–1.

Extrinsic fluorescence titration experiments were performed using all six peptides. Owing to the molecular weight size difference between GroEL and peptides, and the potential number of peptide binding sites, fluorescence titration experiments were actually performed in reverse, keeping peptide concentrations fixed and titrating with increasing GroEL until fluorescence saturation was achieved (Figure 7.20).

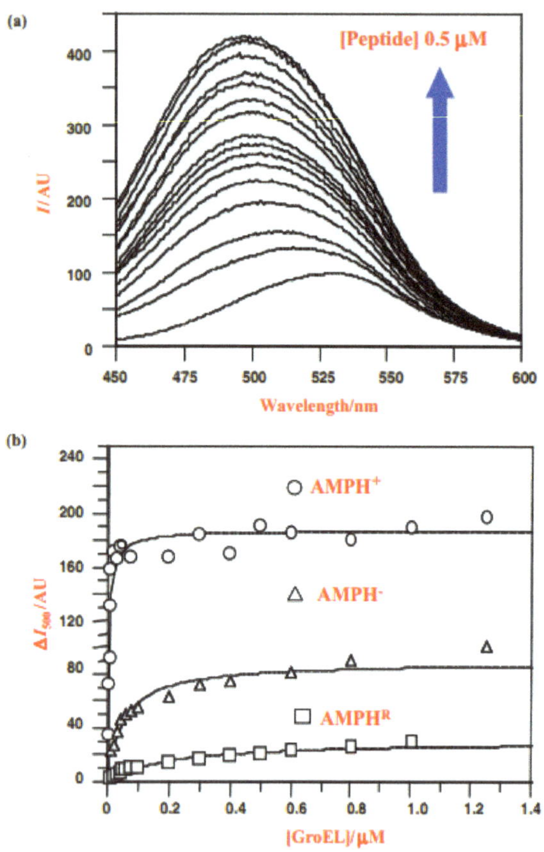

Figure 7.20. *Peptide association with GroEL.*

Source: Andrew Miller, Julian Tanner, Essentials of Chemical Biology Structure and Dynamics of Biological Macromolecules; ISBN 978-0-470-84530-1.

From such experiments values of K_d were determined, with the discovery that the most ⊠-helical AMPH series peptide (AMPH+) was much the most effective GroEL 'substrate,' with a binding interaction affinity (K_d 5 nM) as strong as if not stronger than known binding interactions between GroEL and other protein 'substrates.' Extrinsic fluorescence binding titration experiments under a range of different conditions were also able to show the importance of the hydrophobic effect in peptide–GroEL association (Figure 7.21). Similar titration experiments were used to show that peptides and GroES compete for the same binding sites on GroEL, in agreement with the known GroEL/GroES mechanism (Figure 7.21). Finally, the sum total of all these experiments was the main conclusion that GroEL has a preference to bind regions or subdomains in an unfolded protein substrate that have a propensity to form polar neutral and/or cationic, amphiphilic secondary structural elements in solution. This robust conclusion appears to provide the required explanation for the general–specific molecular recognition and binding behavior of GroEL.

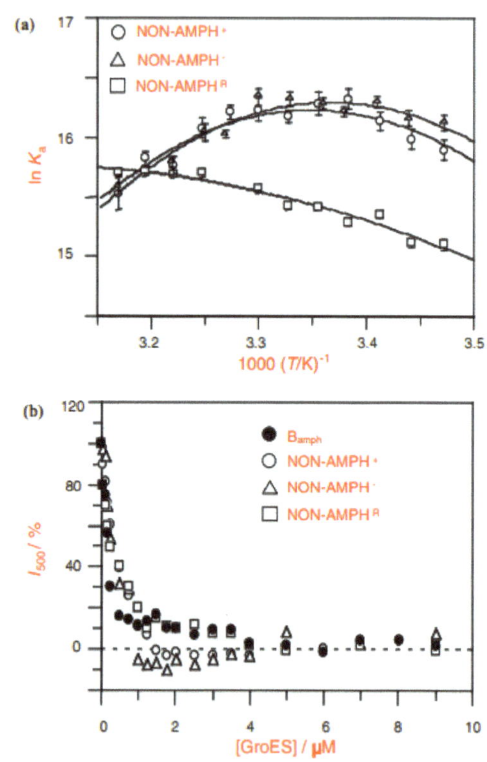

Figure 7.21. *Forces and competition involved in binding GroEL peptide substrates.*

Source: Andrew Miller, Julian Tanner, Essentials of Chemical Biology Structure and Dynamics of Biological Macromolecules; ISBN 978-0-470-84530-1.

CASE STUDY

MOLECULAR RECOGNITION IN DRUG DESIGN

- Introduction

Molecular recognition is a fundamental concept in biochemistry and molecular biology, referring to the specific interaction between two or more molecules through non-covalent bonding. This process is crucial for many biological functions, including enzyme-substrate interactions, DNA replication, and cellular signaling. In drug design, molecular recognition is key to developing compounds that can specifically bind to target proteins, leading to therapeutic effects.

- Background

The case study focuses on the development of a small-molecule inhibitor for a specific protein target involved in a disease pathway. This example will illustrate the principles and techniques used in molecular recognition to achieve selective and potent drug candidates.

- Target Selection: Protein Kinase Inhibitor

Protein kinases are enzymes that modify other proteins by chemically adding phosphate groups. Abnormal kinase activity is associated with various diseases, including cancer. Therefore, targeting kinases with small-molecule inhibitors is a promising therapeutic strategy.

- Steps in Drug Design Using Molecular Recognition

1. **Target Identification and Validation**
   i. **Identification**: Through genomic and proteomic studies, a specific kinase, e.g., BRAF V600E, is identified as a key driver in melanoma.
   ii. **Validation**: Biological assays confirm that inhibiting BRAF V600E reduces cancer cell proliferation.
2. **Structural Analysis of the Target**
   i. **X-ray Crystallography/NMR**: The 3D structure of BRAF V600E is determined using X-ray crystallography.
   ii. **Binding Site Identification**: The ATP-binding pocket of BRAF V600E is identified as the target site for inhibitor binding.
3. **Ligand Design and Optimization**
   i. **Initial Screening**: A library of small molecules is screened for binding affinity to the BRAF V600E kinase.

ii. **Lead Compound Identification**: Compounds with high binding affinity and specificity are identified as lead compounds.

iii. **Structure-Based Drug Design**: Using the 3D structure of the kinase, lead compounds are optimized to improve binding affinity, specificity, and pharmacokinetic properties.

4. **Molecular Dynamics Simulations**

o **Binding Affinity Prediction**: Computational simulations predict the binding affinity and stability of the inhibitor-kinase complex.

o **Optimization**: Iterative cycles of synthesis and testing are conducted to refine the molecular structure for optimal interaction with the target.

5. **Biological Evaluation**

i. *In Vitro* **Assays**: The optimized compounds are tested *in vitro* for their ability to inhibit BRAF V600E activity and reduce cancer cell viability.

ii. *In Vivo* **Studies**: Successful compounds from *in vitro* studies are tested in animal models for efficacy and toxicity.

6. **Clinical Trials**

i. **Phase I**: Safety and dosage are evaluated in healthy volunteers.

ii. **Phase II**: Efficacy and side effects are tested in a small patient group.

iii. **Phase III**: Large-scale testing confirms efficacy, monitors side effects, and compares the new drug to standard treatments.

- Results and Discussion
    - **Lead Compound**: Vemurafenib is identified as a potent BRAF V600E inhibitor.
    - **Mechanism of Action**: Vemurafenib selectively binds to the ATP-binding site of BRAF V600E, inhibiting its kinase activity and blocking the downstream signaling pathway involved in cell proliferation.
    - **Clinical Efficacy**: Vemurafenib shows significant improvement in survival rates of melanoma patients with the BRAF V600E mutation.
    - **Side Effects**: Common side effects include rash, joint pain, and fatigue. Long-term monitoring is necessary for managing adverse effects.
- Conclusion

Molecular recognition plays a pivotal role in the rational design of small-molecule inhibitors for therapeutic targets. The case study of Vemurafenib demonstrates the successful application of molecular recognition principles in drug design, leading to significant advancements in the treatment of melanoma. Continuous research and development in this field hold promise for the discovery of new drugs for various diseases.

CLASS ACTIVITY

Describe the various mechanisms by which molecular recognition occurs in biological systems. Highlight the roles of shape complementarity, electrostatic interactions, hydrogen bonding, and hydrophobic effects. Provide examples of specific biomolecular interactions where each mechanism is predominant.

SUMMARY

- Molecular recognition refers to the specific interaction between two or more molecules through non-covalent interactions such as hydrogen bonds, van der Waals forces, electrostatic interactions, and hydrophobic effects.
- Molecular recognition and binding events found in biology are ubiquitous, diverse and pivotal, but there are common themes and principles. In order to impart a flavor of this, we will take a brief look at a number of interlocking but diverse examples of biological molecular recognition and binding events.
- Brain tissue is comprised substantially of neuronal cells and glial cells. The former are the active component of brain tissue responsible for the transmission of sensory, emotional and motor information between brain regions and into the periphery (body) in the form of electrical pulses.
- The immune system can seem impossibly complex and indeed the study of the immune system (immunology) is without doubt one of the most complex arenas of biology, comprising a veritable feast of molecular recognition and binding events.
- The moment that we lift the constraint of independent binding sites, the number of possible interlinked equilibria, each with its own stoichiometric equilibrium binding constants, might be expected to jump alarmingly together with the complexity of the analysis.
- GroEL is a paradox in molecular recognition and binding. A wide range of different unfolded protein substrates binds to GroEL with K_d values in the micromolar–nanomolar range. Such binding strength should clearly be characterized as specific, and yet GroEL has a promiscuous capacity with GroES to assist the folding/refolding of a diverse range of protein substrates.

REVIEW QUESTIONS

1. What roles do molecular recognition and binding play in the specificity and regulation of cellular processes?
2. Compare and contrast the lock-and-key model with the induced fit model in the context of molecular binding.
3. Explain how molecular recognition is crucial for the adaptive immune system, particularly in the function of antibodies.
4. Describe the role of acetylcholine in neurotransmission and its interaction with acetylcholine receptors.
5. Explain how techniques such as X-ray crystallography and NMR spectroscopy contribute to our understanding of molecular recognition and binding.
6. Define single-site, single-affinity binding and provide an example of this type of binding interaction in biological systems.
7. What is independent multiple-site, equal affinity binding, and how does it differ from single-site binding?

REFERENCES

1. Baron, R., & McCammon, J. A. (2013). Molecular recognition and ligand association. *Annual Review of Physical Chemistry, 64*, 151–175.
2. Breiten, B., Lockett, M. R., Sherman, W., Fujita, S., Al-Sayah, M., Lange, H., et al. (2013). Water networks contribute to enthalpy/entropy compensation in protein-ligand binding. *Journal of the American Chemical Society, 135*(41), 15579–15584.
3. Crick, F. H., & Orgel, L. E. (1964). The theory of inter-allelic complementation. *Journal of Molecular Biology, 8*, 161–165. doi:10.1016/s0022-2836(64)80156-x.
4. Finzel, C., Clancy, L. L., Holland, D. R., Muchmore, S. W., Watenpaugh, K. D., & Einspahr, H. M. (1989). Crystal structure of recombinant human interleukin-1β at 2.0 A resolution. *Journal of Molecular Biology, 209*, 779–791. doi:10.1016/s0022-2836(89)80032-4.
5. Heal, J. R., Bino, S., Roberts, G. W., Raynes, J. G., & Miller, A. D. (2002). Mechanistic investigation into complementary (antisense) peptide mini-receptor inhibitors of cytokine interleukin-1. *ChemBioChem, 3*, 76–85. doi:10.1002/1439-7633(20020104)3:1<76: aid-cbic76>3.0.co;2-u.
6. Hughes, S. J., Tanner, J. A., Miller, A. D., & Gould, I. R. (2006). Molecular dynamics simulations of LysRS: an asymmetric state. *Proteins, 62*, 649–662. doi:10.1002/prot.20808.
7. Keller, M., Tagawa, T., Preuss, M., & Miller, A. D. (2002). Biophysical characterization of the DNA binding and condensing properties of adenoviral core peptide μ (mu). *Biochemistry, 41*, 652–659.
8. Mobley, D. L., & Dill, K. A. (2009). Binding of small-molecule ligands to proteins: "what you see" is not always "what you get." *Structure, 17*(4), 489–498. doi: 10.1016/j.str.2009.02.010.
9. Schmidtchen, F. P. (2010). Hosting anions. The energetic perspective. *Chemical Society Reviews, 39*(10), 3916–3935. doi:10.1039/C0CS00038H.
10. Vigers, P., Caffes, P., Evans, R. J., Thompson, R. C., Eisenberg, S. P., & Brandhuber, B. J. (1994). X-ray structure of interleukin-1 receptor antagonist at 2.0-A resolution. *Journal of Biological Chemistry, 269*, 12874–12879. doi:10.1016/s0021-9258(17)37459-8.

CHAPTER 8

Kinetics and Catalysis

LEARNING OBJECTIVES

After studying this chapter, you will be able to:

- Discuss about catalysis in chemical biology
- Explain steady-state kinetic schemes

KEY TERMS FROM THIS CHAPTER

Bioassay	Biocatalysis
Biocatalysts	Catalysis
Chromogenic reagents	Chromophore
Chymotrypsin	Equilibrium
Fluorescence spectroscopy	Kinetics
Neurotransmission	Radiometric steady-state
Ribozyme analyzes	Steady-state analysis
Tricarboxylic acid	

INTRODUCTION

Kinetics and catalysis are pivotal to the study of chemical biology, providing insights into the rates of biochemical reactions and the mechanisms by which these reactions are accelerated. Chemical biology, which merges principles of chemistry and biology to understand and manipulate biological systems and relies heavily on understanding how biochemical reactions occur and how they can be controlled. Kinetics, the study of reaction rates and the factors influencing them, and catalysis, the acceleration of reactions by catalysts, are fundamental to deciphering the complexities of biochemical pathways and processes.

Kinetics in chemical biology focuses on the rate at which biochemical reactions occur and how these rates are influenced by various factors such as temperature, pH, concentration of reactants, and the presence of inhibitors or activators. Reaction kinetics are crucial for understanding metabolic pathways, signal transduction, and enzyme functionality. By studying the rate laws and mechanisms of reactions, scientists can predict how biological systems respond to changes and design experiments to alter these responses in a controlled manner.

Catalysis is the process by which catalysts, often enzymes in biological systems, accelerate chemical reactions without being consumed in the process. Enzymes are biological catalysts that play a critical role in maintaining life by enabling biochemical reactions to proceed at rates necessary for cellular function. Enzymatic catalysis involves the formation of an enzyme-substrate complex, which lowers the activation energy of the reaction, facilitating a faster conversion of substrates into products. Understanding catalysis is essential for developing inhibitors or activators that can modulate enzyme activity, a key aspect in drug design and metabolic engineering.

The integration of kinetics and catalysis provides a comprehensive framework for studying and manipulating biological processes. It allows scientists to unravel the intricate dynamics of cellular reactions, leading to advancements in fields such as drug discovery, biotechnology, and synthetic biology. By exploring how biochemical reactions are regulated and how they can be modified, chemical biology paves the way for innovative solutions to complex biological challenges, driving progress in medicine, agriculture, and environmental science.

8.1. CATALYSIS IN CHEMICAL BIOLOGY

Some theories concerning the origins of life suggest that molecular complexity sufficient to allow for self-organization is a prerequisite for the creation of organisms. But self-organization by itself cannot bring life. Rather, the ability to speed up or catalyze chemical interconversions must coexist with self-organization. The ability to catalyze chemical reactions is referred to as catalysis. A catalyst is a substance that can accelerate a chemical reaction in both forward and reverse directions without changing permanently during the process. Catalysts are commonly used in catalysis.

As a result, a biocatalyst is a catalyst with biological significance. In comparison to non-catalyzed reactions, biocatalysts typically accelerate biological chemical reactions with relative rate enhancements of 105 to 1,010. The range and diversity of known biocatalysts is simply astounding! Biocatalysts are unquestionably an absolute fundamental for both the origin of life and the propagation of life. Catalysis is universal to all cells of all organisms.

Most biocatalysts are proteins, or enzymes, and occasionally RNA nucleic acids, or ribozymes. For a plethora of distinct and significant reasons, they catalyze an astounding diversity of reactions. In metabolism, the synthesis of adenosine 5 -triphosphate (ATP), the preferred "form" of chemical energy stored in all cells, is coupled with the progressive breakdown and/or reorganization of covalent bonds of primary metabolites, such as glucose. Enzymes play a central role in this process, which generates and stores chemical potential.

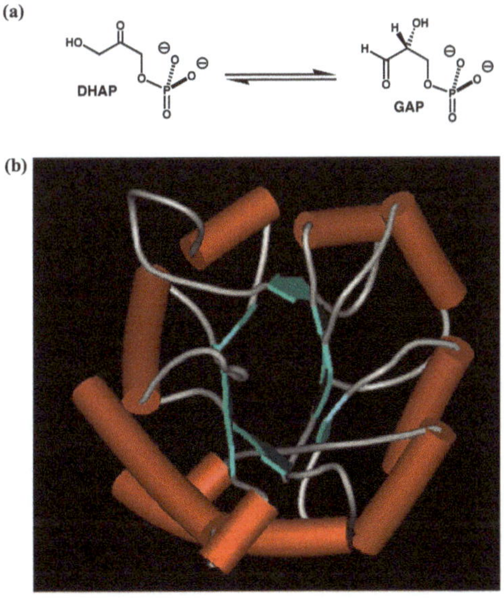

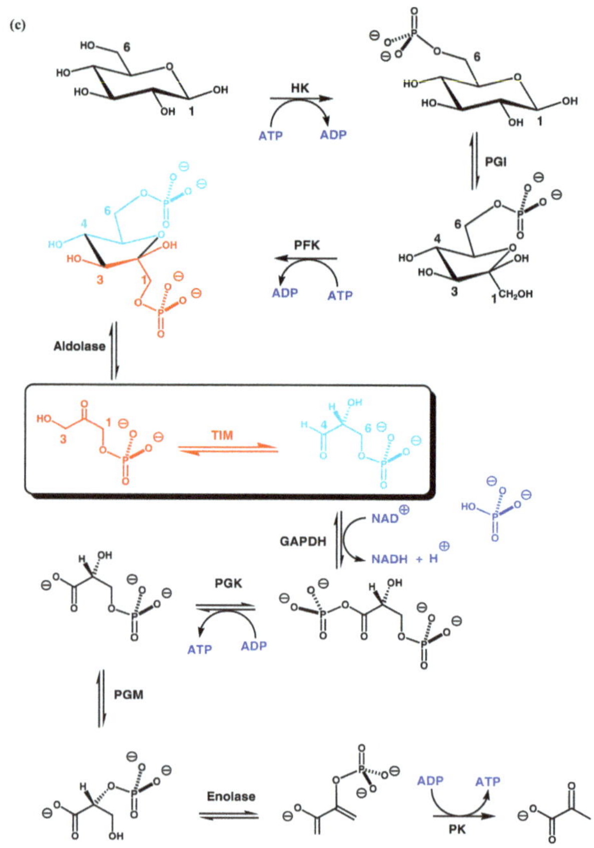

Figure 8.1. *Triose phosphate isomerase (TIM).*

Note: (a) Reversible reaction catalyzed by TIM where DHAP is dihydroxy acetone phosphate, and GAP is glyceraldehyde-3-phosphate; (b) schematic display structure of TIM (chicken muscle) (top view) (pdb: 1tim); (c) chemical illustration of glycolysis – a primary catabolic pathway able to yield two molecules of ATP/glucose consumed, made possible only through the catalytic activity of TIM. Enzyme abbreviations are as follows; HK, hexokinase; PGI, phosphoglucose ismerase; PFK, phosphofructokinase; GAPDH, glyceraldehyde-3-phosphate dehydrogenase; PGK, phosphoglycerate kinase; PGM, phosphoglycerate mutase; PK, pyruvate kinase.

Source: https://nattabiochem.wordpress.com/wp-content/uploads/2009/08/essentials_chemical_biology.pdf.

For example, in the catabolic pathway known as glycolysis, triose phosphate isomerase (TIM) catalyzes the seemingly harmless interconversion between dihydroxyacetone phosphate and glyceraldehyde-3-phosphate. Surprisingly, though, TIM is now regarded as a "perfect enzyme," enabling interconversion to occur at a rate that is precisely equal to the rate at which substrate reaches the active site of the enzyme. In fact, the glycolysis pathway would not be able to produce a net gain of two ATP molecules for every glucose molecule consumed in the absence of TIM.

Similar to this, the dimeric enzyme malate dehydrogenase (MDH) catalyzes the simple reduction of a carbonyl functional group in oxalic acid to malic acid, utilizing the cofactor nicotinamide adenine dinucleotide hydride, reduced form (NADH). However, this interconversion establishes closure of the tricarboxylic acid cycle (TCA cycle), which takes metabolites from glycolysis and yields a net gain of reducing cofactor molecules for each complete rotation through the TCA cycle. Other catabolic enzymes exist, such as chymotrypsin, which breaks down polypeptides into oligopeptides in the gut so that they can pass through the wall of the gut and enter the bloodstream. Alternatively, though through a completely different mechanism, ribonuclease A (RNAse A) performs the same function for RNA polynucleotides as chymotrypsin does for polypeptides.

The role of enzymes extends beyond catabolism (i.e., breakdown), but they also contribute to anabolism (i.e., assembling) of monomeric building blocks necessary for the assembly of biological macromolecules. Examples of this include the enzymes aspartate transaminase (aspAT), glutamate α-decarboxylase, and alanine α-racemase (AlaR), which all require the cofactor pyridoxal phosphate (PLP).

Numerous enzymes frequently serve protective purposes outside of cells. Lysozyme, for example, is an enzyme that is produced externally in tears and serves to catalyze the hydrolysis of O-glycosidic links in complex polysaccharides of bacterial cell walls. This process weakens and destroys the bacterial cell walls, kills the bacteria, and prevents infection of the surface of the eye. One example of a bacterial enzyme that has evolved to inactivate the antibiotic chloramphenicol by controlled acetylation is chloramphenicol acetyl transferase (CAT). Then there is the amazing di-metal enzyme family known as superoxide dismutase (SOD), which catalyzes the remarkably efficient disproportionation of superoxide radicals into oxygen and hydrogen peroxide. When there is an incomplete reduction of oxygen to water during respiration, which happens surprisingly often, superoxide radicals are produced.

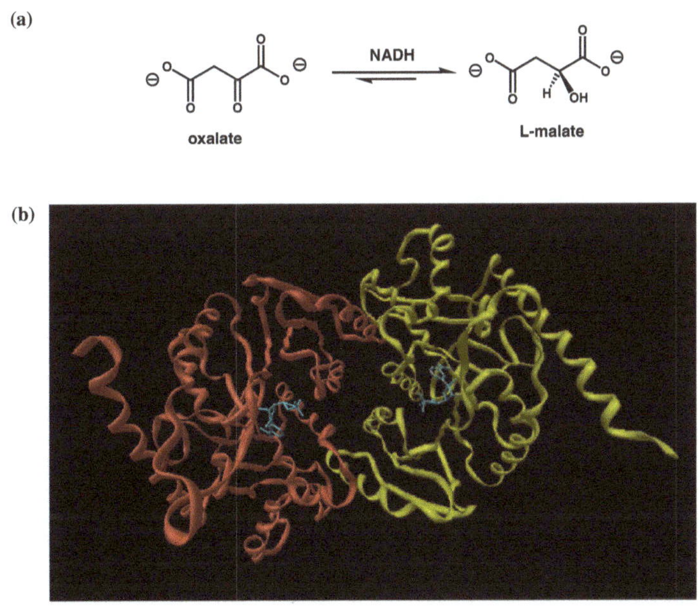

(c)

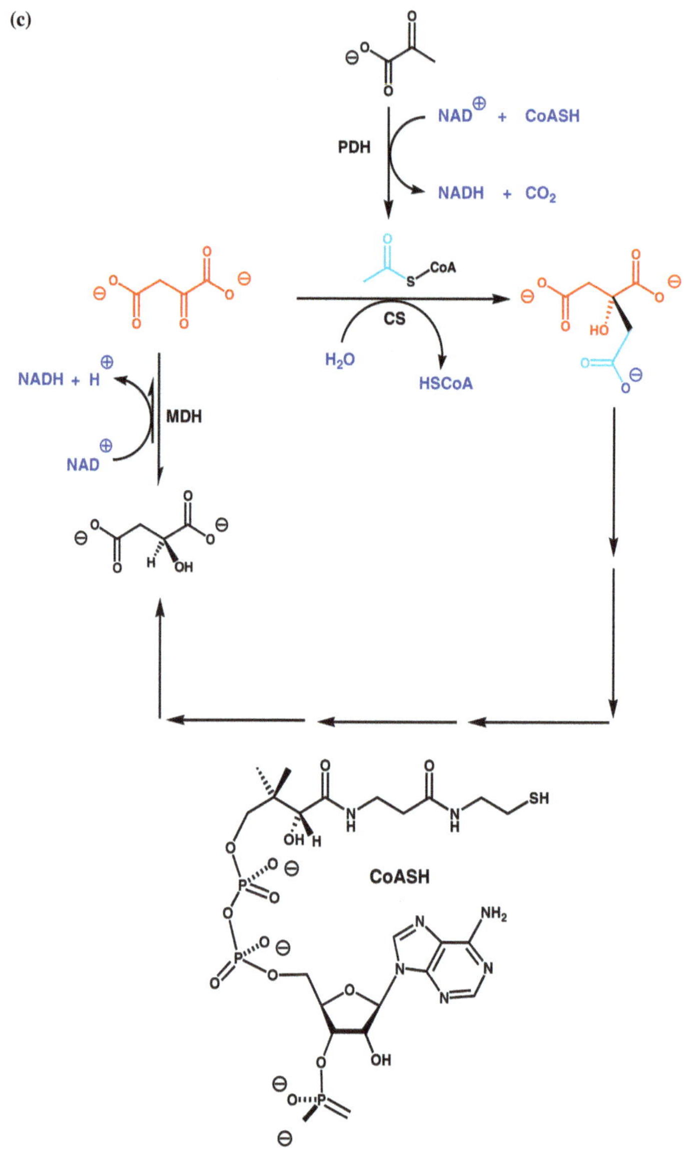

Figure 8.2. *Malate dehydrogenase (MDH).*

Note: (a) Reversible reaction catalyzed by MDH where NADH is nicotinamide adenine dinucleotide, reduced form; (b) ribbon display structure of MDH (porcine heart) (side view) (pdb:4mdh). The homo-dimeric protein consists of two polypeptides chains (yellow and red), with nicotinamide adenine dinudeotide (NAD+) in both independent catalytic sites illustrated in a ball and stick (blue) representation; (c) chemical illustration of the tricarboxylic acid cycle (TCA) to demonstrate the importance of MDH catalysis in cycle closure. Enzyme abbreviations are: PDH, pyruvate dehyrogensase; CS, citrate synthase.

Source: https://nattabiochem.wordpress.com/wp-content/uploads/2009/08/essentials_chemical_biology.pdf.

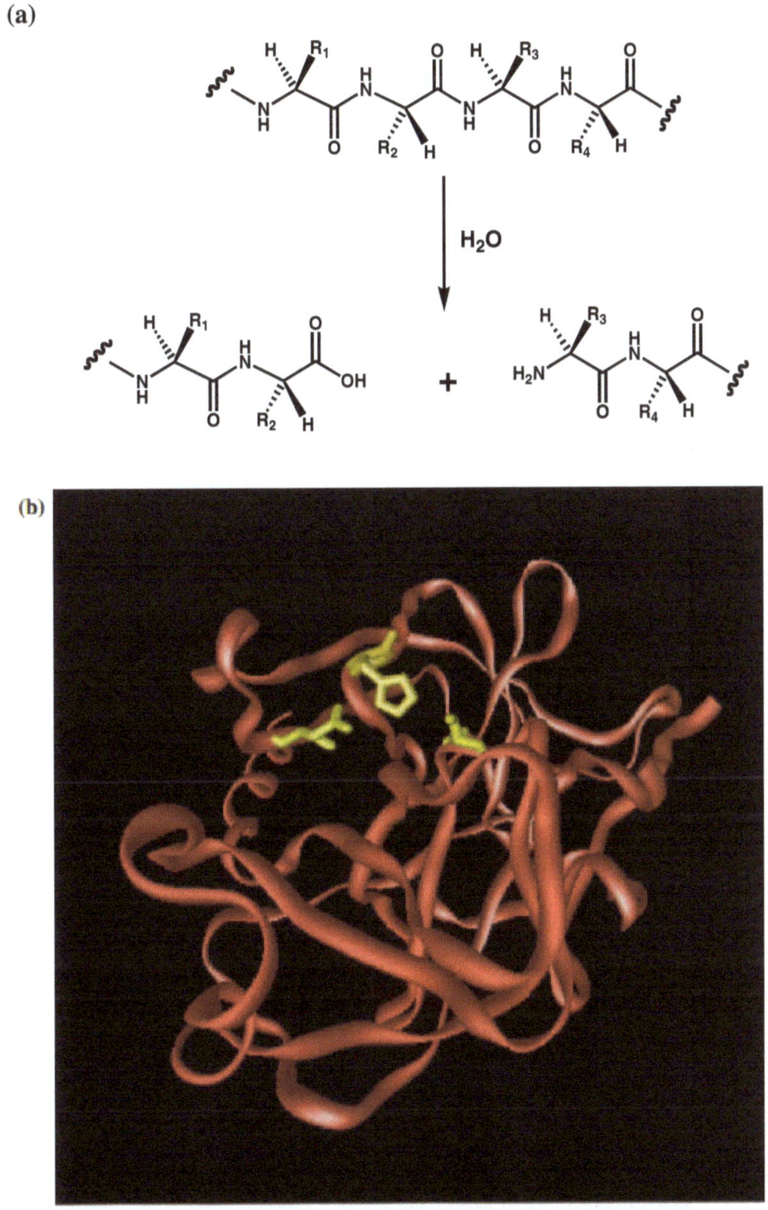

Figure 8.3. *α-chymotrypsin.*

Note: (a) hydrolytic reaction catalyzed by α-chymotrypsin, where amino acid residue side chain R2 is hydrophobic or aromatic in character; (b) ribbon display structure of αchymotrypsin (bovine pancreatic) (pdb: 4cha) in which key active site residues D102, H57 and S195 (left to right) involved in biocatalysis are shown (yellow) rendered in a tube display style.

Source: https://nattabiochem.wordpress.com/wp-content/uploads/2009/08/essentials_chemical_biology.pdf.

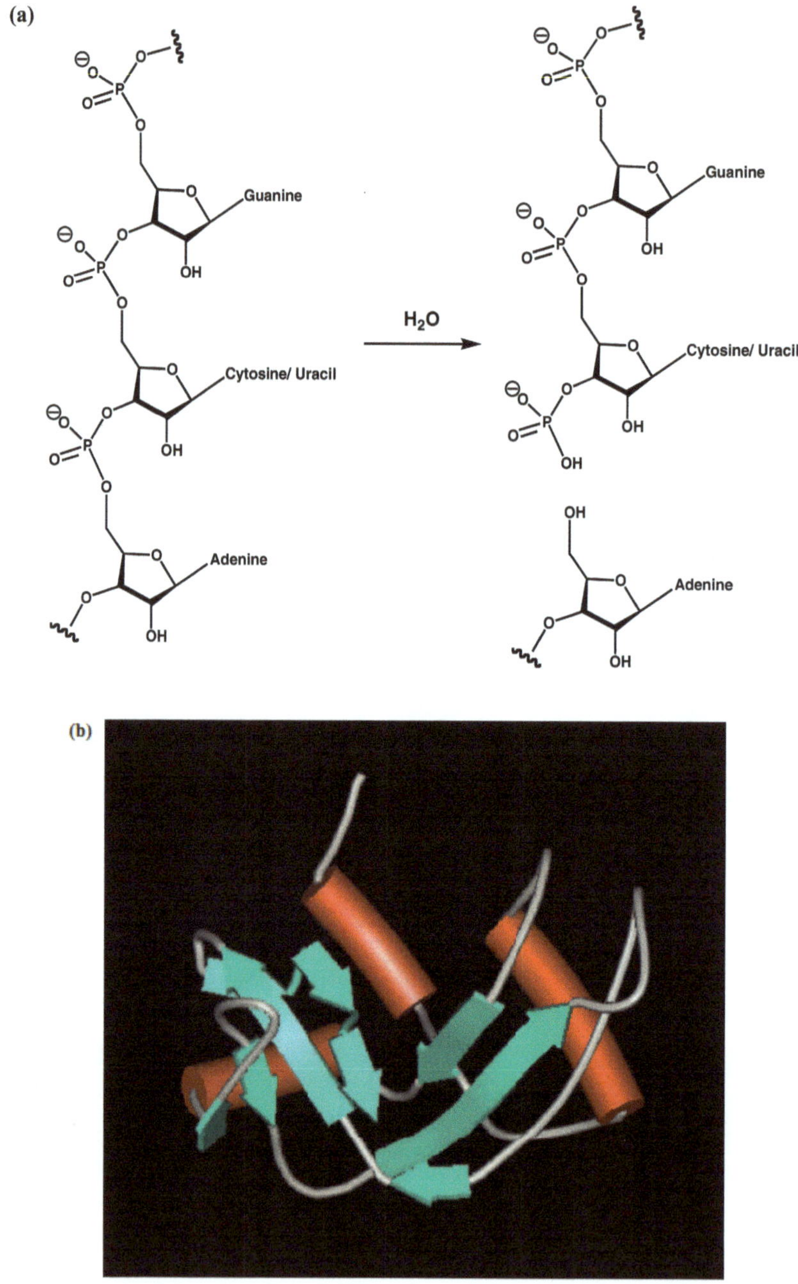

Figure 8.4. *Ribonuclease A.*

Note: (a) Hydrolytic reaction catalyzed by ribonuclease A (RNAse A), note that hydrolytic cleavage occurs at the 3 -side of a pyrimidine nucleoside; (b) cartoon display structure of RNAse A (bovine pancreatic) (side view) (pdb: 7rsa).

Source: https://nattabiochem.wordpress.com/wp-content/uploads/2009/08/essentials_chemical_biology.pdf.

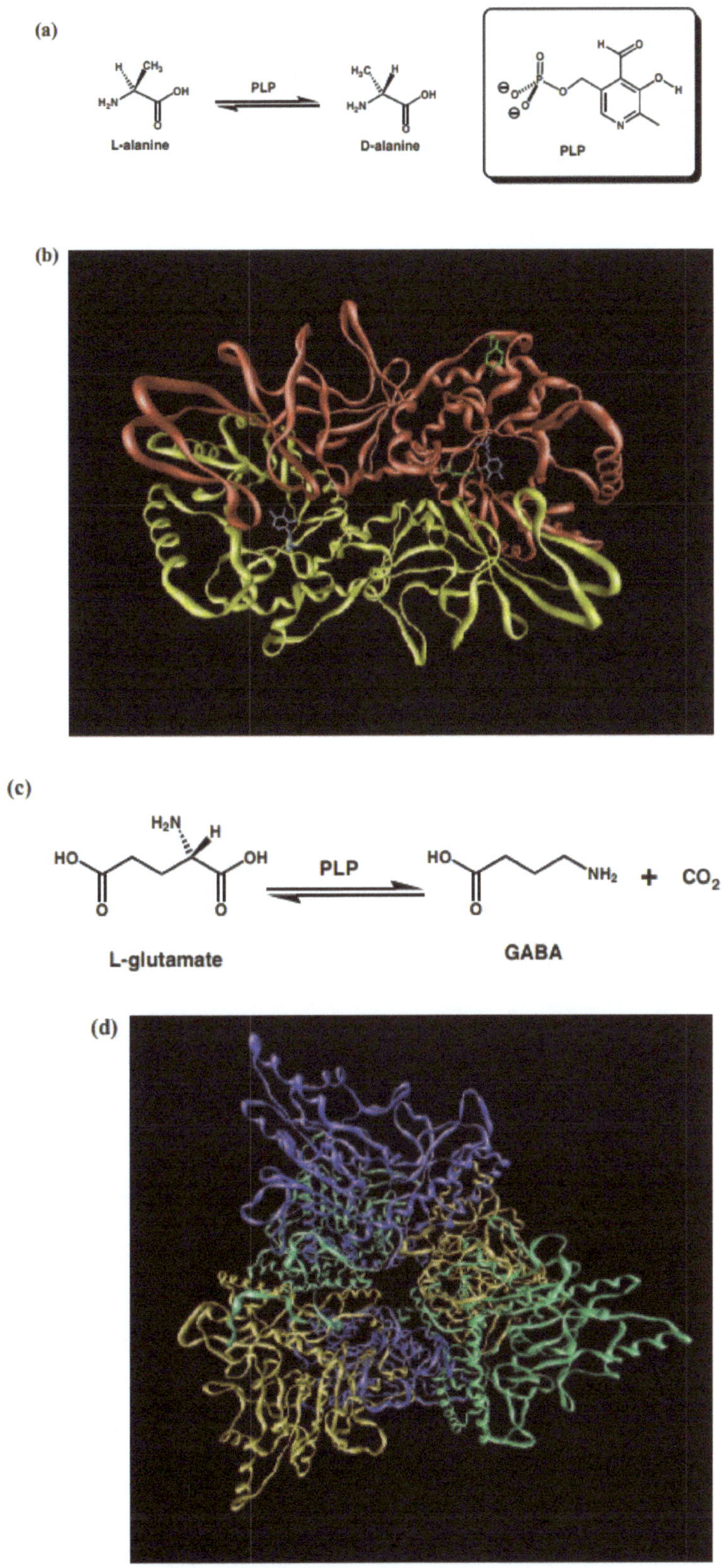
(a)
L-alanine
PLP
D-alanine
PLP
(b)
(c)
L-glutamate
PLP
GABA
+ CO_2
(d)

(e)

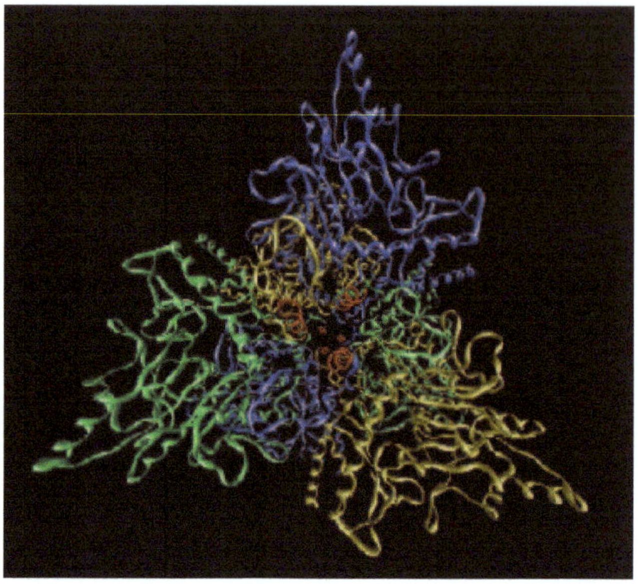

(f)

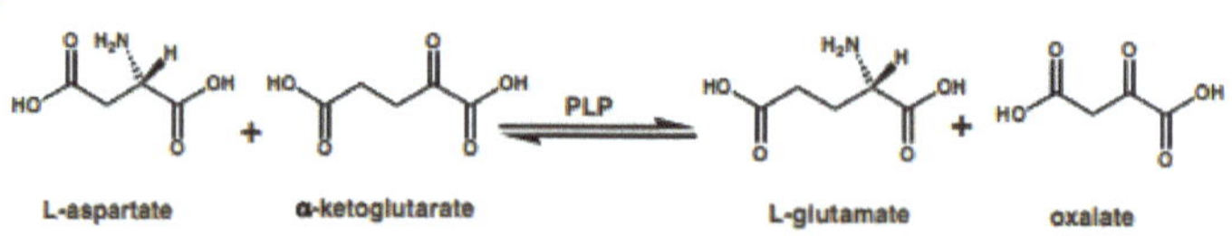

(g)

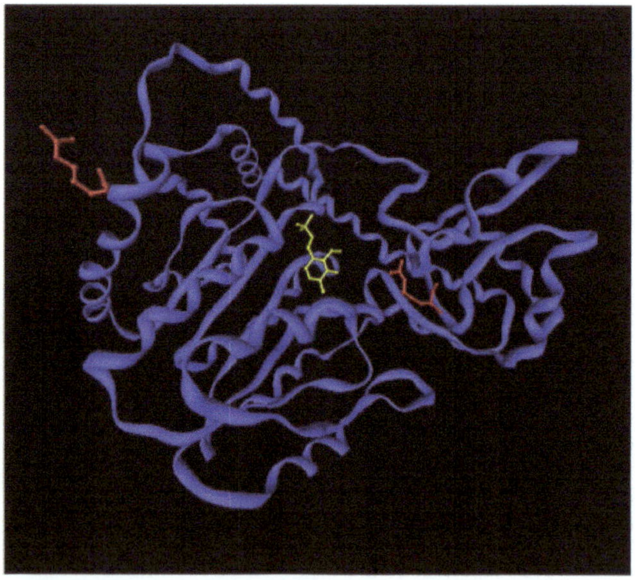

Figure 8.5. *Alanine racemase.*

Note: (a) Racemization reaction catalyzed by alanine racemase (AlaR), note that requires the cofactor pyridoxal phosphate (PLP) in order to take place; (b) ribbon display structure of homo-dimeric AlaR (Bacillus Stearothermophilus) (pdb: 2sfp) showing both polypeptide chains (yellow and red). PLP appears bound in both independent catalytic sites rendered by ball and stick display (blue); in one catalytic site, key catalytic residues Y265 and K39 (top to bottom) are shown (green), also rendered by ball and stick display (green). Glutamate α-decarboxylase. (c) α-decarboxylation reaction catalyzed by glutamate α- decarboxylase that requires the cofactor pyridoxal phosphate (PLP) to take place; (d) ribbon display structure of T-state (neutral pH) homo-hexameric glutamate α-decarboxylase (GadB) (Escherichia coli) showing all polypeptide chains (two in green, yellow and blue) (pdb: 1pmo); (e) ribbon display structure of R-state (acidic pH) homo-hexameric GadB (Escherichia coli) showing all polypeptide chains (two in green, yellow and blue) and six N-terminal α-helices, (1– 14, red) that are key sites for protonation and chloride-ion binding induced cooperativity (pdb: 1pmm). Aspartate transaminase. (f) Transamination reaction catalyzed by aspartate transaminase (aspAT) that requires the cofactor pyridoxal phosphate (PLP) to take place; (g) ribbon display structure of homo-dimeric aspAT (Escherichia coli) (pdb: 2aat) showing one of two polypeptide chains. PLP appears bound in independent catalytic site rendered by ball and stick display (yellow); key substrate binding residues R386 and R292 (right to left) are shown (red), also rendered by ball and stick display.

Source: https://nattabiochem.wordpress.com/wp-content/uploads/2009/08/essentials_chemical_biology.pdf.

If SOD doesn't deal with these radicals quickly, they will kill all of the biological macromolecules and macromolecular lipid assemblies in a cell. Let's talk about carbonic anhydrase (CA), which has been called one of the most effective biocatalysts of all; it catalyzes the straightforward interconversion of carbonic acid into water and carbon dioxide. In fact, some immune system cells intentionally release superoxide radicals in a controlled manner to kill invasive microorganisms. Simply put, the protective effect of this enzyme is due to its ability to maintain blood pH at approximately 7, preventing the acidification that would otherwise occur when blood levels of dissolved carbon dioxide rise as a result of glucose metabolism and TCA cycle activity in cells. Last but not least, we have already met acetyl choline esterase (AChE), without which excess of the neurotransmitter acetyl choline would not be eliminated by hydrolysis from pertinent synaptic clefts, resulting in uncontrolled neurotransmission linked to neurological disorders like epilepsy and seizures. Blood acidification or acidosis is incredibly deadly.

The reader interested in chemical biology should also be aware that some enzymes are multifunctional and that some biocatalysts are not enzymes. The enzyme LysU provides an intriguing case study for the first class.

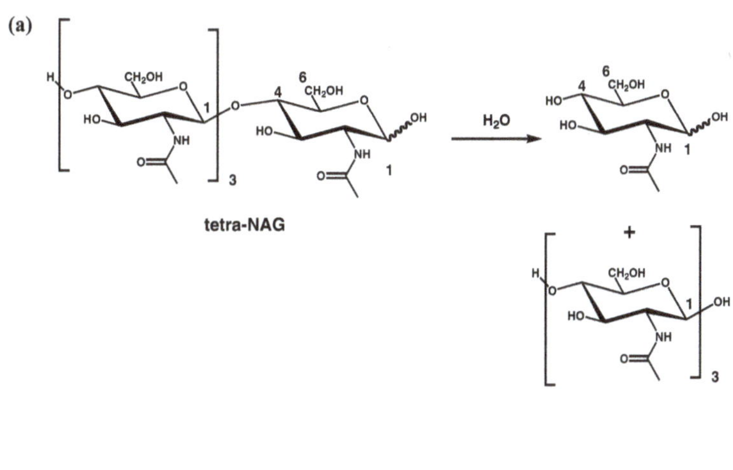

(b)

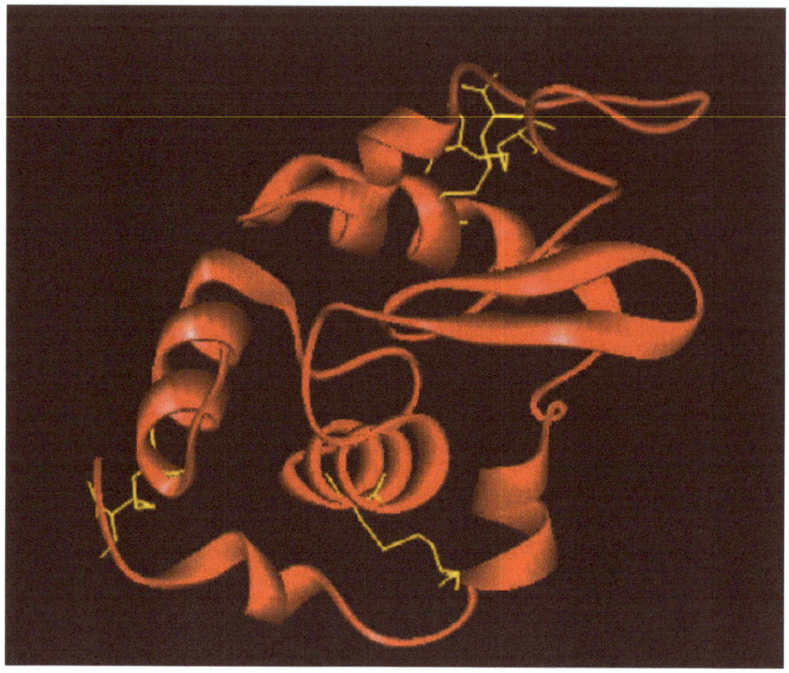

Figure 8.6. *Lysozyme.*

Note: (a) Hydrolysis reaction catalyzed by lysozyme: (b) ribbon display structure of lysozyme (hen-egg white) (pdb: 6lyz) wherein the four disulfide bonds and their corresponding cysteine residues are rendered in a stick display style (yellow). The catalytic site is a broad cleft through the polypeptide.

Source: https://nattabiochem.wordpress.com/wp-content/uploads/2009/08/essentials_chemical_biology.pdf.

First of all, this is a lysyl tRNA synthetase enzyme, which ought to be able to couple the naturally occurring amino acid L-lysine to suitable cognate tRNAs that have complementary anti-codon sequences to lysine codons. However, this enzyme's function changes to catalyze the sequential biosynthesis of diadenosine-5,'5'''-P^1,P^4-tetraphosphate (Ap_4A) and diadenosine-5,'5'''-P^1,P^3-tetraphosphate (Ap_3A) when zinc ions, Zn^{2+}, are present.

(a)

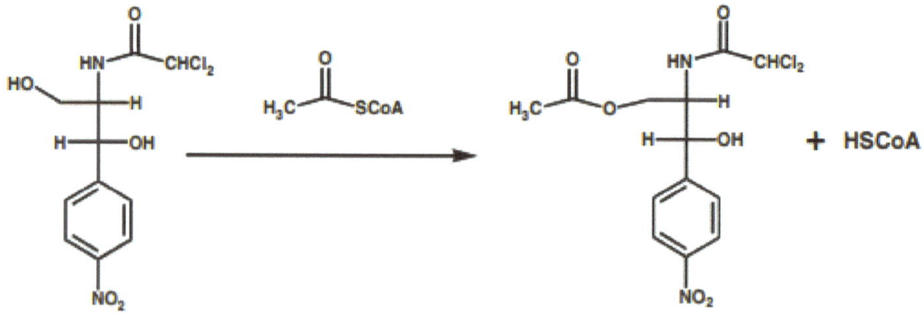

(b)

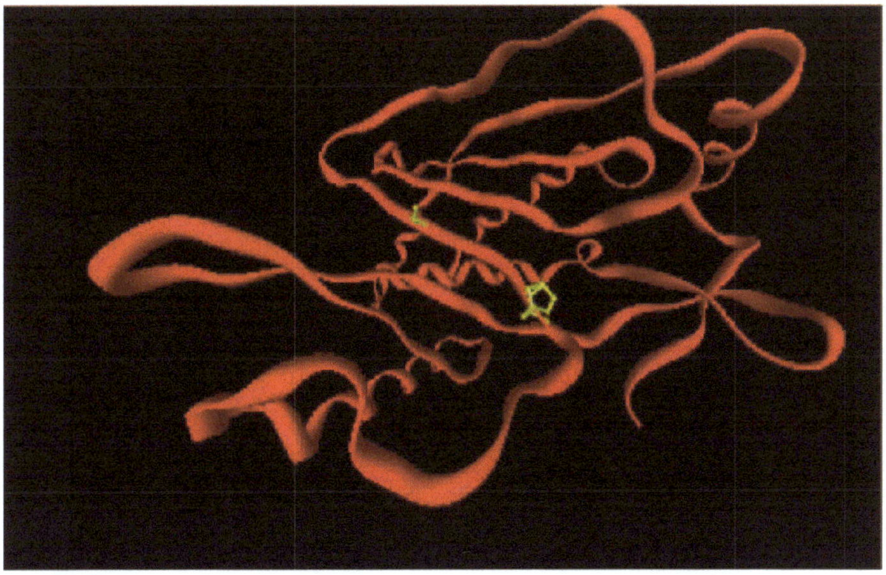

Figure 8.7. *Chloramphenicol acetyl transferase.*

Note: (a) Acylation reaction catalyzed by chloramphenicol acetyl transferease (CAT); (b) ribbon display structure of CAT (Type III, Escherichia coli) (pdb: 1cla) wherein two key catalytic residues H195 and S148 (right to left) are displayed (yellow), in ball and stick representation.

Source: https://nattabiochem.wordpress.com/wp-content/uploads/2009/08/essentials_chemical_biology.pdf.

The fact that RNA itself can function as a biocatalyst, or ribozyme, was a major finding a few years ago. This finding sparked a lot of theories suggesting that ribozymes may have originated before enzymes. Consequently, a molecular theory of evolution that places the RNA world ahead of the protein world has generated a lot of discussion.

8.1.1. Simple Principles in Biocatalysis

The area of a particular biocatalyst that is dedicated to biocatalysis is represented by the active or catalytic site. Catalytic sites typically occupy a very small area/volume of the overall structure and are often surprisingly small. A substrate molecule must be recognized and bound to a catalytic site in the initial step of biocatalysis. This molecular recognition and binding process is exactly the same as those processes that have been outlined. In the second stage of biocatalysis, a bound substrate needs to be released into the bulk solution after being biocatalyzed into a bound product.

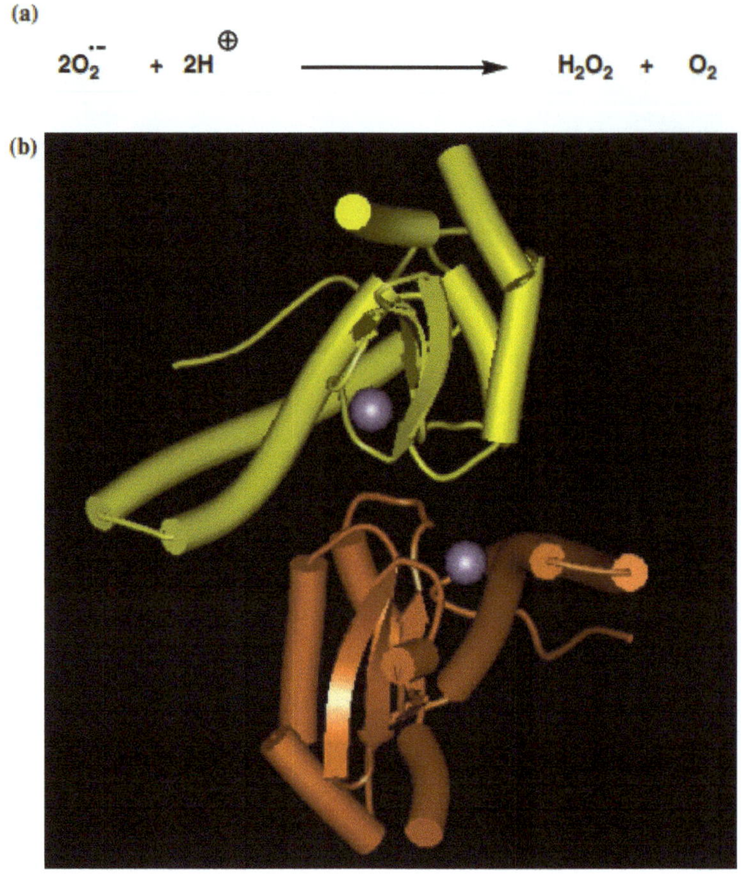

Figure 8.8. *Super oxide dismutase.*

Note: (a) Disproportionation reaction catalyzed by super oxide dismutase (SOD); (b) cartoon display structure of homo-tetrameric MnSOD (manganese containing human mitochondrial) (pdb:1n0j) illustrating two out of the four identical polypeptide chains (orange, yellow) and manganese ions ($Mn^{2+/3+}$) (purple) rendered as van der Waals spheres for complete clarity.

Source: https://nattabiochem.wordpress.com/wp-content/uploads/2009/08/essentials_chemical_biology.pdf.

But first, a readily available biocatalyst of interest made possible by techniques must exist to study biocatalysis. Structure is always a great aid in understanding function. Subsequently, an analysis methodology and a robust theoretical framework are required to interpret biocatalysis data and elucidate the critical mechanisms of biocatalysts that enable biocatalysis. This is the reason we will start this chapter with a thorough explanation of how to collect and analyze biocatalytic data using different biocatalysis models. After that, we'll examine the biocatalysis theories that shed light on how biocatalysts can accelerate chemical reactions so successfully—reactions that are essential to life.

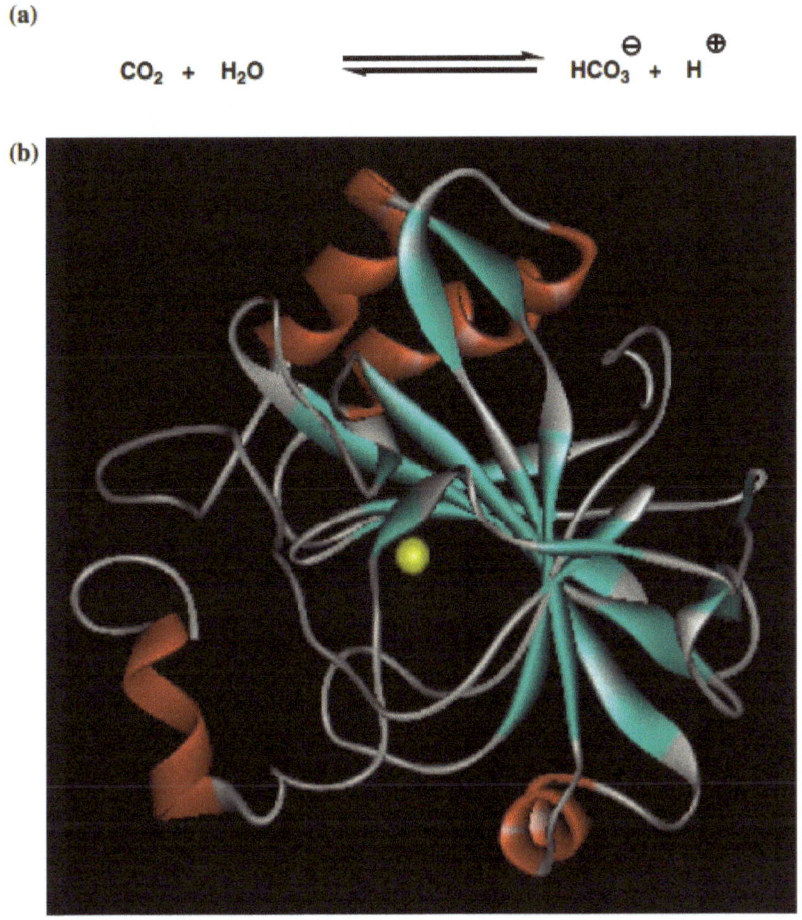

Figure 8.9. *Carbonic anhydrase.*

Note: (a) Reversible hydration reaction catalyzed by carbonic anhydrase (CA): (b) ribbon display structure of monomeric CA (Type I, human erythrocyte) (pdb: 2cab) illustrating α-helices (red) and β-sheet (blue) plus a central zinc ion (Zn^{2+}) (yellow) rendered as van der Waals sphere for complete clarity.

Source: https://nattabiochem.wordpress.com/wp-content/uploads/2009/08/essentials_chemical_biology.pdf.

As a comprehension aid, we will refer to the biocatalyst examples that have already been presented to the chemical biology reader throughout this chapter. For each and every biocatalyst under study, the path to true understanding in biocatalysis is actually convoluted and takes time. It should come as no surprise that there is still a long way to go before we fully comprehend all known biocatalysts, let alone all those that are unknown.

8.1.2. Steady-State Kinetics in Biocatalysis

Kinetics is the study of catalysis and reaction rates. Any given biocatalyst's catalytic activity is usually assessed in steady-state and/or pre-steady-state circumstances.

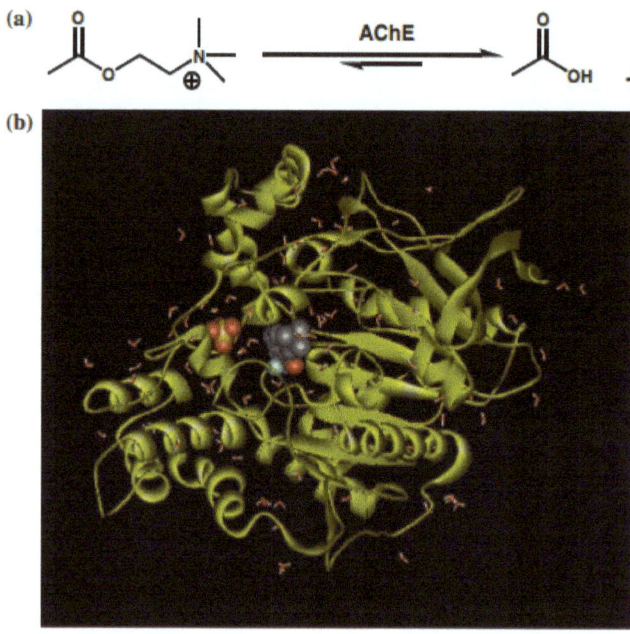

Figure 8.10. *Acetylcholine esterase.*

Note: (a) Hydrolysis reaction catalyzed by acetyl choline esterase (AChE): (b) ribbon display structure of AChE (Torpedo californica) (pdb: 1amn) illustrating crystallographic water molecules (oxygen: red; hydrogen: white) plus sulfate ion and substrate analog bound in catalytic site. Both these are rendered as CPK structures (red: oxygen; yellow: sulfur; gray: carbon: blue: nitrogen).

Source: https://nattabiochem.wordpress.com/wp-content/uploads/2009/08/essentials_chemical_biology.pdf.

To comprehend the surface mechanistic behavior of a specific biocatalyst that is in stasis and able to carry out numerous catalytic turnovers, steady-state kinetic analyzes are utilized. According to stasis, rate measurements (i.e., data that describe the rates at which a specific substrate is converted into a product are limited to brief time intervals during which the concentrations of reactants and substrates do not fluctuate greatly. As a result, stasis also suggests that kinetic analysis occurs when different biocatalyst species, substrates, and/or products are in a metastable equilibrium. Although a relatively blunt tool for the examination of specific mechanistic steps connected to a particular biocatalytic process occurring in the catalytic site of a particular biocatalyst, steady-state analysis is highly representative of biocatalyst behavior under conditions of normal metabolism. However, steady-state kinetic analyzes are routinely carried out because they typically call for relatively simple equipment and straightforward procedures to implement.

8.1.3. Steady-State Bioassays

The variety of biocatalysts available can be reflected in the range of steady-state kinetic analyzes. A bioassay is typically used to measure rate enhancements caused by a single biocatalyst.

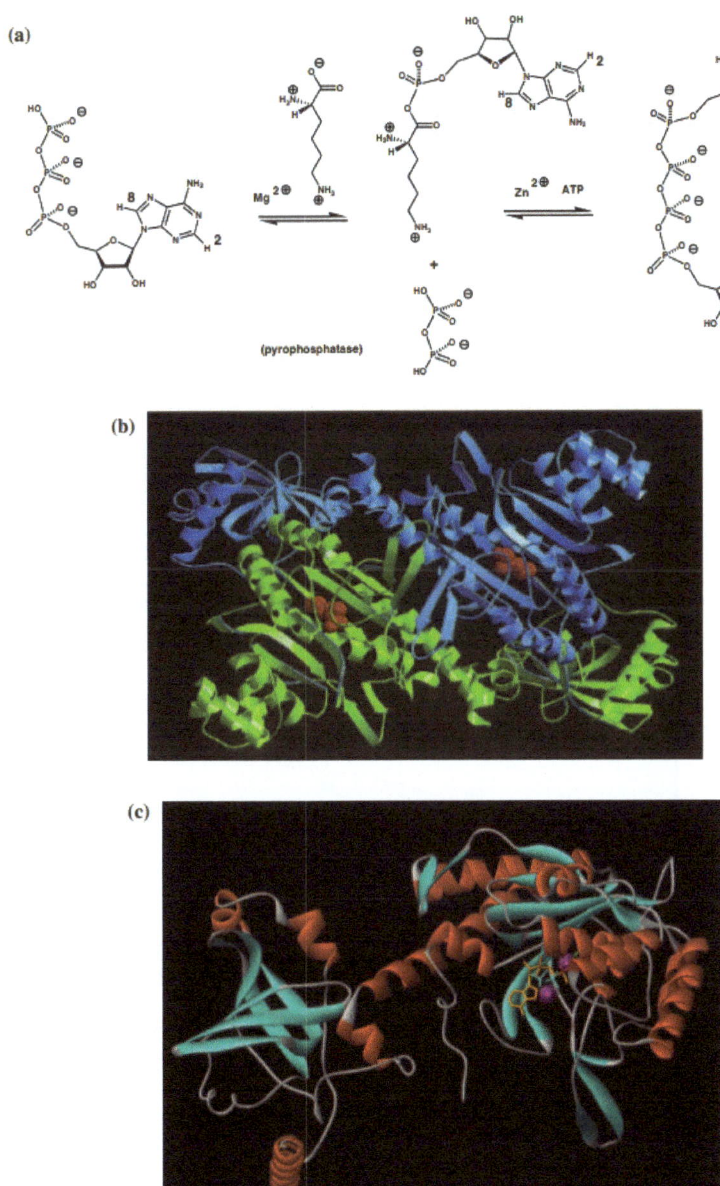

Figure 8.11. *Lysyl tRNA synthetase.*

Note: (a) Ap4A synthesis reaction catalyzed by lysyl tRNA synthetase isozyme (LysU): (b) ribbon display structure of homo-dimeric LysU (Escherichia coli) (pdb: 1lyl) showing both polypeptides (blue and green) and catalytic site bound L-lysine (red) rendered as CPK structures: (c) cartoon display *structure of LysU monomer illustrating α-helices (red) and β-sheets (blue) with ATP (orange) and L-lysine (green) bound in the catalytic site with two Mg^{2+} ions (purple). The Mg^{2+} ions are rendered as van der Waals spheres while ATP and L-lysine are rendered as ball and stick structures (pdb: 1e1t).*

Source: https://nattabiochem.wordpress.com/wp-content/uploads/2009/08/essentials_chemical_biology.pdf.

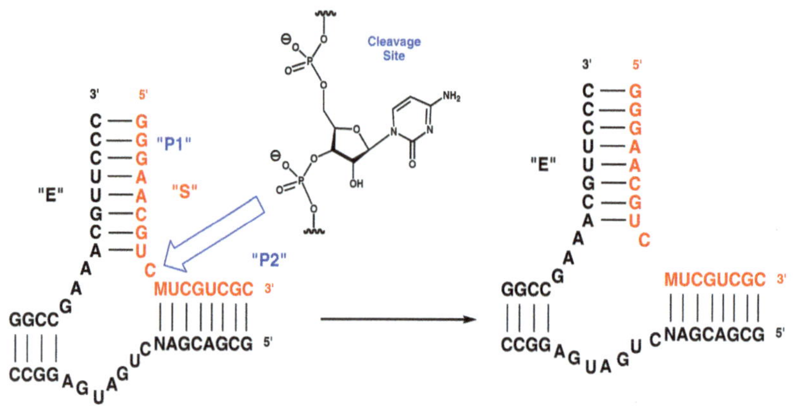

Figure 8.12. *Ribozyme. Hydrolysis reaction catalyzed by ribozyme that consists of an enzyme "E" strand and a substrate "S" strand. Hydrolysis takes place below a cytidine nucleotide, converting "S" into two product strands "P1" and "P2" that dissociate from the "E" strand.*

Source: https://nattabiochem.wordpress.com/wp-content/uploads/2009/08/essentials_chemical_biology.pdf.

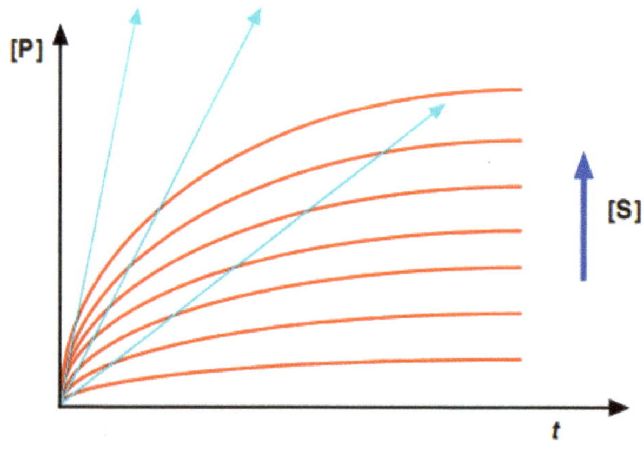

Figure 8.13. *Schematic diagram to illustrate the concept of initial rates. In the presence of a fixed concentration of biocatalyst, variable concentrations of substrate S are converted into product P over a time t. Steady-state analyzes require initial concentrations of substrate, [S] and initial rates of product formation, ν, that derive from the slopes of tangents (blue arrows) to initial slopes as illustrated.*

Source: https://nattabiochem.wordpress.com/wp-content/uploads/2009/08/essentials_chemical_biology.pdf.

Every biocatalyst has at least one distinct bioassay. The variation in the initial rate ν of substrate to product conversion as a function of initial substrate concentration [S] is produced by a bioassay, which maintains a constant concentration of the biocatalyst throughout. Temperature, pH, ionic strength, and even other fixed concentrations of second, third, or even fourth substrates can all be used to determine the rate data itself. Then, using any number of suitable steady-state kinetics equations, this rate data is processed to produce standard biocatalytic parameters (which will be discussed below) as well as a surface mechanism indicator indicating how the biocatalyst catalyzes biocatalysis. These common parameters make it easy to compare the performances of different biocatalysts.

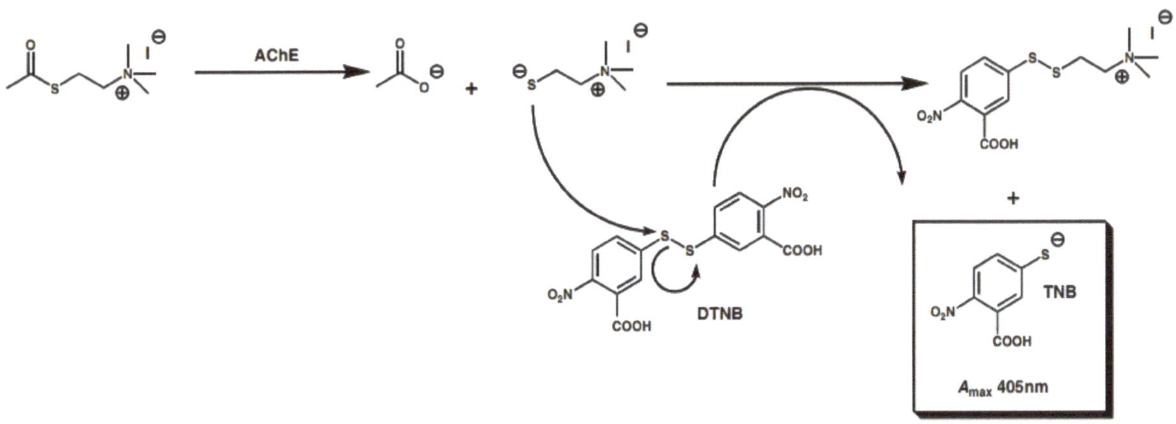

Figure 8.14. *Colourimetric assay. Acetyl choline esterase (AChE) assay system involving thioacetylcholine that is hydrolyzed to thio-choline. This in turn combines with colorless reagent 5,5' – dithio-bis(nitrobenzoic acid) (DTNB) to form yellow-colored 5-thio-2-nitro-benzoic acid (TNB).*

Source: https://nattabiochem.wordpress.com/wp-content/uploads/2009/08/essentials_chemical_biology.pdf.

Photometric assays make up the majority of steady-state bioassays, but there are numerous other options as well, such as 1H-NMR, HPLC, radiometric, oxygen electrode, and gel assays. The primary goal in every scenario is to ascertain how, directly or indirectly, values of ν change as a function of [S] while maintaining a constant concentration of the biocatalyst. In photometric steady-state bioassays, a chromogenic substrate or reagent is used, which undergoes a chemical change during biocatalysis to become a chromophore and absorbs heavily at a specific wavelength.

According to UV-visible spectroscopy, the initial rate of chromophore formation and the initial rate of product formation are generally said to correlate. Chromatogenic reagents are used in both AChE and CA steady-state bioassays.

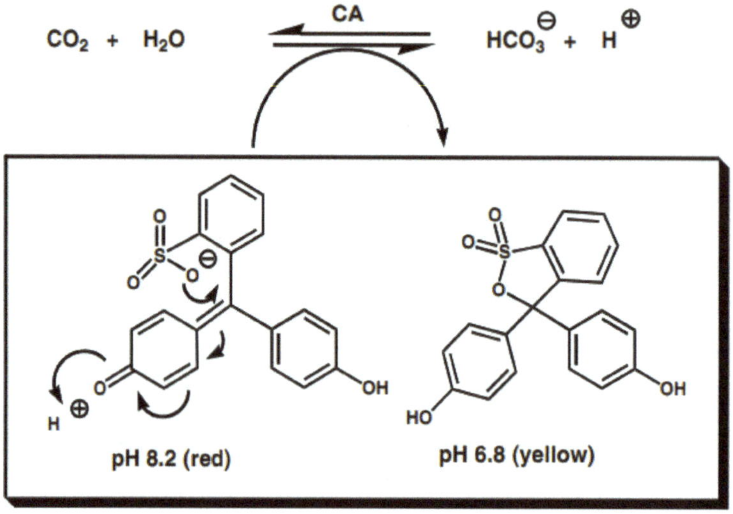

Figure 8.15. *Colourimetric assay. Carbonic Anhydrase (CA) assay system. Changes in pH with enzyme catalysis are followed with great precision and accuracy by pH-sensitive dye that changes absorbance profile with pH-change.*

Source: https://nattabiochem.wordpress.com/wp-content/uploads/2009/08/essentials_chemical_biology.pdf.

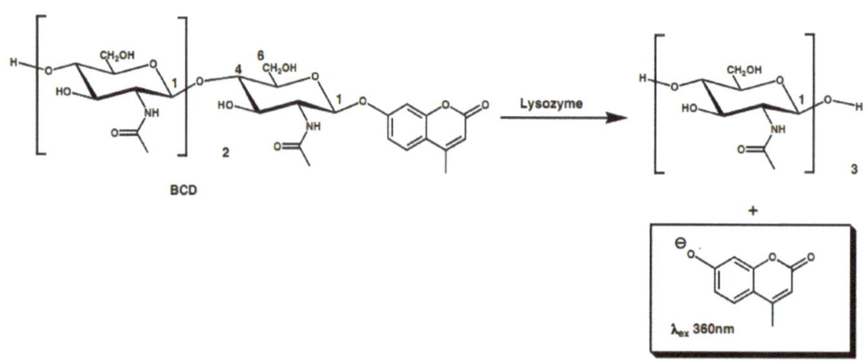

Figure 8.16. *Fluorimetric assay. Lysozyme assay system. The assay is performed under mildly acidic conditions but the appearance of charged fluorophore requires pH-quenching to pH 9.*

Source: https://nattabiochem.wordpress.com/wp-content/uploads/2009/08/essentials_chemical_biology.pdf.

Instead, a chromogenic substrate that can also be detected by fluorescence is used in a steady-state bioassay of lysozyme. As a result, according to fluorescence spectroscopy, the initial rate of product formation correlates with the initial rate of fluorophore formation. When it comes to TIM, the actual reaction that is catalyzed cannot be observed directly using a chromogenic substrate or reagent. However, a coupled assay system can be employed, in which one product is biocatalytically reduced using the reverse chromogenic reagent NADH to produce a follow-on product that is irreversibly changed.

It is believed that the initial rate of TIM-catalyzed dihydroxyacetone phosphate (DHAP) product formation correlates inversely with the initial loss of NADH via oxidation to NAD^+. Since NADH is an intrinsic component of the direct catalytic mechanism in the case of the enzyme, the initial rate of MDH-catalyzed oxalic acid product formation is directly inversely correlated with the initial loss of NADH over time.

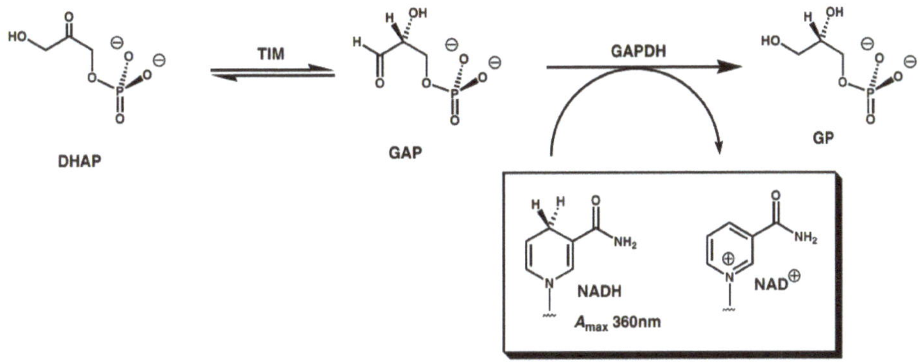

Figure 8.17. Enzyme coupled assay. Triose Phosphate Isomerase (TIM) assay system with glyceraldehyde 3-phosphate dehydrogenase (GAPDH) colorimetric coupled assay for detection purposes. In this version of the TIM assay, dihydroxyacetone phosphate (DHAP) is converted enzymically to glyceraldehyde 3-phosphate (GAP) that is onward converted to glycerol phosphate (GP) by means of the coupled enzyme GAPDH enzyme that uses the reverse-colorimetric reductant NADH.

Source: https://nattabiochem.wordpress.com/wp-content/uploads/2009/08/essentials_chemical_biology.pdf.

When substrate is transformed into product, a change in physical elution time is used in HPLC steady-state bioassays. For example, the fact that the product Ap_4A and substrate adenosine 5 -triphosphate (ATP) elute from an HPLC column at different times is used in a steady-state bioassay for the enzyme LysU. Therefore, the first rate of Ap_4A formation is correlated with the first rate of Ap_4A peak formation or, conversely, with the first rate of ATP peak disappearance. The same HPLC bioassay can also be used to observe the LysU-catalyzed conversion of Ap_4A to Ap_3A. Similar reasoning governs the ^{1}H-NMR steady-state bioassay, where the initial rate of Ap_4A formation is inversely correlated with the initial rate of ATP-purine signal disappearance or with the initial rate of Ap_4A-purine signal appearance as measured by 1D ^{1}H-NMR spectroscopy. When oxygen-evolving or oxygen-consuming biocatalysts are involved, a steady-state oxygen electrode bioassay is typically needed. When it comes to the enzyme SOD, the proper steady-state bioassay necessitates measuring the initial rate of oxygen production in relation to the initial concentration of superoxide using a calibrated oxygen electrode. Radioactive substrates are typically converted into products, and the products are then carefully separated from the substrates in radiometric steady-state bioassays. The product in the CAT bioassay is significantly more hydrophobic than the substrate, allowing for clean extraction into organic solvents.

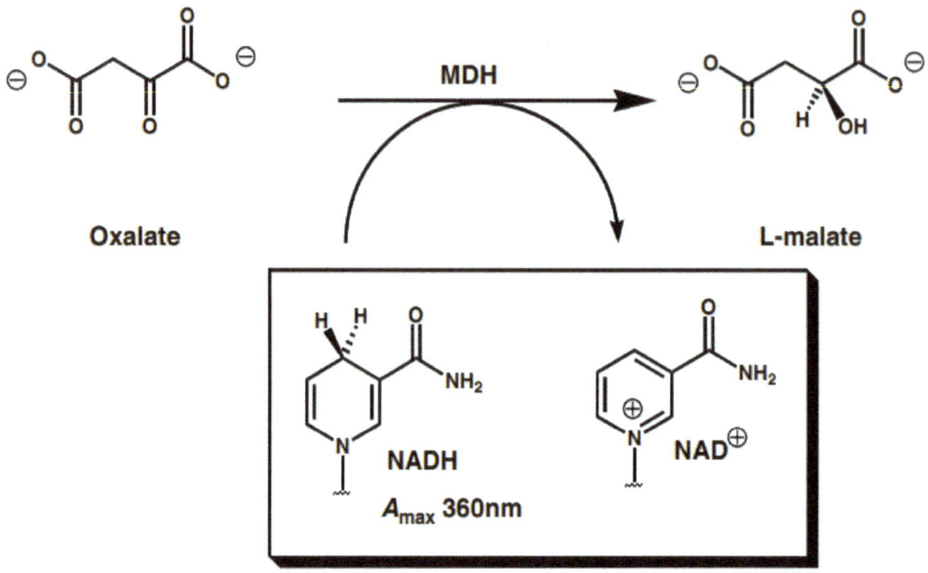

Figure 8.18. *Colourimetric assay. Malate dehydrogenase (MDH) assay system. Oxalate is onward converted to L-malate by means of enzyme MDH that uses the reverse-colorimetric reductant NADH to effect catalytic reduction.*

Source: https://nattabiochem.wordpress.com/wp-content/uploads/2009/08/essentials_chemical_biology.pdf.

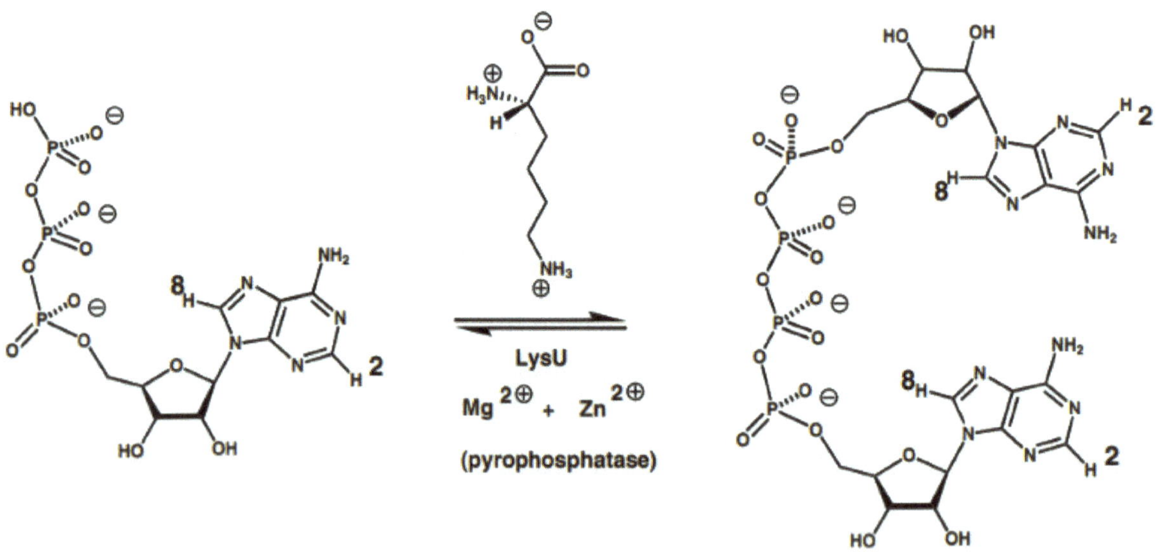

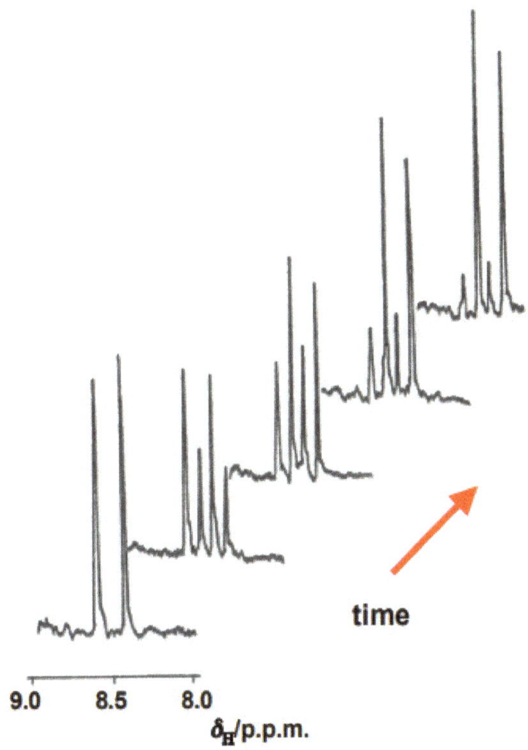

Figure 8.19. *^{1}H-NMR assay. Lysyl tRNA synthetase (LysU) assay system. Conversion of ATP into Ap_4A is observed by 1H-NMR spectroscopy when reaction is run in a 5mm NMR tube. 1H signals of adenine ring are shifted upfield with conversion of ATP (δH 8.55 [H-2] and 8.29 [H-8] p.p.m.) to Ap_4A (δH 8.40 [H-2] and 8.18 [H-8] p.p.m.). Rates of conversion are followed by changes in appropriate 1H-NMR signal peak areas as a function of time.*

Source: https://nattabiochem.wordpress.com/wp-content/uploads/2009/08/essentials_chemical_biology.pdf.

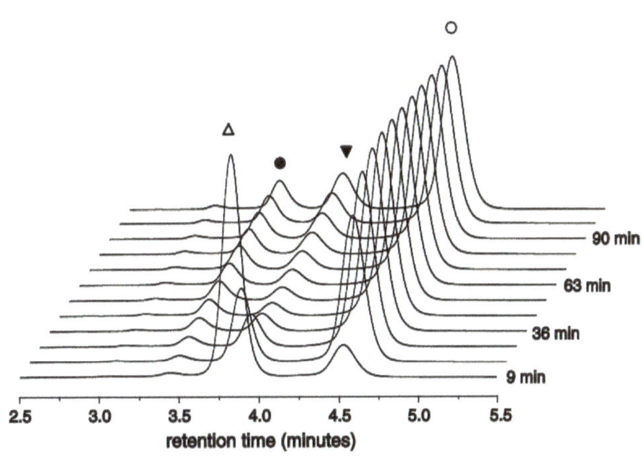

Figure 8.20. *HPLC assay. Lysyl tRNA synthetase (LysU) assay system. LysU converts ATP () into Ap_4A (°). Rates of conversion are followed by changes in appropriate HPLC peak area as a function*

of time. LysU has recently been shown to further catalyze the onwards formation of Ap_3A (▼) from Ap_4A () via an ADP (●) intermediate. This second stage reaction is clearly visible by HPLC assay.

Source: https://nattabiochem.wordpress.com/wp-content/uploads/2009/08/essentials_chemical_biology.pdf.

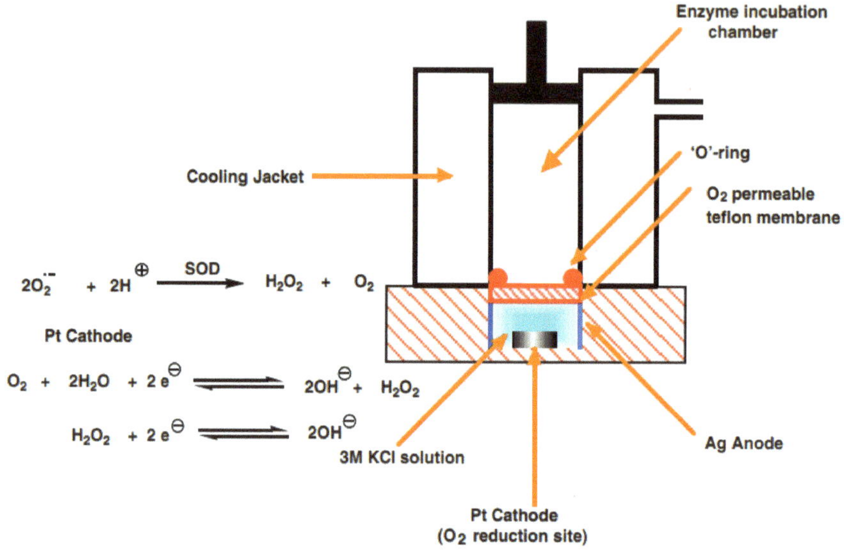

Figure 8.21. *Oxygen electrode assay. Superoxide dismutase (SOD) assay in which enzyme generates H_2O_2 that is further reduced at the Pt-cathode at the expense of oxygen. Current generated through reduction is calibrated with respect to oxygen produced by SOD.*

Source: https://nattabiochem.wordpress.com/wp-content/uploads/2009/08/essentials_chemical_biology.pdf.

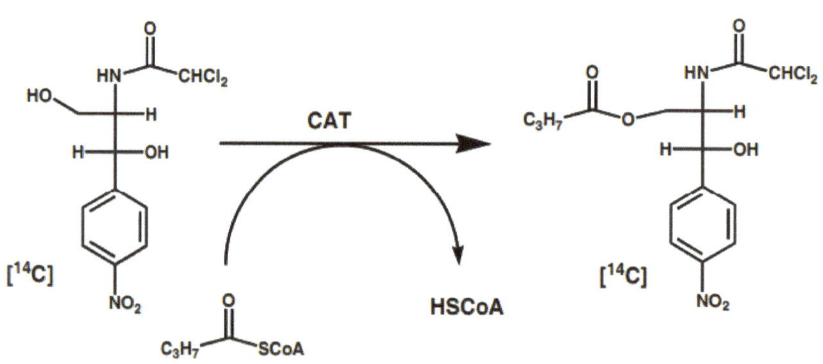

Figure 8.22. *Radiographic assay. Chloramphenicol Acetyl Transferase (CAT) assay system. Butyrylation of Chloramphenicol generates a more hydrophobic product that is separated from substrate by extraction into organic solvent.*

Source: https://nattabiochem.wordpress.com/wp-content/uploads/2009/08/essentials_chemical_biology.pdf.

Thus, it is claimed that the initial formation of the product and the [^{14}C] accumulation in the organic extract are correlated. Lastly, because substrate and product polynucleotides differ in molecular weight, gel steady-state bioassays serve as the foundation for ribozyme analyzes. In this instance, the initial decrease in substrate band intensities or the initial increase in product band intensities determine the initial formation of the product, respectively.

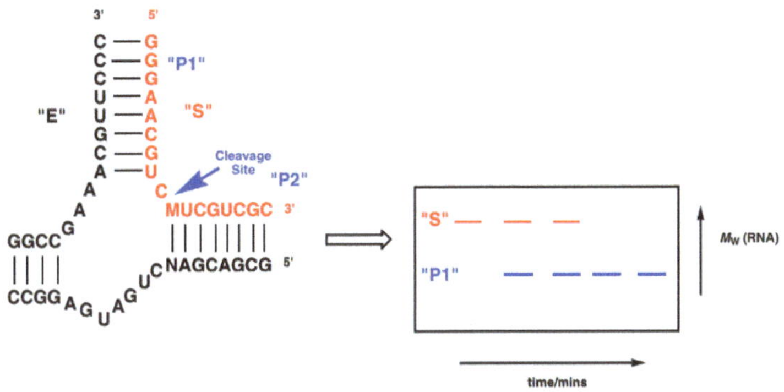

Figure 8.23. *Gel assay. Ribozyme assay system. One strand of the ribozyme acts as the biocatalyst E and the other strand acts as the substrate S. S is specifically cleaved at the indicated site generating two product oligonucleotides P1 and P2. The conversion of S into a larger molecular weight product P1 can be followed by gel electrophoresis with time.*

Source: https://nattabiochem.wordpress.com/wp-content/uploads/2009/08/essentials_chemical_biology.pdf.

8.2. STEADY-STATE KINETIC SCHEMES

Every steady-state kinetics equation is the result of meticulous examination of several steady-state kinetic schemes, all of which aim to identify the most likely surface mechanism through which a single biocatalyst could catalyze a reaction. The goal of the final steady-state kinetics equations is to establish a graphical relationship between ν and [S] that varies based on variations in the biocatalysis surface mechanism. Therefore, the most likely surface mechanism can be determined along with the most accurate biocatalytic parameter determination using a steady-state equation that most closely accounts for the observed relationship between ν and [S] data. As many steady-state kinetics equations as there are kinetic schemes is no more than it can be, but there is a limit to how much more complexity can be understood. Because of this, we will only provide the reader of chemical biology with those systems and formulas that are currently in use and that we believe to be generally the most significant. For the purpose of analyzing the steady-state kinetic behavior of those biocatalyst examples, the most suitable kinetic scheme and steady-state equations will be cited whenever possible.

8.2.1. Simple Steady-State Kinetics and Michaelis–Menten Equation

The Uni-Uni kinetic scheme, in which Uni denotes a single substrate and the second Uni denotes the evolution of a single product, is the most basic kinetic scheme that can exist. This extremely basic kinetic scheme's primary characteristic is the existence of an intermediate chemical species called ES, which is a complex (basically non-covalent) between biocatalyst E and substrate molecule S. For reasons that will become clear in a moment, this is sometimes called the Michaelis complex. With such a basic mechanistic scheme, ν can be expressed as

$$\nu = k_2 [ES] \tag{1}$$

For the purposes of the following analysis, biocatalysis is assumed to be irreversible and each biocatalyst E possesses only a single catalytic site. Hence, if the total concentration of biocatalyst is $[E]_0$, then this must be the sum of free enzyme, [E], and Michaelis complex, [ES], concentrations as indicated in

$$[E]_0 = [E] + [ES] \tag{2}$$

$$E + S \underset{k_{-1}}{\overset{k_1}{\rightleftharpoons}} ES \xrightarrow{k_2} E + P$$

$$K_s$$

Scheme 8.1

If Eqn. (1) is then divided through by Eqn. (2) we arrive at:

$$\frac{\nu}{[\mathrm{E}]_0} = \frac{k_2[\mathrm{ES}]}{[\mathrm{E}] + [\mathrm{ES}]} \qquad (3)$$

In order to solve Eqn. (3), expressions for [E] and [ES] are required that are obtained by applying Briggs–Haldane steady-state principles. According to these principles, biocatalysis rapidly attains a condition of stasis under which all biocatalyst species are at a constant equilibrium concentration. In other words [E] and [ES] are constant with time. Stasis is reflected by:

$$k_1[\mathrm{E}][\mathrm{S}] = k_{-1}[\mathrm{ES}] + k_2[\mathrm{ES}] \qquad (4)$$

Which rearranges to give the steady-state expression:

$$[\mathrm{ES}] = \frac{k_1}{k_{-1} + k_2}[\mathrm{E}][\mathrm{S}] \qquad (5)$$

Hence, substituting for [ES] in Eqn. (3) followed by rationalization of terms, we arrive at:

$$\nu = \frac{k_2[\mathrm{E}]_0[\mathrm{S}]}{\left(k_2 + k_{-1}/_{k_1}\right) + [\mathrm{S}]} \qquad (6)$$

This equation may be further simplified by substituting for the bracketed aggregate of rate constants by the constant K_m, Where;

$$K_m = \left(k_2 + k_{-1}/_{k_1}\right) = k_2/k_1 + K_s \qquad (7)$$

The result is:

$$\nu = \frac{k_{cat}[\mathrm{E}]_0[\mathrm{S}]}{K_m + [\mathrm{S}]} \qquad (8)$$

where K_m is known as the Michaelis constant. Compared with Eqn. (6), there is one further change in that rate constant k_2 has been relabeled as k_{cat}, which is known as the catalytic constant or turnover number. Eqn. (8) is known as the Michaelis–Menten or Henri–Michaelis–Menten equation after the pioneers of biocatalysis who originally derived this equation.

The Michaelis–Menten equation classically describes a rectangular hyperbola and has the form shown in Figure 8.24. In this graphical form K_m can be described in simple terms as the value of [S] corresponding to an initial rate of catalysis equivalent to $V_{max}/2$, where V_{max} is the maximum initial rate of catalysis that is possible.

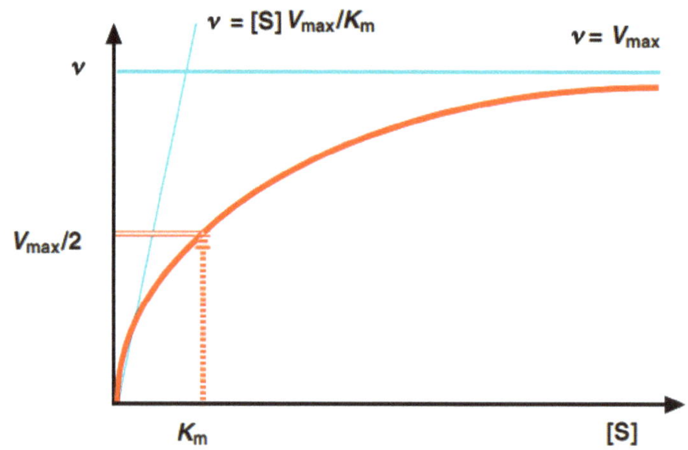

Figure 8.24. *Basic steady-state biocatalyst kinetic profile.*

Source: https://nattabiochem.wordpress.com/wp-content/uploads/2009/08/essentials_chemical_biology.pdf.

Studies on molecular recognition also reveal a rectangular hyperbola resembling this one, which illustrates the idea that, in the presence of enough substrate, every active site in a biocatalyst should be continuously and effectively saturated with reacting substrate over time. The rectangular hyperbolic graphical representation and the Michaelis-Menten equation are fundamental to steady-state biocatalysis. As we shall see, the form of this equation is often maintained regardless of the complexity of the kinetic model, even though it was obtained above through a steady-state analysis of the most basic kinetic model for biocatalyst action. The explanation for this is straightforward: biocatalysis invariably involves saturation at high values of [S]. As a result, a saturating function of some kind is required to characterize the catalytic behavior of a particular biocatalyst, and the Michaelis–Menten equation provides nearly the most straightforward description of a saturating function. There are two important limits to Eqn. (8). In the first instance, when [S] >> K_m, then Eqn. (8) reduces to

$$\nu = k_{cat}[E]_0 = V_{max} \quad (9)$$

Alternatively, when K_m >> [S] then Eqn. (8) reduces to

$$\nu = \left(\frac{k_{cat}}{K_m}\right)[E]_0[S] \tag{10}$$

However, at low [S] almost all the biological catalyst is free of bound substrate. Therefore, $[E]_0$ may be substituted for by [E], the concentration of free catalyst in solution. This results in the following equation:

$$\nu = \left(\frac{k_{cat}}{K_m}\right)[E][S] \tag{11}$$

where (k_{cat}/K_m) represents the specificity constant for a given substrate of a given biological catalyst. In practice, Eqn. (11) applies over a wider range of [S] values than those obeying the inequality that Km >> [S], which is extremely useful for reaction kinetic analyzes.

8.2.2. Interpretation of k_{cat} and K_m

The definition of Km can be found either in terms of an aggregate of rate constants or in terms of the value of [S], which corresponds to an initial rate of catalysis equal to $V_{max}/2$. With reference to Eqn. (7), we can provide a definition of Km that is far more nuanced and comprehensive. The Km reduces to K_s, an ES complex equilibrium dissociation constant, at the point where $k_{-1} >> k_2$. Therefore, K_m can be broadly interpreted as an equilibrium dissociation constant, either real or apparent, that gives a precise indication of how much biocatalyst is "locked up" in binding to a substrate, reaction intermediate, or even the final product during a catalytic process.

It is evident from Eqn. (9) that k_{cat}, also known as the catalytic constant or turnover number, is a first order rate constant that indicates the maximum catalytic rate in a biocatalyst per unit time at saturating concentrations of substrate per catalytic binding site. On the other hand, k_{cat} is more accurately defined as the first order rate constant for the decomposition of an ES complex with respect to Eqn. (1). Consequently, k_{cat} can be viewed as a first order rate constant in general since it is a function of the first order rate constants for the transformation or breakdown of all complexes that are created during a catalytic process between a substrate, reaction intermediate, or even a product and the biocatalyst.

8.2.3. Determination of k_{cat} and K_m

Parameters k_{cat}, V_{max} and K_m may be determined from real catalytic rate data by means of a single or double reciprocal plot. The single reciprocal plot or Hanes plot is based upon the following algebraically rearranged version of Eqn. (8):

$$\frac{[S]}{\nu} = \frac{K_m}{V_{max}} + \frac{[S]}{V_{max}} \tag{12}$$

The data should fit into a straight line when the initial rate data, ν, and the initial substrate concentration data, [S], are plotted using Eqn. (12). The plot's typical error distortion is only seen at extremely low values of ν and [S]. Compared to the Hanes plot, the double-reciprocal plot, also known as the Lineweaver–Burk plot, has been applied more frequently. This popularity is a little out of place, though.

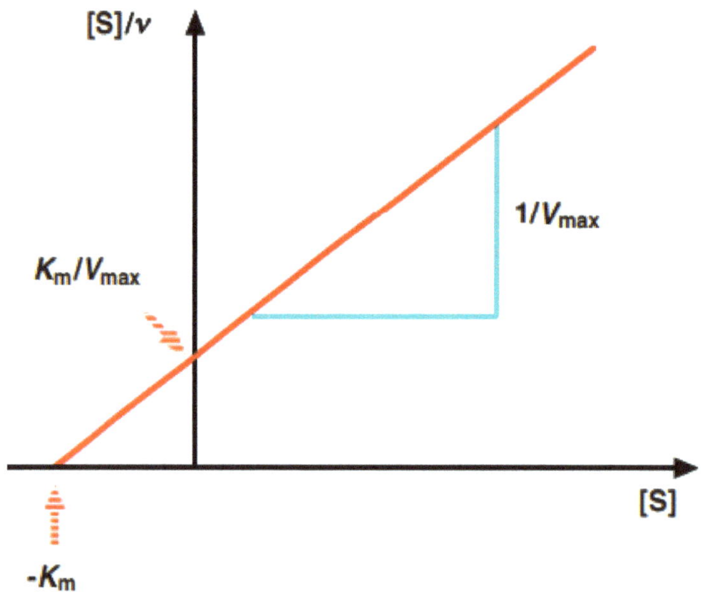

Figure 8.25. *Linear, single-reciprocal Hanes plot.*

Source: https://nattabiochem.wordpress.com/wp-content/uploads/2009/08/essentials_chemical_biology.pdf.

A Lineweaver–Burk plot is based upon this alternative rearranged version of Eqn. (8):

$$\frac{1}{\nu} = \frac{K_m}{V_{max}}\frac{1}{[S]} + \frac{1}{V_{max}} \tag{13}$$

The initial rate and concentration data in this instance also produce a straight line plot, but it suffers from gross systematic error distortion, with ν and [S] decreasing with time. Because of this, we generally do not advise using Lineweaver-Burk plots unless conducting multisubstrate kinetics experiments.

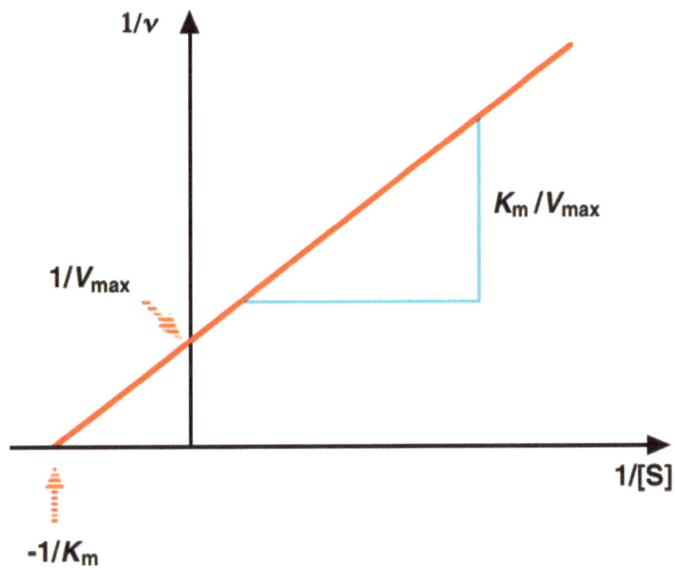

Figure 8.26. *Linear, double-reciprocal Lineweaver-Burk plot.*

Source: https://nattabiochem.wordpress.com/wp-content/uploads/2009/08/essentials_chemical_biology.pdf.

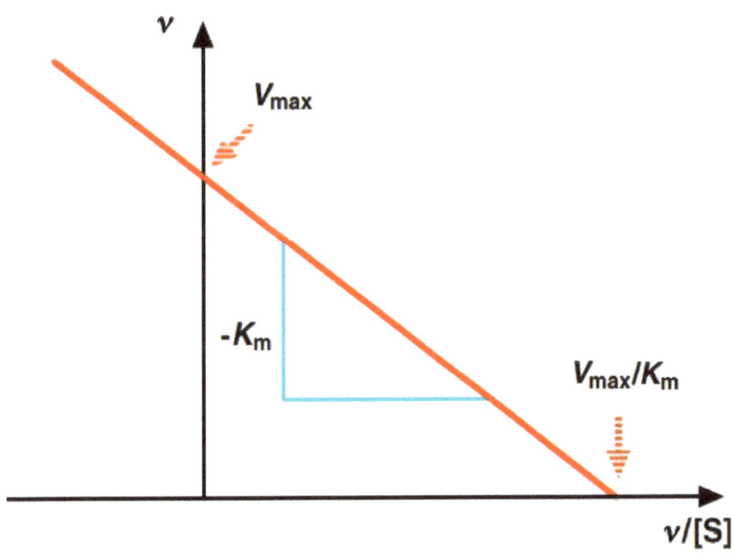

Figure 8.27. *Linear, single-reciprocal Eadie-Hofstee plot.*

Source: https://nattabiochem.wordpress.com/wp-content/uploads/2009/08/essentials_chemical_biology.pdf.

Another version of kinetic data presentation is the Eadie– Hofstee plot, which is a single-reciprocal plot that is based upon the following rearranged version of Eqn. (8):

$$\nu = V_{\max} - K_{\mathrm{m}}\frac{\nu}{[\mathrm{S}]} \tag{14}$$

In this case, initial rate and concentration data also give a straight line plot, but one also suffering from gross systematic error distortion with increasingly low values of ν and [S].

8.2.4. Effect of Steady-State Inhibitors

"Substrate-like" molecules known as biocatalyst inhibitors interact with a specific biocatalyst and obstruct the process of biocatalysis. Competitive inhibition, non-competitive inhibition, or uncompetitive inhibition are the three main modes of action of inhibitors. Each case has a unique mode of inhibition, necessitating the use of a unique steady-state kinetic scheme to account for them all. As a result, as we will demonstrate below, every mode of inhibition is characterized by a unique steady-state kinetic equation that results in a unique graphical output of ν versus [S] data. If the type of inhibition is unknown, these significant variations in the graphical output can be used to diagnose it.

8.2.4.1. Competitive Inhibition

Inhibitor I binds to the biocatalyst's active site and directly competes with the substrate to inhibit the biocatalyst when it functions as a competitive inhibitor. The competitive inhibitor alters the kinetic scheme in the manner depicted in Scheme 8.2 when viewed through the prism of the Uni-Uni kinetic scheme, which was utilized to derive the Michaelis–Menten Eqn. (8).

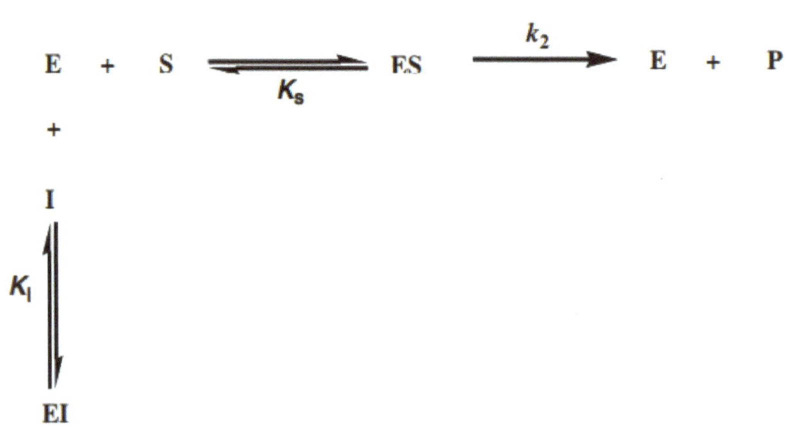

Scheme 8.2

Applying a Briggs–Haldane steady-state approach to this kinetic scheme yields the following steady-state kinetic equation:

$$\nu = \frac{k_2[\mathrm{E}]_0[\mathrm{S}]}{K_\mathrm{m}\left(1+\frac{[\mathrm{I}]}{K_\mathrm{I}}\right)+[\mathrm{S}]} \qquad (15)$$

where KI is an equilibrium dissociation constant for the biocatalyst–inhibitor complex, EI, that defines the strength of interaction between biocatalyst and inhibitor. Eqn. (15) can be rearranged into a double-reciprocal form as follows:

$$\frac{1}{\nu} = \frac{K_\mathrm{m}\left(1+\frac{[\mathrm{I}]}{K_\mathrm{I}}\right)}{V_\mathrm{max}}\frac{1}{[\mathrm{S}]}+\frac{1}{V_\mathrm{max}} \qquad (16)$$

In the event that a particular inhibitor I acts by competitive inhibition, then plots of 1/ν versus 1/[S] data obtained at different fixed concentrations of [I] should obey Eqn. (16) and take on the visual appearance shown in Figure 8.28. This graphical depiction of 1/ν versus 1/[S] data not only confirms the existence of competitive inhibition but allows for the determination of K_I as well.

8.2.4.2. Non-Competitive Inhibition

If I functions as a non-competitive inhibitor for the biocatalyst, it binds to an allosteric site rather than the active site. This site is found in both the free biocatalyst and the biocatalyst-substrate complex, and is different from the active site. The classic non-competitive inhibitor does not impact substrate binding, and the resulting ESI complex is not catalytically active.

Hence, if we repeat the analysis then the appropriate kinetic scheme becomes as in Scheme 8.3, and the modified steady-state equation becomes:

$$\nu = \frac{\left[k_2[\mathrm{E}]_0\Big/\left(1+\frac{[\mathrm{I}]}{K_\mathrm{I}}\right)\right][\mathrm{S}]}{K_\mathrm{m}+[\mathrm{S}]} \qquad (17)$$

Once again, Eqn. (17) can be rearranged into a double-reciprocal form as follows:

$$\frac{1}{\nu} = \frac{K_\mathrm{m}}{V_\mathrm{max}}\left(1+\frac{[\mathrm{I}]}{K_\mathrm{I}}\right)\frac{1}{[\mathrm{S}]}+\frac{1}{V_\mathrm{max}}\left(1+\frac{[\mathrm{I}]}{K_\mathrm{I}}\right) \qquad (18)$$

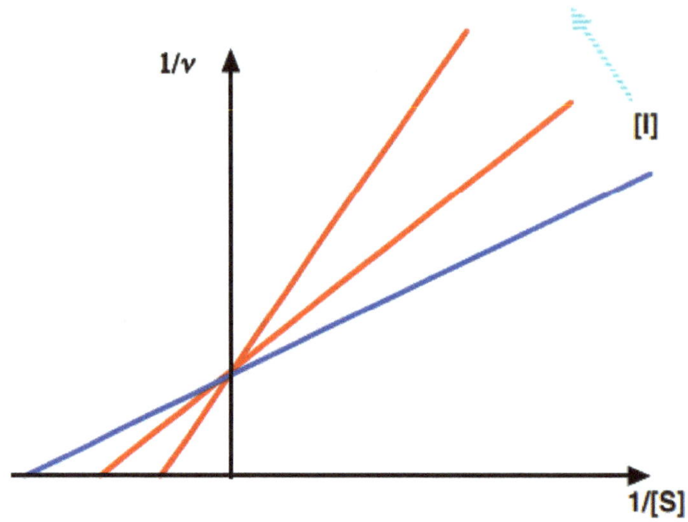

Figure 8.28. *Successive Lineweaver-burke plots typical of competitive inhibition of a bio-catalyst that operates mechanistically through the simplest Uni-Uni kinetic Scheme (Scheme 8.1).*

Source: https://nattabiochem.wordpress.com/wp-content/uploads/2009/08/essentials_chemical_biology.pdf.

In the event that a particular inhibitor I acts by non-competitive inhibition, plots of 1/v versus 1/[S] data obtained at different fixed concentrations of [I] should obey Eqn. (18) and take on the visual appearance shown in Figure 8.29. Clearly, when I acts as a non-competitive inhibitor then the only practical effect is to reduce the maximal effective rate of catalysis, V_{max}.

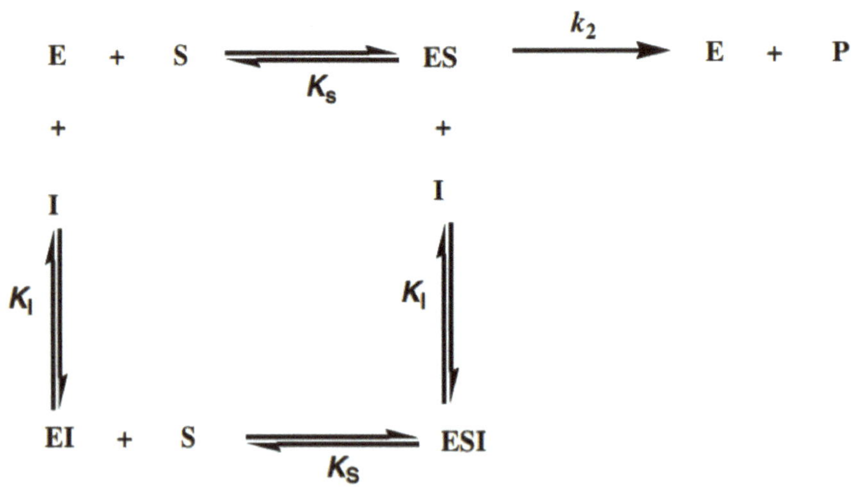

Scheme 8.3

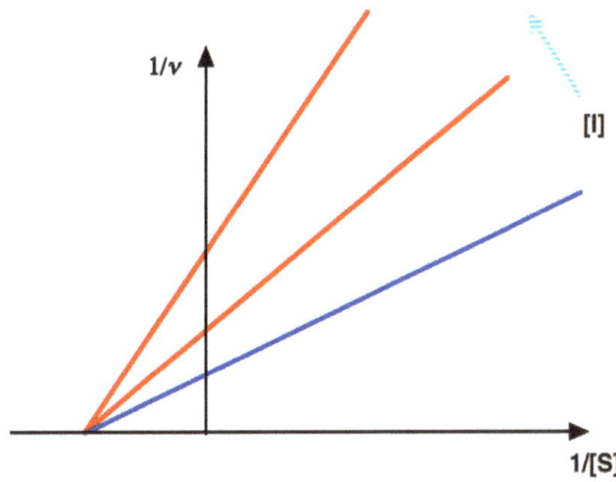

Figure 8.29. *Successive Lineweaver-Burke plots typical of non-competitive inhibition of a biocatalyst that operates mechanistically through the simplest Uni-Uni kinetic Scheme (Scheme 8.1).*

Source: https://nattabiochem.wordpress.com/wp-content/uploads/2009/08/essentials_chemical_biology.pdf.

8.2.4.3. Uncompetitive Inhibition

If I acts as an uncompetitive inhibitor of the biocatalyst, then the inhibitor acts by binding to an allosteric site revealed only in the biocatalyst–substrate complex. In this case the kinetic scheme becomes modified again (Scheme 8.4), and the modified steady-state equation becomes:

$$v = \frac{\left[k_2[\mathrm{E}]_0 \Big/ \left(1 + \frac{[\mathrm{I}]}{K_\mathrm{I}}\right)\right][\mathrm{S}]}{\left[K_\mathrm{m} \Big/ \left(1 + \frac{[\mathrm{I}]}{K_\mathrm{I}}\right)\right] + [\mathrm{S}]} \tag{19}$$

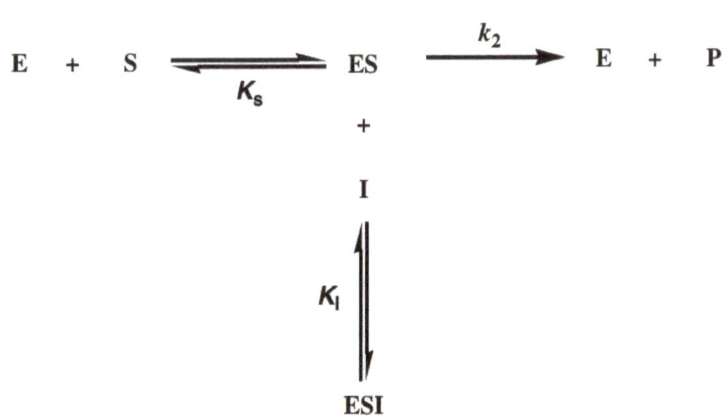

Scheme 8.4

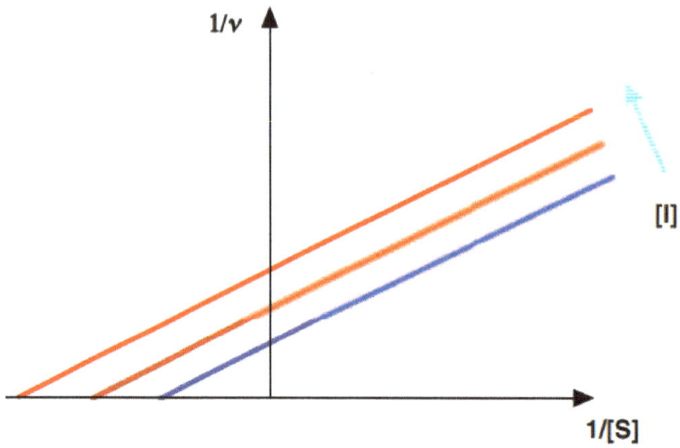

Figure 8.30. *Successive Lineweaver-Burke plots typical of un-competitive inhibition of a bio-catalyst that operates mechanistically through the simplest Uni-Uni kinetic Scheme (Scheme 8.1).*

Source: https://nattabiochem.wordpress.com/wp-content/uploads/2009/08/essentials_chemical_biology.pdf.

which rearranges to double-reciprocal form:

$$\frac{1}{\nu} = \frac{K_m}{V_{max}} \frac{1}{[S]} + \frac{1}{V_{max}} \left(1 + \frac{[I]}{K_I}\right) \qquad (20)$$

When a particular inhibitor I acts by uncompetitive inhibition, then plots of 1/ν versus 1/[S] data obtained at different fixed concentrations of [I] should obey Eqn. (20) and take on the visual appearance shown in Figure 8.30. In this case, when I acts as a non-competitive inhibitor, there are practical reductions in both the maximal effective rate of catalysis V_{max} and K_m in equal proportion.

8.2.5. Applicability of Michaelis–Menten Equation

The irreversible Uni-Uni kinetic scheme (Scheme 8.1) and the Michaelis-Menten Eqn. (8) are only truly applicable to irreversible biocatalytic processes in which a single substrate interacts with a single biocatalyst consisting of a single catalytic site. Therefore, the Uni-Uni kinetic scheme is only really directly applicable to the steady-state kinetic analysis of TIM biocatalysis with reference to the biocatalyst examples given, Eqn. (8). Moreover, even this assertion is only true provided that all biocatalytic initial rate values are calculated without any product. Similarly, the steady-state kinetic analysis for the inhibition of TIM is really the only context in which the Uni-Uni kinetic schemes for competitive, uncompetitive, and non-competitive inhibition are really directly applicable. The main reason for the widespread application of Eqn. (8) and the irreversible Uni-Uni kinetic scheme for the steady-state analysis of various biocatalytic processes is their ease of use, as it allows for the easy interpretation of measured k_{cat} and KM parameters. All that needs to be done is modify the catalysis conditions so that steady-state kinetic measurements of initial rates can be performed under pseudo-irreversible Uni-Uni kinetic conditions, in which there is no product and all substrates except one are significantly excessed. In addition, if a biocatalyst has multiple catalytic sites, steady-state analyzes ought to be carried out in a situation where only one of the sites is active at any given time, or alternatively, in a situation where the catalytic sites can function entirely independently of one another. In the latter instance, a value per catalytic site—also referred to as the turnover number per catalytic site—should be obtained by dividing the measured values of k_{cat} by the total number of catalytic sites.

Since water is clearly an abundant substrate, steady-state kinetic analyzes of RNAse A, lysozyme, and CA can therefore be easily accommodated within Eqn. (8) and the irreversible Uni-Uni kinetic scheme to a reasonable approximation. The same could be said for steady-state kinetic analyzes of acetylcholine hydrolysis catalyzed by AChE and polypeptide hydrolysis catalyzed by bovine pancreatic α-chymotrypsincat. Eqn. (8) and the Uni-Uni kinetic scheme may accommodate reactions that are catalyzed by PLP-dependent AlaR, GadB, and aspAT, or that are assisted by CAT and LysU, provided that the proper steady-state kinetic conditions are met. Although this method may seem straightforward, more thorough and meticulous steady-state kinetic analyzes using more intricate kinetic schemes are necessary in order to demonstrate and/or validate the mechanisms involved in biocatalysis.

CASE STUDY

WILLS CATALYSTS: COMMERCIALIZED SYSTEMS FOR ENANTIOSELECTIVE PRODUCTION OF PHARMACEUTICAL INTERMEDIATES

- Summary

A process for the commercial production of a family of Warwick-invented organometallic catalysts has been developed and patented by Johnson Matthey (JM). The catalysts — which have been sold internationally to several fine chemical and pharmaceutical companies in kilogram quantities, capable of producing tons of product — are in widespread industrial use for synthesis and scale-up.

New Ru(II) tethered catalysts have been developed and their mechanisms and wider applications have been reported. Catalytic hydrogenation is a pivotal chemical transformation which underpins the synthesis of numerous high-value target molecules, materials and intermediates. Wills tethered catalysts — invented at Warwick and commercialized by collaborators and competitor businesses — have been shown to have significant commercial advantages over the existing systems.

The Wills catalyst was tested by Johnson Matthey Catalysis and Chiral Technologies (JM CCT) for a scalable synthetic route to the catalysts, which was later demonstrated on multi-100s gram scale and kg-scale at Alfa Aesar; a Johnson Matthey company. A patent was filed in 2009.

- Details of the Impact

Since 2009 JM CCT — through Alfa Aesar, SigmaAldrich and Strem — has sold four Wills tethered catalysts in the family, either individually or as a part of a `kit' (catalog prices £368-£560 per gram). JM CCT also uses the material in collaborative projects with client companies worldwide.

AstraZeneca (Sweden & UK), one of the largest pharmaceutical companies in the world, describes in a process patent the use of (*S*,*S*)-teth-TsDpen-RuCl (a JM CCT trade name for a Wills catalyst) in the catalytic asymmetric synthesis of a class of anti-asthmatic bronchodilator drugs.

Synthon BV (Netherlands) report the use of a Wills catalyst in the key chirality-inducing reduction step in their synthesis of the anti-asthmatic drug montelukast (singulair), a drug which in 2010 was the fourth most prescribed drug in the US (24.7 M). They conclude, "It has now been discovered that the use of [Wills catalyst] can provide a more suitable process for the asymmetric transfer hydrogenation reaction."

Archimica GMBH (Germany) patented the use of a Wills tethered catalyst in the key chirality-generating step of their synthesis of the third generation antiepileptic drug Eslicarbazepine.

Boehringer Ingelheim (Germany), one of the largest pharmaceutical companies in the world, describe their route to certain chemokine inhibitors via the reduction of an early-stage intermediate using a Wills tethered catalyst.

Lek Pharmaceuticals (Slovenia), part of the Sandoz group, has reported an improved process for the preparation of intermediates on route to non-steroidal selective estrogen receptor modulators such as lasofoxifene using the Wills catalyst.

In collaboration with Eli Lilly (USA), one of the largest pharmaceutical companies in the world, JM CCT have developed an achiral version of the Wills tethered catalysts and have applied this to racemic reductions.

Additional examples from a rapidly growing list further demonstrate the international commercial reach of the Wills catalysts, in e.g., China (Hunan Fangsheng Pharma, patent CN102978253A), USA (Ambit Biosciences, patent US2012053193A1) and Italy (Zach Systems/Zambon Chemicals, patent WO2012120086).

JM CCT recognized the commercial need for the use of dihydrogen as the primary reducing agent and will develop this area with Wills through a TSB-funded collaborative grant *Development of the future generation of catalysts for asymmetric reduction* (TSB ref: 101330, Mar 2013, £216k plus £108K from JM CCT), extending the range of catalysts and their applications. In a further development, a new synthetic route to the Wills catalysts and novel variants is the subject of a patent application filed by Warwick (UK patent application 1219716.6, 02 Nov 2012; PCT/GB2013/05286901, 01 Nov 2013); licensing negotiations are ongoing.

CLASS ACTIVITY

Students write a brief reflection on what they learned and how the principles of kinetics and catalysis are relevant to their field of study or interests.

SUMMARY

- Biocatalysts are overwhelmingly proteins (enzymes) and sometimes RNA nucleic acids (ribozymes). They catalyze an amazing diversity of reactions for myriads of different important reasons.
- Enzymes are not only involved in catabolism (i.e., breaking down), but also play a role in anabolism (i.e., building up) of monomeric building blocks required for biological macromolecular assembly.
- The active or catalytic site of any biocatalyst represents the region of structure devoted to biocatalysis in any given biocatalyst.
- The study of reaction rates and catalysis is known as kinetics. Typically, the catalytic effect of any one biocatalyst is evaluated under steady-state and/or pre-steady-state conditions.
- Steady-state kinetic analyzes can be as diverse as the range of biocatalysts themselves. Rate enhancements brought about by any one biocatalyst are usually assessed by means of a bioassay.
- Photometric steady-state bioassays make use of some form of chromogenic substrate or reagent that becomes chemically altered into a chromophore during biocatalysis and as a result absorbs significantly at a particular wavelength of absorption.
- HPLC steady-state bioassays make use of a change in physical elution time when substrate is converted to product.
- The simplest possible kinetic scheme is the Uni-Uni kinetic scheme, where Uni indicates one substrate and the second Uni indicates the evolution of only a single product.

REVIEW QUESTIONS

1. Explain the simple principles in biocatalysis.
2. What is steady-state kinetics in biocatalysis?
3. Write short note on steady-state bioassays.
4. What are the effects of steady-state inhibitors?
5. Illustrate the Michaelis–Menten equation.

REFERENCES

1. Beard, D. A., & Qian, H. (2008). *Chemical Biophysics: Quantitative Analysis of Cellular Systems*. Cambridge Texts in Biomedical Engineering. Cambridge University Press.
2. Gillespie, D. T. (1976). A general method for numerically simulating the stochastic time evolution of coupled chemical reactions. *Journal of Computational Physics, 22*(4), 403–434.

3. Gorban, A. N., & Yablonsky, G. S. (2015). Three waves of chemical kinetics. *Mathematical Modeling of Natural Phenomena, 10*(5), 1–5. doi:10.1051/mmnp/201510501.

4. Karsenti, E. (2008). Self-organization in cell biology: A brief history. *Nature Reviews Molecular Cell Biology, 9*, 255–262. doi:10.1038/nrm2357.

5. Ptáček, P., Opravil, T., & Šoukal, F. (2018). A brief introduction to the history of chemical kinetics. In P. Ptáček & T. Opravil (Eds.), *Introducing the Effective Mass of Activated Complex and the Discussion on the Wave Function of this Instanton* (pp. 129–140). London, UK: IntechOpen. doi:10.5772/intechopen.78704.

6. Resaco, D. C., Gao, F., Morgan, F., Novak, I. L., Schafft, J. J., & Stepchenko, B. M. (2012). Virtual cell comp tools for modelling in cell biology. *Wiley Interdisciplinary Reviews: Systems Biology and Medicine*, 129–140. doi:10.1002/wsbm.165.

CHAPTER 9

Mass Spectrometry and Proteomics

LEARNING OBJECTIVES

After studying this chapter, you will be able to:

- Describe the mass spectrometry in chemical biology
- Learn about key principles in mass spectrometry
- Understand about proteomics
- Describe the challenge of proteomics

KEY TERMS FROM THIS CHAPTER

Electron ionization	Electrospray
Electrospray ionization	Fast atom bombardment
Functional genomics	Hormones
Human genome project	Ion implanters
Ion source	Ionization
Ionization source	Mass analyzer
Mass spectrometry	Particle accelerators
Polypeptides	Protein purification

INTRODUCTION

In many related scientific fields, including chemistry, biochemistry, pharmacy, and medicine, mass spectrometry is an essential analytical tool. Mass spectrometry is used in combinatorial library analysis, biomolecule sequencing, and single-cell or spacecraft exploration. Mass spectrometry plays a major role in a number of processes, including the structure-elucidation of unknown substances, environmental and forensic analytes, and quality control of drugs, foods, and polymers (Király et al., 2016).

One analytical technique that can be used to determine the mass-to-charge ratio (m/z) of one or more molecules in a sample is mass spectrometry. It is also frequently possible to ascertain the exact molecular weight of the sample's constituent parts using such measurements. A technique for determining a compound's molecular mass, mass spectrometry indirectly contributed to the identification of isotopes.

Large-scale protein research is known as proteomics (Anderson et al., 1998; Blackstock et al., 1999). With a variety of roles, including the synthesis and replication of DNA, the enzymatic digestion of food, and the development of the structural fibers of muscle tissue, proteins are essential components of all living things. Moreover, there are other protein types such as hormones that carry vital signals throughout the body and antibodies that shield an organism against infection.

The whole collection of proteins that a system or organism produces or modifies is called its proteome. An increasing amount of proteins can be identified thanks to proteomics. This changes in response to different demands, or stresses, that an organism or cell encounters over time.

The interdisciplinary field of proteomics has benefited immensely from the genetic data of several genome projects, such as the Human Genome Project. It is a crucial part of functional genomics and covers the investigation of proteomes from the general level of protein composition, structure, and activity.

Mass spectrometry and protein purification are frequently specifically mentioned when the term "proteomics" is used to describe the large-scale experimental analysis of proteins and proteomes. The most effective technique for analyzing proteomes in both large samples made up of millions of cells and individual cells is mass spectrometry.

9.1. MASS SPECTROMETRY IN CHEMICAL BIOLOGY

Originating in the physical sciences, mass spectrometry has quickly emerged as one of the most potent methods for examining biological systems. The development of soft ionization techniques has enabled mass spectrometry to advance, allowing for in-depth analyzes of the primary structures of biological macromolecules and amphiphilic lipids. Since mass spectrometry can provide molecular mass and sequence information from minuscule amounts of material (femtomole–picomole, even attomole levels), it can be considered the gold standard method for identifying the primary structures of polypeptides, nucleic acids, and complex carbohydrates.

Recombinant technologies are not even necessary to obtain such information. Therefore, even though mass spectrometry cannot directly provide three-dimensional information, it can still yield an impressive amount of primary structural information that can be combined with other data to try and offer helpful insights into the structure of biological macromoleculars.

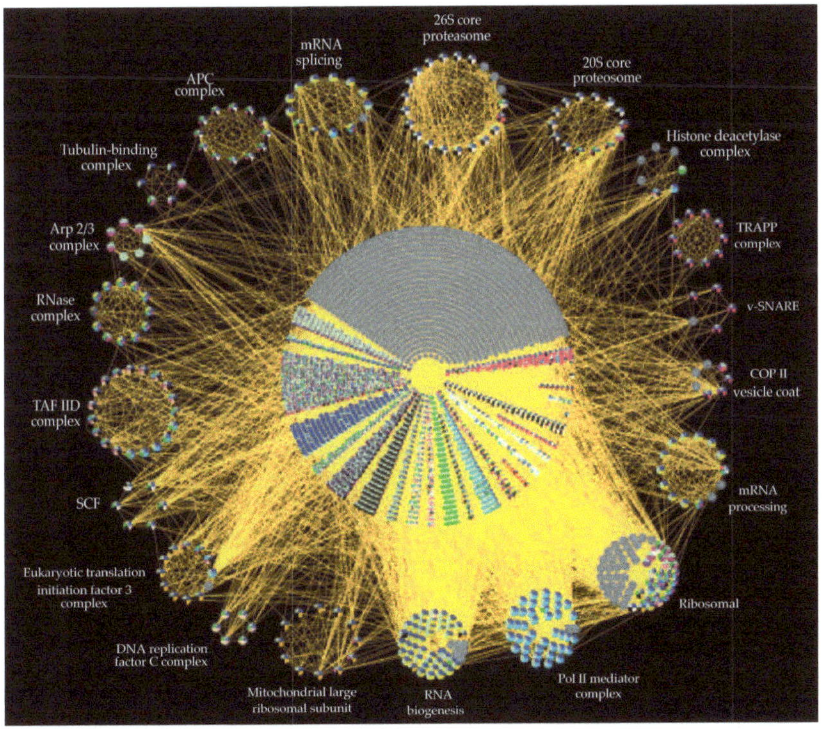

Figure 9.1. *Visualization of combined datasets of protein-protein interaction in yeast.*

Source: Andrew Miller and Julian Tanner (2008). Essentials of Chemical Biology. John Wiley & Sons, Ltd. ISBN: 978–0-470–84530–1.

But these days, proteomics—the study of all the interactions and presumed functions of all the proteins expressed from a genome—is also recognized to benefit greatly from mass spectrometry. The development of proteomics has been largely driven by chemical biology and multidisciplinary thinking, from the creation of appropriate chromatographic systems to map out protein networks to the development of suitable mass spectrometric methods for large-scale protein identification (Figure 9.1). Put another way, mass spectrometry allows us to define intricate, functional, three-dimensional networks of interacting proteins and polypeptides within cells in addition to providing us with primary structures. Simply put, mass spectrometry gives us the ability to conduct in-depth analyzes of the structures and functions of specific biological macromolecules of interest. It also enables us to investigate protein-protein interactions within multi-protein complexes and, eventually, networks of these interactions, which dictate not only molecular function but also cellular function.

9.2. KEY PRINCIPLES IN MASS SPECTROMETRY

In mass spectrometry, individual molecules are ionized in the gas or vapor phase, differentiated, and then detected based on the mass/charge (m/z) ratio. Usually, z is equals one, meaning that the ratio corresponds to mass. Because of this, the main application of mass spectrometry was to accurately ascertain molecular weight based on the identification of the molecular ion ($[M]^+$). Fortunately, spontaneous decomposition reactions of ions in gas phase can produce fragment ions ($[M - x]^+$) that can be utilized to determine structural elements. An ionization source, a mass analyzer, and a detector are the three fundamental parts of a mass spectrometer in terms of instruments (Figure 9.2). In order to control the molecules' flight, the ionization source must first send them into the gas phase as ionic species. In order for the ions to avoid colliding or interacting with other species, the flight occurs in a high vacuum. The ions are separated for detection based on their m/z ratio using a mass analyzer. The detector records the ions' abundance (a function of signal) as well as their m/z (a function of physical property and/or time).

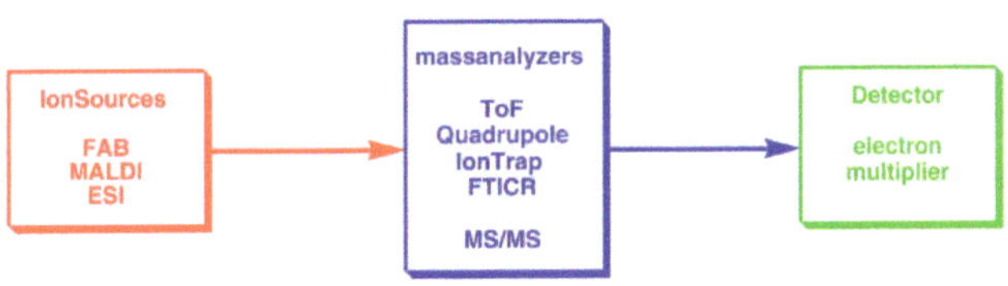

Figure 9.2. *The basic components of a mass spectrometer.*

Source: Andrew Miller and Julian Tanner (2008). Essentials of Chemical Biology. John Wiley & Sons, Ltd. ISBN: 978–0-470–84530–1.

In the long history of chemistry, mass spectrometry has been used to analyze the molecular weights of small molecules using traditional ionization techniques like chemical ionization and electron bombardment. However, neither technique was really useful for mass analyzing biological macromolecules or lipid amphiphiles. However, the mass analysis of oligo- and polypeptides entered the field of mass spectrometry with the advent of fast atom bombardment (FAB) mass spectrometry in the early 1980s. The development of electrospray ionization (ESI) and matrix-assisted laser desorption ionization (MALDI) mass spectrometry thereafter opened up a wide range of possibilities for the mass analysis of almost all biological macromolecules and lipid amphiphiles. Because of this, the reader interested in chemical biology will naturally focus on these three primary methods of soft ionization and how they are combined with suitable mass analyzers and detector units to create mass spectrometers suitable for characterizing biological macromolecules and amphiphilic lipids.

9.2.1. Ionization Sources

A device that produces atomic and molecular ions is called an ion source (Bernhard Wolf 1995). Mass spectrometers, optical emission spectrometers, particle accelerators, ion implanters, and ion engines all require ions, which are created by ion sources. In mass spectrometry, electron ionization is frequently used, especially for organic molecules.

9.2.1.1. Traditional Techniques of Ionization

The most popular ionization method for mass spectrometry in synthetic chemistry is electron ionization (EI). Electrochemical Ionization (EI) is a two-step process that involves first vaporizing the sample of a target molecule and then bombarding it with an electron stream. An electron will be displaced in the event that the colliding electron's energy surpasses the molecule's ionization energy, producing a positively charged molecule known as an ion $[M]^+$ (radical cation). Since an electron has a very small mass, the mass of the molecules and the observed mass of $[M]^+$ must be correlated. Overuse of collision energy promotes spontaneous fragmentation and fragment ion $[M - x]^+$ formation, which provides information that, at the very least, partially reveals the structure of the target molecule (analyte). Although it still has some major limitations, EI is still a useful technique for the analysis of small molecules (<1000 Da). Firstly, fragmentation frequently occurs to such an extent that the molecular ion is invisible. Second, this ionization technique is too high energy (i.e., hard) for mass analysis of biological macromolecules or lipid amphiphiles; it cannot analyze thermally labile or non-volatile compounds.

In situations where the ionization process was not as difficult as it was with EI, chemical ionization (CI) was created as a technique to supplement it. Using this method, an electron beam is used to ionize a reagent gas at a partial pressure that is much higher than that of a sample analyte molecule of interest in the vapor phase (Figure 9.3). After that, the ionized reagent gas combines with the target analyte molecule to form stable molecule–ion adducts, which can be measured during analysis and the adduct mass taken into consideration. For this purpose, a variety of reagent gases can be used, including hydrogen, methane, water, methanol, ethanol, or ammonia. The choice of reagent gas can somewhat control the degree of fragmentation. While EI can offer more detailed structural information, CI has the advantage of controlled fragmentation. However, this ionization technique is still hard and has drawbacks similar to EI for non-volatile compounds, so it is only appropriate for the analysis of compounds with a molecular weight of less than 1,000 Da.

9.2.1.2. Desorption Ionization Techniques – FAB and MALDI

The ionization process used in nearly all desorption ionization techniques is the same: the target analyte molecule is mixed with or dissolved in a matrix prior to being exposed to high energy particles or light, producing a high concentration of solvated gas phase neutral or ionic species slightly above the point of impact. Vaporization happens before thermal breakdown following the beam impact, allowing ions to "escape" the impact region before fragmentation starts. Field desorption (FD) and plasma desorption were two of the initial desorption ionization methods. An extraordinarily powerful electric field was used in FD to directly desorbed ions into the vapor phase. This process lowers the barrier to electron removal from the molecule through quantum mechanical tunneling, leaving behind a radical cation molecular ion $[M]^{+}$. Despite being successful in mass analyzing non-volatile compounds, FD spectrometry was not widely used because of the challenges associated with creating repeatable emitter electrodes that allowed for ionization. It was clear that softer, more adaptable ionization methods were required.

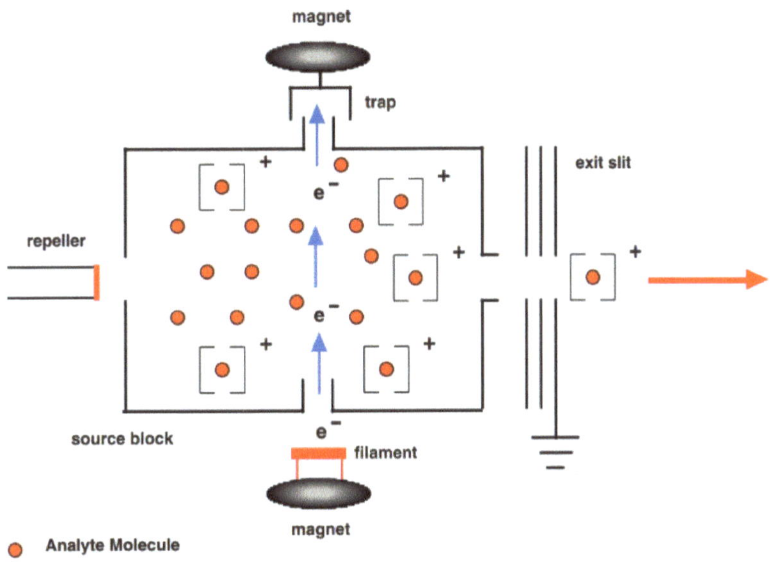

Figure 9.3. *Schematic of chemical ionization. A reagent gas is present in the source block and is initially ionized by a beam of electrons, which then interacts with sample analyte molecules also in the vapor phase to form stable molecular ion adducts that are repelled into the mass analyzer. The mass of the reagent gas adduct must be corrected during the analysis.*

Source: Andrew Miller and Julian Tanner (2008). Essentials of Chemical Biology. John Wiley & Sons, Ltd. ISBN: 978–0-470–84530–1.

The FAB technique, which is also a desorption technique, was the first of these adaptable soft ionization techniques to be developed. FAB strikes the target analyte molecule dissolved in a liquid matrix (usually glycerol) using a rapid stream of argon or xenon atoms (Figure 9.4). By bombarding inert gas atoms with electrons, which accelerate the ions through a potential difference, a fast stream of inert gas atoms is produced. After that, a cloud of extra neutral gas atoms transforms the rapidly moving ions into atoms by electron transfer, neutralizing the ions (residual ions are deflected by a potential field before they are able to strike the matrix). Many ions, including the molecular ions $[M + H]^{+}$ or $[M - H]^{-}$, are sputtered out of the sample as a result of the fast moving large atoms colliding with the matrix. The protonated molecular ions and cationic fragment ions are collected and mass analyzed if the FAB is operated in the positive ion mode. Only the deprotonated molecular ions and anionic fragment ions will be subjected to mass analysis if the negative ion mode is employed. Na^{+} and K^{+} molecular ion adducts (i.e., $[M + Na]^{+}$ or $[M + K]^{+}$) are also commonly observed in the positive ion mode.

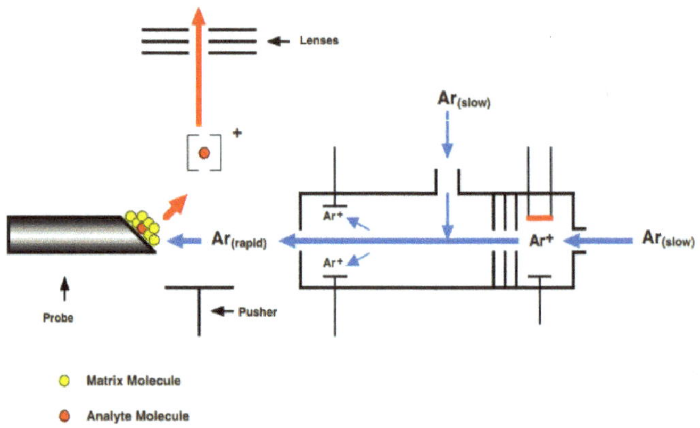

Figure 9.4. *Schematic of Fast Atom Bombardment. A stream of inert atoms (argon or xenon) strike a probe target on which analyte molecules of interest are dissolved in a liquid matrix (frequently glycerol). The fast stream of inert atoms is created by electron bombardment of inert atoms to create ions that are accelerated by a potential difference. Kinetic energy is transferred from ions to more inert atoms leading to the stream of fast atoms that act to sputter molecular and fragment ion species into the vapor phase for mass analysis.*

Source: Andrew Miller and Julian Tanner (2008). Essentials of Chemical Biology. John Wiley & Sons, Ltd. ISBN: 978–0-470–84530–1.

Regardless of mode, fragmentation is also seen, and it is typically well-balanced in relation to the intensity of the molecular ions. For this reason, FAB mass spectrometry can also be utilized to learn more about the structure of the target analyte molecule. The fact that minutes can pass between the acquisition of FAB mass spectra and the maintenance and even increase of molecular ion and fragment ion intensities is one of the primary causes of this.

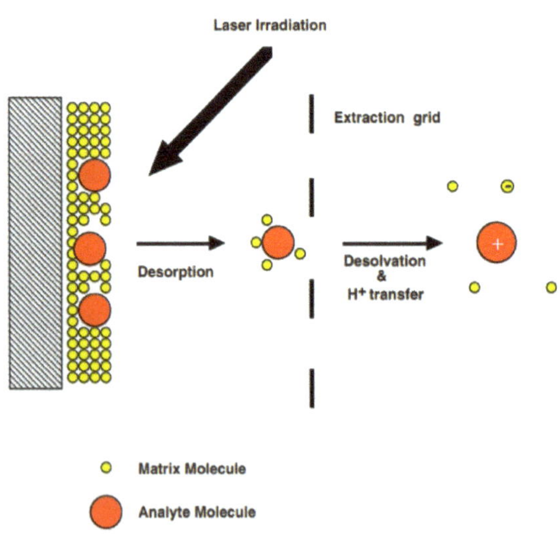

Figure 9.5. *Schematic of Matrix Assisted Laser Desorption Ionization (MALDI). The laser is fired at the sample analyte of interest mixed with a crystalline matrix that readily absorbs laser energy and allows the sputtering of analyte into the vapor phase in association with matrix molecules of*

solvation. Desolvation and proton transfer leads to naked molecular ion species, ready for mass analysis.

Source: Andrew Miller and Julian Tanner (2008). Essentials of Chemical Biology. John Wiley & Sons, Ltd. ISBN: 978–0-470–84530–1.

The matrix's nature is the cause of this. In FAB mass spectrometry, the matrix is crucial because it absorbs the excess energy from the atomic bombardment and restocks the matrix surface with relevant analyte molecules as the collision process proceeds. FAB mass spectral matrices are most often used for viscous, non-volatile liquids like glycerol, but other matrices can also be helpful in certain situations. For the mass analysis of smaller biological macromolecules (or fragments thereof) of less than 4000 Da, such as oligodeoxy- or oligonucleotides, short polysaccharides, and short polypeptides, FAB mass spectrometry has proven to be very helpful.

An important technical advancement on FAB is represented by MALDI. In MALDI, laser radiation is used in place of the fast atom beam, and the target analyte molecule is distributed throughout a crystalline matrix as opposed to a liquid one. Crystals are placed on a solid surface after the analyte has been pre-mixed with a matrix material (Figure 9.5). Pulses of nanosecond lasers sublime both the matrix and the target analyte molecule. The majority of the laser energy is then absorbed by the matrix, which keeps the analyte from breaking down too much during sublimation and ionization, guaranteeing a sizable molecular ion population in the vapor phase.

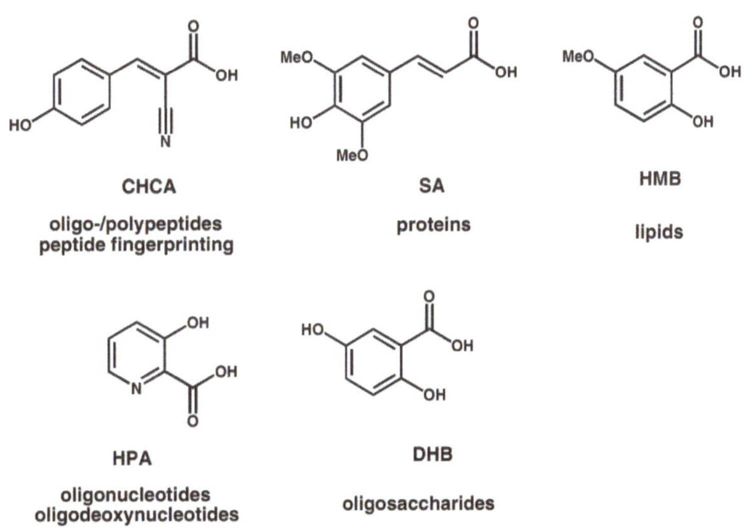

Figure 9.6. *Summary of main MALDI matrices matched with the analytes most widely mass analyzed with the assistance of each of these different matrix materials.*

Source: Andrew Miller and Julian Tanner (2008). Essentials of Chemical Biology. John Wiley & Sons, Ltd. ISBN: 978–0-470–84530–1.

In MALDI, the crystalline matrix fulfills two additional crucial supporting roles. First, the analyte is actually solvated by the matrix, which lessens the analyte molecule's tendency to aggregate and encourages the formation of molecular ions in the vapor phase. Second, certain applications have identified matrices with particular characteristics (Figure 9.6). α-cyano-4-hydroxycinnamic acid (CHCA) is the most widely used matrix for peptide fingerprinting and mass analysis of oligo-/polypeptides. For protein mass analysis, hydroxypicolinic acid (HPA), for oligonucleotides and oligodeoxynucleotides, di-2,5-hydroxybenzoic acid (DHB), for oligosaccharides, and 5-hydroxy-2-methoxybenzoic acid (HMB) for lipids are frequently utilized as the preferred matrix. When using the proper version of MALDI mass spectrometry, MALDI is a soft enough and versatile enough ionization technique to allow the observation of the molecular ions of even very large biological macromolecules, approaching 10^6 Da.

9.2.1.3. Spray Ionization Techniques – Thermospray and Electrospray

Many of the issues surrounding the passage of liquid solutions into the gas phase are resolved by spray ionization techniques. These methods are the easiest to integrate with a liquid chromatography system because they all use liquid phase samples. These methods don't operate in a vacuum; instead, they operate at room pressure.

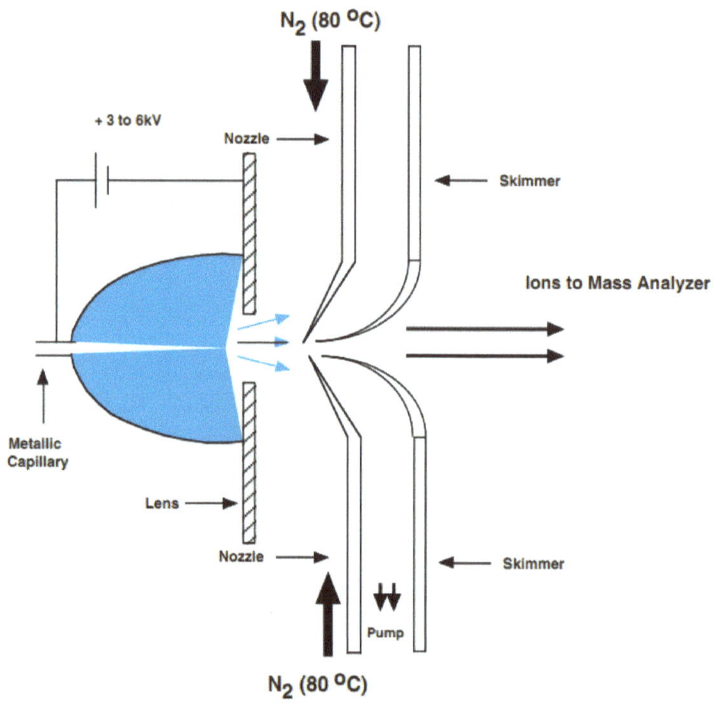

Figure 9.7. *Schematic of Electrospray ionization. A "spray" of droplets is formed that evaporate until the destabilizing electrostatic forces cause the droplets to "explode" releasing multi-charged molecular ion species for mass analysis.*

Source: Andrew Miller and Julian Tanner (2008). Essentials of Chemical Biology. John Wiley & Sons, Ltd. ISBN: 978-0-470-84530-1.

Thermospray ionization (TI) was among the earliest spray ionization methods to be documented. Although this method is illustrative, it is not useful for biological macromolecules of amphiphilic lipid mass analysis because it is best suited to volatile compounds of less than 1000 Da. In TI, a solution containing the target analyte molecule is passed through a heated capillary; as a result of the liquid's quick expansion and cooling, a superheated mist containing a mixture of small droplets and analyte molecules in the vapor phase is ejected from the capillary's end. The tendency of sprayed droplets to become charged facilitates the development of molecular ions for subsequent mass analysis. For the analysis of small organic molecules such as oligosaccharides, dinucleotides, dideoxynucleotides, and oligopeptides, thermospray is a helpful ionization technique. The large molecule upgrade of TI is called ESI. The atmospheric pressure chemical ionization (APCI) method was developed into ESI. When using ESI, a solution containing the target analyte molecule is run through a capillary that has a potential difference of 3–16 kV applied to it. This creates a strong electrostatic field around the capillary tip, which helps to generate molecular ions (Figure 9.7). The electrostatic field causes a "spray" of droplets to form as the solution exits the capillary's end. The stability of the droplet is then determined by two forces. Initially, droplet integrity is preserved by droplet surface tension. Second, the droplet disintegrates due to electrostatic forces in the analyte. Destructive electrostatic forces therefore rise as droplets evaporate and eventually cause the droplets to "explode," releasing fragments of ions and molecules for mass analysis. Analyte ions may be multiply charged in the ESI technique, which is unusual. Biological macromolecules, like proteins, for example, often generate molecular ion series with the structure $[M + zH]^{z+}$ or $[M - zH]^{z-}$. The cationic molecular ion series can be observed in the positive ion mode and the anionic molecular ion series in the negative ion mode because, similar to FAB, ESI allows for the observation of both positive and negative ion modes. A primary benefit of multiple ion series is that true molecular ion weights can be orders of magnitude larger even though actual m/z values are within the mass analysis's range. Because of this, mass analysis of very large biological macromolecules—molecular weights of at least 10^6 Da—is now possible thanks to ESI.\

9.2.2. Mass Analyzers in Mass Spectrometry

The front end of the contemporary mass spectrometer used by chemical biology researchers consists of soft ionization devices for FAB, MALDI, and ESI. It is obvious that these instruments must be connected to mass analyzers in order to resolve and subsequently detect fragment and molecular ions. In order to disturb ion motion in vacuo through arc trajectories that were a function of ion mass, velocity, and net charge, the conventional EI mass spectrometer used various combinations of electric and magnetic sectors. Because these mass analyzers were bulky and cumbersome, mass analysis equipment was extremely specialized. These days, much less complicated mass analyzers that significantly reduce instrument specialization in favor of more complex experiments are possible thanks to the availability of soft ionization devices. The reader studying chemical biology should be most familiar with the four main categories of mass analyzers. These analyzers include the Fourier transform ion cyclotron resonance (FTICR), quadrupole, ion trap, and time of flight (TOF). In some ideal configurations, these mass analyzer devices are connected either separately or simultaneously to a soft ionization device.

9.3. PROTEOMICS

A novel kind of "omics" called proteomics has emerged quickly, particularly in the field of therapeutics. Marc Wilkins coined the term "proteome" in 1995 (Agrawal et al. 2013). The study of protein interactions, functions, compositions, and structures as well as how they affect cellular processes is known as proteomics. Compared to genomics, proteomics offers a deeper understanding of the structure and function of the organism. However, because protein expression varies with time and environmental factors, it is far more complex than genomics (Holman et al. 2013). Human proteins are thought to number close to 1 million, and many of them have modifications like post-translational modifications (PTMs). Nonetheless, the human genome is thought to code for between 26000 and 31000 different proteins (Chandramouli & Qian, 2009). Proteomics encompasses a range of methods, such as two- and one-dimensional gel electrophoresis (2-DE), stable isotope labeling with amino acids in cell culture, multidimensional protein identification technology, isobaric tagging for relative and absolute quantitation, and isotope-coded affinity tags. Additionally, there are gel-free high-throughput screening technologies available. Tissues, organelles, and cells can be treated with shotgun proteomics, 2D difference gel electrophoresis (2D-DIGE), and protein microarrays. High-throughput processing techniques include label-free quantification of high mass resolution liquid chromatography (LC)-tandem mass spectrometry (MS), large-scale western blot assays, and multiple reaction monitoring assays. Proteomics has been divided into two categories in the past 10 years: protein interaction mapping and protein expression mapping. The former technique quantifies the expression of the proteome in cells, bodily fluids, or tissues by combining 2-DE and MS. Understanding the PTMs of expressed proteins in various environmental or disease states can be achieved through protein expression mapping. Protein-protein interaction mapping identifies the interaction partners for the encoded proteins in each cell and at the proteome-wide level by using the yeast two-hybrid system in conjunction with mass spectrometry.

Proteomics is a multi-step process that requires strict control at each stage to prevent extra-biological influences from affecting the expression and interaction of proteins. The most crucial stage is sample preparation, which solubilizes all of the sample's proteins and gets rid of any inhibitory substances that could interfere, like lipids. For results to be dependable, accurate, and repeatable, the sample must be properly prepared. The most popular technique for isolating and separating proteins is PAGE. Proteins are separated via 1-DE, 2-DE, and high-performance LC (HPLC). Using 1-DE, proteins are separated according to their molecular mass. Since proteins are soluble in sodium dodecyl sulfate (SDS), protein solubility is rarely a problem.

Moreover, 1-DE can resolve proteins with molecular masses ranging from 10 kDa to 300 kDa and is simple to use and repeatable. 1-DE gel is most frequently used to characterize proteins after they have been purified because of its low resolving power. 2-DE, however, can be applied to more complicated protein mixtures, like a basic cell lysate. Proteins in 2-DE are identified by their molecular mass and net charge.

Proteomics is able to examine a protein's expression at various scales, which permits evaluation of particular quantitative and qualitative cellular reactions associated with that protein. Measurements of qualitative and quantitative proteomes are made at the transcriptomic, genomic, and post-transcriptional levels. Qualitative proteomics uses 2-DE, 2D-LC, and microarrays, depending on the circumstances, to track changes in protein expression and mixture composition. Furthermore, it can furnish details regarding the molecular mechanisms underlying illnesses and draw comparisons between two groups, for example, patients and healthy controls. Moreover, quantitative proteomics can offer profound understanding of cellular processes, disease mechanisms, and biomarker discovery. Quantitative proteomics uses a number of novel techniques, such as metabolic stable-isotope labeling or post-extraction labeling, either alone or in conjunction with affinity labeling. By classifying cations based on their mass-to-charge ratio, MS distinguishes between various compounds.

9.3.1. Types of Proteomics

Expression proteomics, functional proteomics, and structural proteomics are the three primary subtypes of proteomics.

9.3.1.1. Expression Proteomics

A novel method for studying the quantitative and qualitative expression of proteins is called expression proteomics. Its goal is to identify the changes in protein expression that occur between two states, like patients and controls. It can also recognize novel proteins in signal transduction and proteins specific to diseases. Experiments using expression proteomics are typically used to examine the patterns of protein expression in various cell types. To find variations in the amounts of proteins, for instance, a sample of tumor tissue is compared to a sample of normal tissue. Using 2-DE and MS techniques, differences in protein expression that exist or are absent in tumor tissue relative to normal tissue are found.

9.3.1.2. Structural Proteomics

Structural proteomics uses X-ray crystallography and nuclear magnetic resonance spectroscopy to ascertain the three-dimensional structure and structural complexity of functional proteins. It details every interaction between proteins in the mixture, including those with membranes, cell organelles, and ribosomes. One application of structural proteomics is the investigation of the nuclear pore complex.

9.3.1.3. Functional Proteomics

This kind of proteomics ascertains the interactions between protein partners by examining the molecular mechanisms and protein functions within the cell. It specifically looks into how a protein that isn't known interacts with other members of a particular protein complex that's involved in a given process. This could point to the protein's biological function. Comprehensive descriptions of cellular signaling pathways can also result from the determination of protein-protein interactions *in vivo*.

9.3.2. Proteomics Workflow

Proteomics allows for the application of both top-down and bottom-up workflows. Another name for the bottom-up approach is peptide-based proteomics. Here, trypsin breaks down the protein and separates it using a particular column. Then, MS. analyzes the peptides. Based on the fractionation step, the bottom-up approach can be divided into two categories. In the first method, the proteins are separated from the gel using 2-DE. After that, the proteins are broken down into peptides that MS can recognize. The term "shotgun" proteomics refers to the second strategy. In top-down proteomics, whole proteins or polypeptides are immediately assessed by MS. The molecular mass of proteins is sometimes calculated by using electrospray ionization (ESI) followed by matrix-assisted laser desorption/ionization (MALDI). MS. Top-down proteomics can identify proteins with a molecular mass of > 200 kDa. In this case, protein digestion happens without fractionation, and LC is used to separate the peptides identified by MSDoT. Each strategy has different benefits and drawbacks. Because only tiny and irregular fractions of total peptides are recovered in the recovered sample, the bottom-up approach results in low percentage coverage of the protein sequence. A significant percentage of alternative splice variants and PTMs are thus missed as a result. Top-down proteomics, on the other hand, preserves all of a protein's properties and allows for the recovery of practically all known alterations and correlations. Furthermore, the outcomes of the gradual exclusion of protein digestion are maintained in top-down proteomics. The main obstacle in top-down proteomics is that proteins are less soluble than short peptides. SDS washing is necessary for certain highly soluble proteins in the membrane, but ESI does not allow the use of SDS. Analytical flow and sample preparation are part of proteomics workflows. These latter consist of protein identification, validation, and separation.

- **Sample Preparation:** Proteomics experiments rely heavily on precise sample preparation as well as a well-thought-out pre-analytical workflow. Proteomics does not have a standard method for preparing samples. The number of proteins in the sample, the sample's complexity, and the goals of the study all influence each method. In sample preparation, the most important step is to extract the proteins from the mixture. Tissue disruption should come first in the extraction process, then organic solvents and detergents to optimize protein extraction and solubilization. By using lyophilization, organic solvents and detergents can be eliminated. Comparable protein detection rates have been obtained in the past using detergent-based techniques for the extraction of 2,2,2-trifluoroethanol (TFE) macro-scale (> 100 µg) materials and nano-scale (30 µg)-based lysis.

- **Separation and Isolation of Protein:** Proteins are separated and isolated from the mixture using gel-based and chromatography-based techniques.
- **Gel-Based Approach:** PAGE is the most effective method for detecting and isolating proteins. 1-DE and 2-DE can be used for separation. Additionally, two 2D variations used in gel electrophoresis are 2D-DIGE and SDS-PAGE.
- **1-DE:** Proteins ranging in molecular weight from 10 kDa to 300 kDa can be separated using 1-DE. In order to combine secondary and non-disulfide-linked tertiary structures with a negative charge proportional to their volume, it employs SDS, a detergent. This makes it possible to determine molecular weights. Protein purification tests, sample purity checks, and the determination of molecular weights for proteins that are unknown can all be accomplished with SDS-PAGE.
- **2-DE:** Because protein molecules differ in their molecular weight and isoelectric point, 2-DE is more effective at differentiating proteins than 1-DE. Because the protein is divided into two distinct dimensions, it also has a higher resolution than 1-DE. Whereas the molecular mass and isoelectric point are the bases for protein separation in 2-DE, net charge is the basis for protein separation in 1-DE. As a result, this technique can identify various protein forms, including PTMs and phosphorylation. 2-DE can be used to resolve some proteins that result from various proteolysis processes and alternative mRNA splicing. 2-DE has numerous uses, such as cell map proteomics and protein expression profiling. Using protein expression profiling, tissues with and without disease can be compared. Protein complexes, microbes, and cellular organelles can all benefit from 2-DE protein mapping. 2-DE can assist with protein cataloging, and an online database can be established. Nevertheless, 2-DE is not able to identify proteins with low molecular weights or those that are separated only by size and isoelectric point.
- **Chromatography-Based Approach:** Protein-based chromatography can be purified using ion-exchange chromatography (IEC), size exclusion chromatography (SEC), and affinity chromatography. Moreover, the enzyme-linked immunosorbent assay and western blotting are employed to identify particular proteins.
- **IEC:** Proteins can be purified based on their charges using IEC. Other methods do not allow for the separation of proteins based on their charge nature, but this technique does. By changing the buffer's pH, it is possible to easily use the charge that the target molecule accepted. The IEC method can continue in fluctuating buffer conditions and is inexpensive.
- **SEC:** Different compounds can be separated using SEC based on their size (hydrodynamic volume), which is determined by how well they enter the pores of the stationary phase. This method isn't as helpful as other proteomics methods, though. Gel permeation chromatography (GPC), which uses organic solvents for polymer analysis, and gel filtration, which

uses aqueous solvents, are the two fundamental forms of SEC that are employed.

- **Affinity Chromatography:** Protein separation based on its interaction with an immobilized ligand is known as affinity chromatography. Affinity chromatography reduces protein complexity in both 2-DE and non-2-DE situations. Separation of protein before 2-DE, affinity chromatography of protein before MS, and affinity chromatography of peptides before MS are the three classifications of affinity chromatography.
- **LC:** Large and delicate biomolecules can be analyzed and proteins can be extracted from complex mixtures using the potent LC technique. It can be used to identify the peptides in the mixture when paired with MS. LC can assist researchers in identifying new biomarkers and comprehending the mechanisms underlying protein modification-based carcinogenesis. For instance, some researchers use dried filter-paper blood samples and LC-MS/MS to quickly monitor congenital adrenal hyperplasia.
- **Protein Identification and Characterization:** One of the most important steps in proteomics is protein identification. After the proteins have been separated by chromatography or electrophoresis, MS can be used. Protein microarray and Edman sequencing are two more methods that can be used to identify proteins.
- **Edman Sequencing:** Proteins and peptides can have their amino acid sequences found using Edman sequencing. This method involves the use of chemical reactions to extract and identify amino acid residues that are present at the polypeptide chain's N-terminus. As a result, it is important for determining the quality of therapeutic proteins and biopharmaceuticals.
- **MS:** The most effective analytical tool for quickly enabling protein sequencing is mass spectrometry. It can also be used to determine a protein's molecular weight. This method ionizes protein molecules and uses mass-to-charge ratios to calculate their mass. Three major parts make up a mass spectrometer: an ion source, a detector, and an analyzer. ESI and MALDI are ionization techniques used. To create a crystal lattice in MALDI, the peptides are combined with a chemical matrix and spotted onto a metal multiwall microliter plate. The energy is absorbed by the matrix chemicals and then transferred to the samples. A mass analyzer then finds the peptide ions. The majority of the ions produced by MALDI are singly charged, which aids in m/z calculation. In ESI, the protein sample's power is turned on to produce charged droplets that boost the production of gaseous ions, which are subsequently examined using a mass spectrometer. The advantages of ESI include its high elasticity to combine multiple MS. categories and high reproducibility. Moreover, ESI can be fixed to quadruple, ion traps, Fourier transform ion cyclotron resonance, and time-of-flight (TOF)-MS. Conversely, the drawbacks of ESI include its inability to be used for molecular imaging, the need for

a large number of samples, and the production of multiple peaks as a result of the numerous charged ions that contribute to the complexity of MS/MS spectra.

- **Protein Identification and Validation:** Currently, instruments for database searching include Sequent, Mascot, Comet, and Tandem. However, because most search devices rely on differentiation algorithms and recording functions, creation integration, and data comparison from numerous studies and experiments, they rarely yield matching data. Consequently, more time is required for the identification of peptides through data search. High-quality data reduces the time required for data searches and increases their effectiveness. Furthermore, measuring ion fragments with accurate mass can speed up database searches and yield more precise results.

9.4. THE CHALLENGE OF PROTEOMICS

Research on individual biological macromolecules and macromolecular lipid assemblies has made significant contributions to our understanding of how cells and organisms function internally. It's highly likely that things will stay this way. Removing cellular components and then attempting to characterize their structures and functions without being complicated by the "background noise" of other cellular components can yield significant insights. Having said that, the "background noise" needs to be equally linked to the structure and function of the relevant cellular components. Consequently, the reductionist philosophy of chemistry and the inclusionist philosophies of biology and medicine must both be followed in chemical biology research. As previously mentioned, mass spectrometry plays a crucial role in facilitating the transition from individual to collective molecular structure and function analyzes by means of the study and application of proteomics. This is particularly true for research pertaining to cellular proteins, as there is an increasing understanding that proteins frequently participate in larger multi-protein complexes and do not always function in isolation. Stated differently, there is a far greater prevalence of quaternary association between catalytic and functional units than previously thought. It's still unclear what this realization means in practice.

9.4.1. Early Developments in Proteomics

With the introduction of 2D polyacrylamide gel electrophoresis (PAGE) as a technique for visualizing sets of proteins in the 1970s, attempts to characterize protein systems on a large scale were first described (Figure 9.8). With this technique, protein mixtures were separated first by pI and subsequently by molecular mass in the second dimension. The result was a two-dimensional gel that could be stained and seen as a complex pattern. Generally, comparative experiments were carried out, wherein two tissues, one afflicted and the other healthy, were compared. To look for modifications in the protein spot pattern, patterns were then probed. The issue with the first method was that it was very hard to determine the protein from the spot.

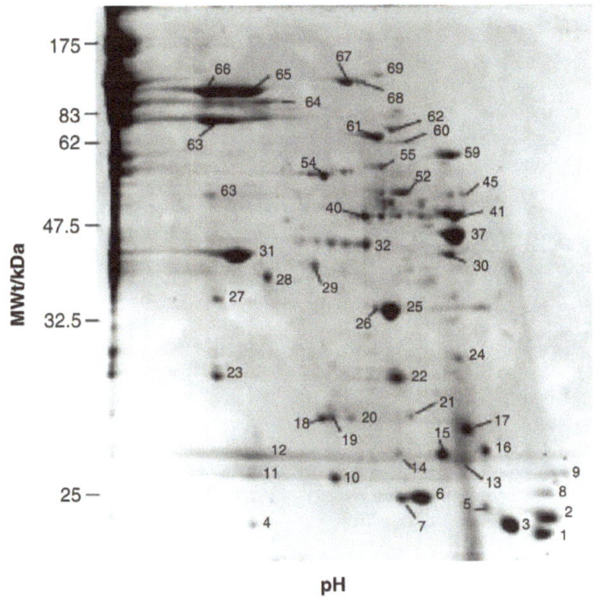

Figure 9.8. *2-D PAGE of protein mixture. Relative abundances of proteins may be visualized by gel staining and each individual spot can be excised and identified by mass spectrometry.*

Source: Andrew Miller and Julian Tanner (2008). Essentials of Chemical Biology. John Wiley & Sons, Ltd. ISBN: 978–0-470–84530–1.

The protein might be subjected to chemical degradation in an attempt to sequence it, but identification of the protein would require prior knowledge of its sequence. These challenges effectively put a stop to 2D electrophoresis in the 1980s, and only recently has it gained traction again thanks to the availability of sizable gene databases and methods for identifying peptides from proteins digested "in gel" by soft-ionization mass spectrometry.

9.4.2. Using 2D Gel Electrophoresis With Mass Spectrometry

Proteomics is traditionally approached using mass spectrometry and 2D gel electrophoresis. Proteins are separated by 2D gel electrophoresis based on orthogonal properties; gels are stained, spots are cut from the gel, and peptide mass fingerprinting combined with sequence data from gene databases is used to identify the spots (Figure 9.9). Just these methods have made it possible to conduct certain kinds of "cataloging" experiments, which enable the examination of patterns on 2D gels and the identification of spots on gel. These methods have been most effective when studying systems that are a little bit simpler than the whole cell scenario, such as the structures of multiprotein complexes or subcellular organelles. Alternatively, 2D gel electrophoresis results can be compared using pattern-matching algorithms in comparative experiments to try and identify differences, such as protein profiles between diseased and normal tissues, which could lead to the discovery of disease biomarkers.

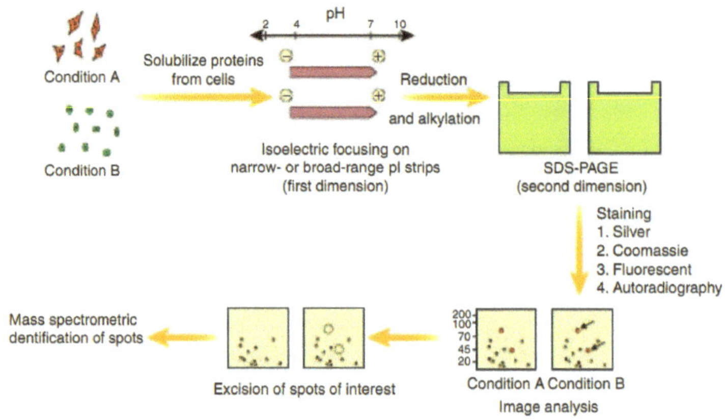

Figure 9.9. *Comparative 2-D electrophoresis/MS approach to proteomics. Proteins in two separate gels can be compared, differentially expressed spots identified, and identities confirmed by mass spectrometry.*

Source: Andrew Miller and Julian Tanner (2008). Essentials of Chemical Biology. John Wiley & Sons, Ltd. ISBN: 978-0-470-84530-1.

Nevertheless, there are some serious issues with applying mass spectrometry and 2D electrophoresis together. Initially, 2D gels are only used to identify high abundance proteins; it is estimated that over 10,000 genes are expressed simultaneously in a single type of human cell. Even within a single cell, there may be millions of different polypeptides if post-translational modification is taken into account. Several 100 spots can be analyzed individually on a 2D gel; this number could increase several times with specialized subcellular fractionation techniques or zoom gels that cover smaller pH ranges. Furthermore, it's possible that very few membrane and basic proteins can be distinguished in 2D gels. Because of this, the majority of proteins found in databases that list proteins in particular tissues are those that are common, like cytoskeletal proteins, chaperones, and matrix-associated proteins. There is a severe underrepresentation of many low-abundance proteins, including transmembrane receptors and signaling proteins. Therefore, it becomes essential to push for more quantitative proteomics.

9.4.3. Isotope-Coded Affinity Tags

The fact that mass spectrometry is not strictly a quantitative technique with respect to molecular abundance is one of the primary obstacles to be overcome in order to realize quantitative proteomics. Doing parallel comparative experiments with one sample unlabeled and the other labeled with an isotope (like ^{18}O or ^{2}H) is one way to get around this issue. A mass spectrometer can be used to measure the relative ratios and observe mass differences between labeled and unlabeled samples (Figure 9.10). Evidently, this quantification technique depends on the equivalent mass "expression" of both samples, even if one isotope is labeled and the other is not. Put another way, both samples should exhibit the same chemical behavior regardless of the labeling. Using isotope-coded affinity tags (ICATs) is the most straightforward method of effecting labeling. This method maintains two identical populations of proteins under various circumstances (e.g., temperature, pH, and ionic strength) but is labeled with distinct isotope-coded affinity tags (ICATs) at the end. Tags are unique to each site (e.g., amine groups, phosphate esters, or sulfhydryl), and the affinity tag of the ICAT tagging agent is also present. Because eight 1H hydrogen atoms are substituted for eight 2H deuterium atoms, the heavier ICAT typically has a mass difference of eight compared to the lighter ICAT. Before affinity purification of the affinity-tagged oligo-/polypeptides of the combined digest, both protein populations are then pooled post-tagging and digested into oligo-/polypeptides. Then, using tandem mass spectrometry, tagged oligo- and polypeptides are directly compared in an effort to identify quantitative variations in molecular ion abundance that can be traced back to variations in the environmental factors that supported the two protein populations that were identical. One of the greatest methods for quantitative and comparative proteomics is still this one.

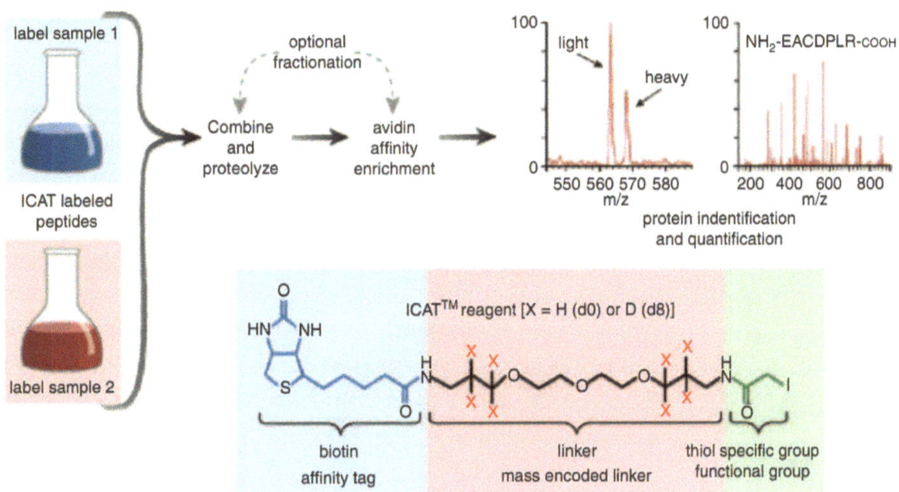

Figure 9.10. *Isotope coded affinity tagging as an approach for comparative proteomics. Two pools of peptides can be differentially labeled with tags of different masses, then the pools combined, enriched by avidin affinity column purification (avidin has a very high affinity for biotin) and relative abundance of particular peptides can be directly compared by mass spectrometry.*

Source: Andrew Miller and Julian Tanner (2008). Essentials of Chemical Biology. John Wiley & Sons, Ltd. ISBN: 978-0-470-84530-1.

9.4.4. Deciphering Protein Networks by Tandem Affinity Purification

The creation of a map of protein–protein interactions in yeast cells is among the most prominent early achievements of proteomics. One of the main methods for creating these maps has been to employ yeast two-hybrid mapping's genetic approach. The theory behind yeast two-hybrid experiments is that certain transcriptional activators need the actions of two proteins: one that binds DNA and the other that activates transcription (Figure 9.11).

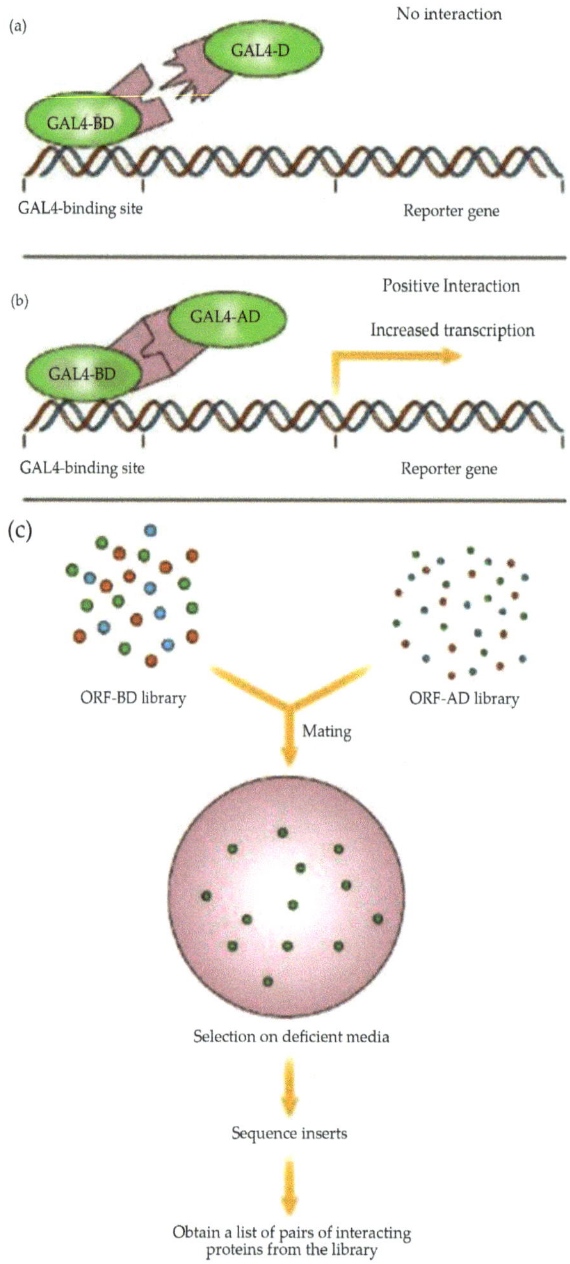

Figure 9.11. *The yeast two-hybrid system. Different open reading frames are fused to a Gal4 DNA-binding domain (BD) or an activation domain (AD). If the encoded proteins interact then a reported gene is transcribed and translated, allowing for identification of protein-protein interactions.*

Source: Andrew Miller and Julian Tanner (2008). Essentials of Chemical Biology. John Wiley & Sons, Ltd. ISBN: 978-0-470-84530-1.

Only when the two proteins are linked to one another does transcription start. Typically, the DNA binding domain of *E. coli* LexA serves as one component of the transcriptional activator complex, which also includes the yeast protein GAL4. By fused the activation domain of one protein and the complementary DNA binding domain of the other, two proteins can be examined for binding interactions; if the two proteins interact, transcription of a reporter gene will occur automatically.

Large-scale execution of these experiments can be achieved in two ways: either by employing an array approach or a library approach. An array approach involves the design of a living array of haploid yeast transformants, or the prey, that are fused to transcriptional activation domains with distinct open reading frames (ORFs). A different yeast transformant, known as the bait, is made by fusing a single protein of interest with the appropriate DNA-binding domain. Diploid yeast strains are created when the bait transformant mates with every transformant on the array; however, transcriptional activation is only observed in those strains that express the appropriate molecular partners to the single protein of interest from the ORFs. Molecular partners of our single protein of interest should be easy to identify, since the array's ORF identities and array addresses are known. All of the prey transformants are combined and fused to the bait transformant in a library approach. This approach suggests that transcriptional activation is linked to a survival trait (e.g., antibiotic resistance), resulting in the expression of the appropriate molecular partners to the single protein of interest from the ORFs by the only diploid yeast strains that survive. Subsequently, the "survival" ORFs of the protein of interest are sequenced to identify its molecular partners individually.

As demonstrated by the yeast two-hybrid method, tandem affinity purification (TAP) is an additional technique for examining protein–protein interactions that can characterize multi-protein complexes as opposed to just the binary partners. An improvement on protein tagging is tandem affinity purification (TAP), which uses two tags to enable the targeted purification of molecular partners interacting with a particular bait protein of interest. Additionally, there are two distinct protease sites: one is situated between Tags 1 and 2, and the other is situated between the two tags and the particular bait protein of interest. First, an extract containing the target bait protein that has been TAP-tagged is run over an affinity column for Tag 1, enabling Tag 1 to bind. Following a wash of the column, Tag 1 is cleaved from Tag 2 by protease, enabling the target protein and its interacting protein partners to elute from the affinity column together. After that, this enriched protein mixture is run through a second affinity column for Tag 2, which enables Tag 2 binding. This time, the target protein and its molecular partners are able to elute together without the interference of non-specific binding interactions because the target protein is protease-cleaved from Tag 2. Mass spectrometry and 2D gel electrophoresis can then be used to identify the bait protein of interest and its partners.

9.4.5. The Challenge of Membrane Proteins in Proteomics

An estimated 20–30% of the proteins that the human genome encodes are membrane proteins. Given that over 70% of drug targets are membrane proteins, it is obvious that understanding membrane proteins is essential. The fact that much fewer membrane proteins are detected by mass spectrometry than are known to exist in any given sample by other techniques is a major issue. Solubility prevents membrane proteins from being observed on conventional 2D gels. First, these proteins (for the first dimension) are not soluble in the buffer. Second, at their isoelectric point pI, solubilized membrane proteins will all too quickly precipitate once more. Solubilizing membrane proteins in organic solvents or non-ionic/zwitterionic detergents is one method of resolving the initial issue. One way to avoid the pI issue is to only use 1D gels. However, because oligo-/polypeptides in digests tend to be particularly hydrophobic, additional issues arise during the digestion steps even if membrane proteins are clearly resolved within a gel (Figure 9.12).

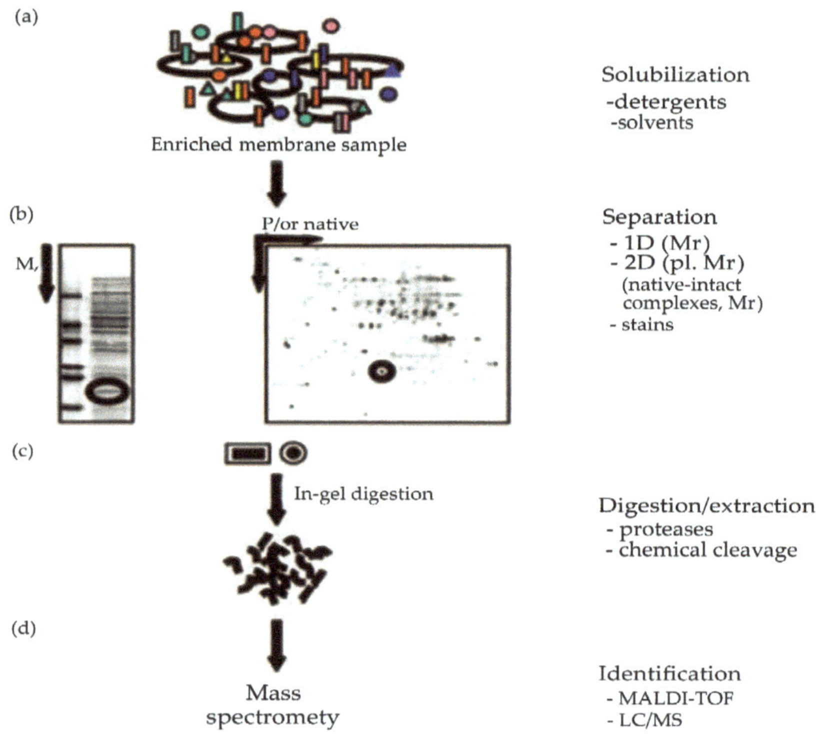

Figure 9.12. *Gel-based membrane proteomics. Membrane proteins can be analyzed by a four-step process of solubilization, separation, digestion/extraction, and finally MS identification.*

Source: Andrew Miller and Julian Tanner (2008). Essentials of Chemical Biology. John Wiley & Sons, Ltd. ISBN: 978-0-470-84530-1.

An alternative to using gels is shotgun proteomics—i.e., in situ digestion of complex mixtures of proteins followed by 2D liquid chromatography mass spectrometry (2D LC-MS) to identify proteins in particular samples (Figure 9.13). One more important consideration is initial solubilization; different detergents, organic acids, and solvents may be employed. Enzymes (e.g., trypsin, which is still active in 60% methanol) or chemicals (e.g., cyanogen bromide) can be used to carry out proteolysis. Protein mass fingerprinting can be used to identify proteins after cleaving the exposed soluble domains with a non-specific protease like proteinase K. The resulting peptides can then be sequenced using LC-MS. In summary, the field of membrane protein proteomics continues to present challenges, but with increased interest in membrane proteins from the pharmaceutical industry, these methods will surely become more popular.

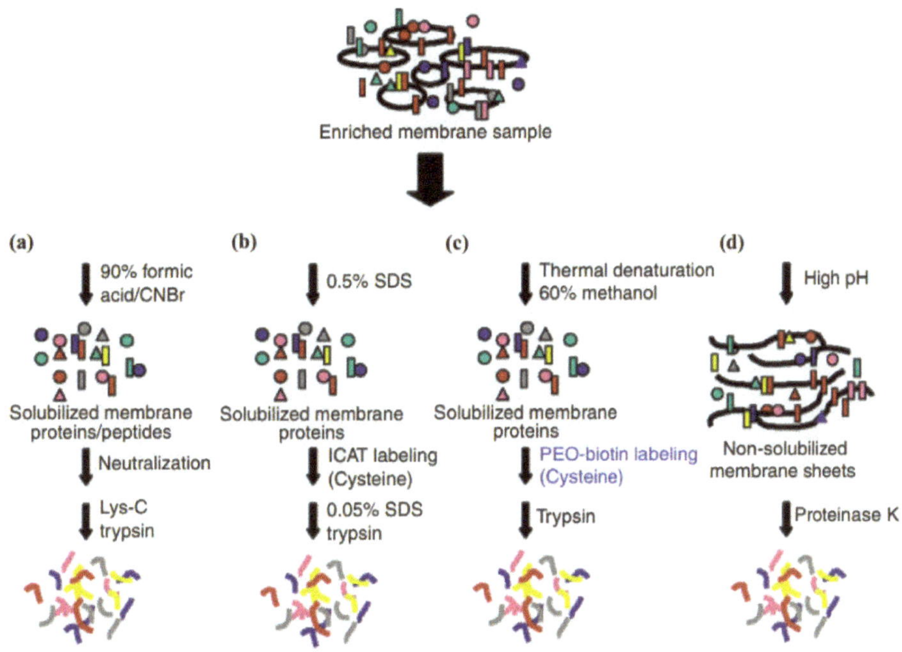

Figure 9.13. *Comparison of four methods to prepare membrane proteins for analysis. Analysis can be by shotgun proteomics or by conventional mass spectrometry.*

Source: Andrew Miller and Julian Tanner (2008). Essentials of Chemical Biology. John Wiley & Sons, Ltd. ISBN: 978–0-470–84530–1.

9.4.6. Proteomics and Post-Translational Modifications

Posttranslational modifications (PTMs) are covalent processing events that add a modifying group, such as acetyl, phosphoryl, glycosyl, or methyl, to one or more amino acids and cause proteolytic cleavage, altering the properties of a protein.

Particularly in eukaryotes, post-translational modifications (PTMs) play a critical role in altering the structure and function of proteins. Large-scale research presents a different set of difficulties. In proteomics investigations, proteins are typically isolated and identified using protein mass fingerprinting, supplemented, if required, by product ion scanning. Regretfully, PTM analysis necessitates a far more in-depth examination of the complete amino acid sequence. This is made more difficult by the fact that different types of proteins of interest can have a wide range of modifications on them, and that modifications are not always homogenous. Even biologists do not fully comprehend the structural and functional implications, so the role of PTMs in cells remains a vast field of research that the reader interested in chemical biology is free to explore.

9.4.6.1. Comprehensive PTM Analysis of a Single Protein

Purification is the initial step in a thorough PTM analysis of a target protein. Gel electrophoresis, using either a 1D or 2D gel, is the most convenient method for achieving this, depending on the initial level of purification. Electrophoresis frequently provides hints about the type of modification. For instance, when a protein is phosphorylated, the phosphorylated and non-phosphorylated states of the protein separate into spots on a 2D gel with different pI values. Accurately identifying the molecular ion is a helpful first step in the process; comparing the measured and expected masses can provide some early insight into potential modifications. Peptides that have been obtained through enzymatic or chemical degradation can be analyzed to obtain detailed information about the site(s) of modification after information about the intact protein has been obtained. Using trypsin in one experiment and endoproteinase Asp-N or Glu-C in another would be a common combination. Unexpected molecular ions can be used to identify modified oligo- or polypeptides in digests. Afterward, product ion scanning can be used to identify the amino acid residues at which PTMs have occurred. Finding acetylation is the simplest, as it frequently occurs on arginine amino acid residues. Although it is typically visible, phosphorylation can be either stable or labile (phosphotyrosine tends to be stable, phosphothreonine less so, and phosphoserine tends to be labile). Sadly, tandem mass spectrometry may have a harder time identifying sulphation and glycosylation modification sites since they can be considerably more labile.

9.4.6.2. PTM Analysis of Protein Populations

Compared to the single-protein approach, the PTM mapping of protein populations calls for a different methodology. For example, phosphorylated proteins can be identified by western blot using antiphospho-amino-acid antibodies after a protein sample mixture has been resolved by 2D-gel electrophoresis. It is also possible to use individual anti-phosphoserine, anti-phosphothreonine, or anti-phosphotyrosine antibodies. A different strategy is to use affinity column purification to precisely target and isolate a subset of proteins from a complex mixture. Anti-phospho-amino-acid antibodies, for example, attached to a resin (e.g., A complex protein mixture's phospho-proteins can be affinity purified using agarose beads. After the enriched population of proteins is eluted from the column, it can be resolved using 2D electrophoresis, and tandem mass spectrometry can be used to identify individual proteins by their phosphorylation sites.

Shotgun proteomics can be used to remove gels, but before mass spectrometry, certain protein modifications of interest must be examined and resolved using chromatography. Phosphopeptides, for instance, can be precisely purified using immobilized-metal affinity chromatography prior to mass spectrometry (IMAC). Prior to mass spectrometry, phosphorylated peptides can be purified using metal ion affinity columns using both iron (as Fe^{3+}) and gallium (as Ga^{3+}). Tandem mass spectrometry can then easily identify their sequences and phosphorylation site(s), which can then be linked back to the identities of the parent proteins that these daughter digest peptides were derived from. Future development in the field of PTM analysis using mass spectrometry is expected to be significant.

ROLE MODEL

MARC WILKINS

Marc R. Wilkins is an Australian scientist who is credited with the defining the concept of the proteome. Wilkins is a Professor in the School of Biotechnology and Biomolecular Sciences at the University of New South Wales, Sydney.

Wilkins coined the term proteome in 1994 whilst developing the concept as a PhD student at Macquarie University, Sydney, Australia, describing it as the 'PROTein complement expressed by a genOME.' The term is a generalization of the concept of the genome to encompass the set of all proteins that can be produced through the genome through alternative splicing and post-transcriptional modification of messenger RNA.

- Career

Wilkins held a post-doctoral fellowship from 1995–1997 at the University of Geneva, Switzerland, working with Prof Denis Hochstrasser and Dr Amos Bairoch. He co-developed many of the protein analysis tools available on the ExPASy web server. He subsequently served as a senior post-doctoral fellow in the Australian Proteome Analysis Facility, which was established by the Australian Government in 1995 as the world's first dedicated proteome research center.

In 1997 he co-edited the first book on proteomics, 'Proteome Research: New Frontiers in Functional Genomics' (Wilkins et al. (eds), Springer Verlag), which sold more than 4,000 copies.

He was a co-founder of the proteomics company Proteome Systems Ltd. It was established in January 1999 and listed on the Australian Securities Exchange in 2004. Dr Wilkins worked in Proteome Systems full-time for 6 years as leader of its bioinformatics Research and Development team then as Head of Proteomics. It was awarded IBM's "Rookie of the Year Award" for companies in the Asia-Pacific region, in 2003. Proteome Systems was renamed Tyrian Diagnostics and ultimately divested its IP. In 2007, Wilkins co-founded the regenerative medicine company, Regeneus. This company has developed treatments for osteoarthritis using adult stem cells, which are now in clinical trials.

Wilkins is currently a Professor of Systems Biology at the University of New South Wales and is the director of the Systems Biology Initiative and the Ramaciotti Centre for Genomics. He has published more than 200 research papers and book chapters. His current research concerns the role of protein methylation in the eukaryotic cell, the use of crosslinking mass spectrometry (XL-MS) for proteome-scale analysis of protein interaction networks (the interactome) and the bioinformatic analysis of next-generation sequencing data. As part of the koala genome consortium, his team did all PacBio long-read sequencing and then Falcon-based assembly to generate the most complete and contiguous marsupial genome to date.

In 2012, Wilkins was awarded the Australian Society for Biochemistry and Molecular Biology's Beckman Coulter Discovery Science Award. This prize is awarded to a member for distinguished contributions to the field of biochemistry and molecular biology. The University of New South Wales awarded Wilkins a Doctor of Science (D. Sc.) in 2018.

Recently, he has enjoyed collaborating on the Human Proteome Project (published 2020) and the Human RNA Atlas Project (published 2021).

CASE STUDY

PROTEOMICS & METABOLOMICS

- How H. Pylori Gets Its Shape

Helicobacter pylori is the primary cause of stomach cancer, which is the third leading cause of cancer deaths worldwide. The helical shape of the bacterium is necessary for efficient stomach colonization. Thus, if researchers can understand the mechanisms by which H. pylori achieves its shape, they may be able to identify novel therapeutic strategies.

Previously, Fred Hutchinson Cancer Center faculty member Dr. Nina Salama and her team used genetic screening to identify a set of proteins that, when deleted, resulted in distinctly non-helical cell shapes.

- The Challenge

Salama and her team wanted to find other proteins that interact with one of the shape-forming proteins they'd identified in their earlier work, called Csd5. Mass spectrometry-based proteomics is the primary technology for identifying protein-protein interactions. However, the Salama Lab is not well-versed in the complexities of the technology or how to obtain optimal results.

- The Approach

The Proteomics & Metabolomics shared resource at Fred Hutch offered the technological expertise the Salama Lab needed to answer their questions about Csd5. Salama's team worked in conjunction with the core to design and carry out immunoprecipitation mass spectrometry, or IP-MS, experiments using Csd5 fusion constructs to identify proteins that interact with Csd5. The core provided high-resolution mass spectrometry experiments and data analysis to identify protein-protein interactors. This research would not have possible without access to an experienced proteomics facility.

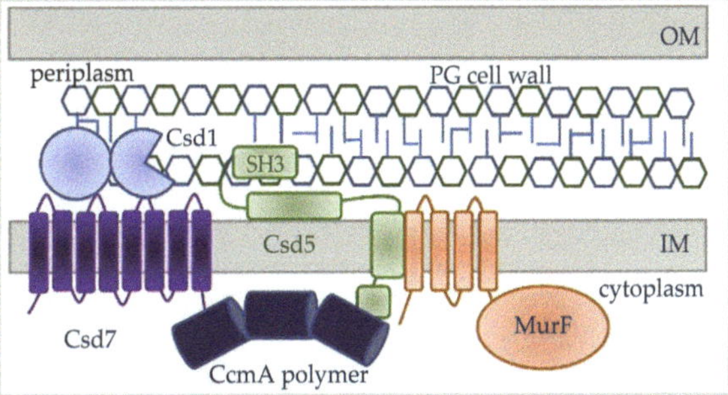

Source: https://www.fredhutch.org/en/research/shared-resources/core-facilities/proteomics-metabolomics/case-study-h-pylori-shape.html.

- The Outcome

These experiments identified MurF, a cytoplasmic proteoglycan synthesis enzyme, and CcmA, a known H. pylori cell-shape protein, and they showed enrichment for nearly all the components of the F1F0 ATP synthase. The team validated these interactions using reciprocal IPs with western blotting detection.

The researchers also conducted IP-MS experiments using domain deletions of Csd5. Those results strongly suggested that the N-terminal and transmembrane domains of Csd5 are required for interactions with MurF, CcmA and ATP synthase.

Finally, the team's IP-MS experiments in mutant strain backgrounds strongly suggested Csd5, MurF and CcmA interact individually and/or together with ATP synthase.

Ultimately, this research has built a model in which Csd5 promotes helical shape as part of a membrane-associated, multiprotein complex that includes interactions with the periplasmic cell wall, a proteoglycan precursor synthesis enzyme, the bacterial cytoskeleton and ATP synthase. Moreover, it has led to additional proteomic work to identify more of CcmA's interaction partners — suggesting that these bacteria have an additional subcomplex for generating a helical cell shape that, together with the Csd5 complex, form a larger "shapeosome."

CLASS ACTIVITY

How proteins interact with each other, which proteins interact, and when and where they interact. Students should know this.

SUMMARY

- Mass spectrometry is an indispensable analytical tool in chemistry, biochemistry, pharmacy, medicine and many related fields of science. Mass spectrometry is employed to analyze combinatorial libraries, sequence biomolecules and help explore single cells or objects from outer space.
- Mass spectrometry is an analytical method useful for calculating the mass-to-charge ratio (m/z) of one or more molecules in the sample. Such measurements may also often be used to determine the precise molecular weight of the sample components. Mass spectrometry is an analytical method to find the molecular mass of a compound and indirectly helps to prove the identity of isotopes.
- Proteomics is the large-scale study of proteins. Proteins are vital parts of living organisms, with many functions such as the formation of structural fibers of muscle tissue, enzymatic digestion of food, or synthesis and replication of DNA.
- Mass spectrometry is a technique with origins in the physical sciences that has fast become one of the most powerful techniques available for probing biological systems. Advances in mass spectrometry made possible by the advent of soft ionization techniques have made possible detailed investigations into the primary structures of biological macromolecules and amphiphilic lipids.
- An ion source is a device that creates atomic and molecular ions. Ion sources are used to form ions for mass spectrometers, optical emission spectrometers, particle accelerators, ion implanters and ion engines. Electron ionization is widely used in mass spectrometry, particularly for organic molecules.
- Electron ionization (EI) is the most widespread technique of ionization used for mass spectrometry in synthetic chemistry. EI consists of two steps: first, the sample of a molecule of interest is passed into the vapor phase before being bombarded, second with a stream of electrons.
- Proteomics is a new type of 'omics' that has rapidly developed, especially in the therapeutics field. The word proteome was created by Marc Wilkins in 1995. Proteomics is the study of the interactions, function, composition, and structures of proteins and their cellular activities.
- Expression proteomics is a novel approach that studies the quantitative and qualitative expression of proteins. It aims to specify the difference in protein expression between two conditions such as patients and controls.
- Nuclear magnetic resonance spectroscopy and X-ray crystallography are used in structural proteomics to determine the three-dimensional structure and structural complexities of functional proteins. It specifies all protein interactions such as membranes, cell organelles, and ribosomes in the mixture. The study of the nuclear pore complex is an example of structural proteomics.
- Tandem affinity purification (TAP) is another method for probing protein–protein interactions that is able to characterize multi-protein complexes rather than just the binary partners as illustrated with the yeast two-hybrid method.

- Posttranslational modifications (PTMs) are covalent processing events that change the properties of a protein by proteolytic cleavage and adding a modifying group, such as acetyl, phosphoryl, glycosyl and methyl, to one or more amino acids.

REVIEW QUESTIONS

1. What is mass spectrometry?
2. Name the three basic components of a mass spectrometer.
3. What is proteomics?
4. Name the three main types of proteomics.
5. What is the post-translational modification?

REFERENCES

1. Agrawal, G. K., Sarkar, A., Righetti, P. G., Pedreschi, R., Carpentier, S., Wang, T., Barkla, B. J., Kohli, A., Ndimba, B. K., Bykova, N. V., Rampitsch, C., Zolla, L., Rafudeen, M. S., Cramer, R., Bindschedler, L. V., Tsakirpaloglou, N., Ndimba, R. J., Farrant, J. M., Renaut, J., & Rakwal, R. (2013). A decade of plant proteomics and mass spectrometry: translation of technical advancements to food security and safety issues. *Mass Spectrometry Reviews, 32*, 335–365.
2. Anderson, N. L., & Anderson, N. G. (1998). Proteome and proteomics: new technologies, new concepts, and new words. *Electrophoresis, 19*(11), 1853–1861. https://doi.org/10.1002/elps.1150191103.
3. Blackstock, W. P., & Weir, M. P. (1999). Proteomics: quantitative and physical mapping of cellular proteins. *Trends in Biotechnology, 17*(3), 121–127. https://doi.org/10.1016/S0167-7799(98)01245-1.
4. Chandramouli, K., & Qian, P. Y. (2009). Proteomics: challenges, techniques and possibilities to overcome biological sample complexity. *Human Genomics and Proteomics, 2009*. https://doi.org/10.4061/2009/239204.
5. Dass, C. (2001). *Principles and Practice of Biological Mass Spectrometry*. Wiley–Interscience.
6. De Hoffmann, E., Charette, J., & Stroobant, V. (1996). *Mass Spectrometry Principles and Applications*. Wiley, Chichester; Masson, Paris.
7. Hamdan, M. H., & Righetti, P. G. (2005). *Proteomics Today: Protein Assessment and Biomarkers Using Mass Spectrometry, 2D Electrophoresis and Microarray Technology*. Wiley–Interscience.
8. Holman, J. D., Dasari, S., & Tabb, D. L. (2013). Informatics of protein and posttranslational modification detection via shotgun proteomics. *Methods in Molecular Biology, 1002*, 167–179.

9. Kinter, M., & Sherman, N. E. (2000). *Protein Sequencing and Identification Using Tandem Mass Spectrometry*. Wiley–Interscience.
10. Király, M., Dalmadiné Kiss, B., Vékey, K., Antal, I., & Ludányi, K. (2016). Mass spectrometry: past and present. *Acta Pharmaceutica Hungarica, 86*(1), 3–11.
11. Laskin, J., & Lifshitz, C. (2006). *Principles of Mass Spectrometry Applied to Biomolecules*. Wiley.
12. Matthiesen, R., & Bunkenborg, J. (2013). Introduction to mass spectrometry-based proteomics. *Methods in Molecular Biology, 1007*, 1–45.
13. Pennington, S. R., & Dunn, M. J. (2001). *Proteomics from Protein Sequence to Function*. Springer-Verlag Telos.
14. Twyman, R. M. (2004). *Principles of Proteomics*. BIOS Scientific.
15. Wolf, B. (1995). *Handbook of Ion Sources*. CRC Press.

CHAPTER

10 Molecular Selection and Evolution

LEARNING OBJECTIVES

After studying this chapter, you will be able to:

- Discuss on origins of molecular evolution, the molecular clock, and the neutral theory
- Describe the molecular breeding; natural selection acting on self-organization
- Understand the directed evolution of protein function
- Identify the directed evolution of nucleic acids
- Explain the catalytic antibodies

KEY TERMS FROM THIS CHAPTER

Amino acid sequence
Bioinformatics
DNA aptamers
DNA molecules
DNA sequences
Genetic drift
Molecular diversity
Molecular selection
Aptamers
Catalytic antibodies
DNA catalysts
DNA repair pathways
DNA/RNA purine
Molecular breeding
Molecular evolution
Mutations

INTRODUCTION

Understanding the dynamics of genetic variation and the mechanisms underlying the diversity of life requires an understanding of molecular selection and evolution. Natural selection favors alleles that confer adaptive advantages in a particular environment based on genetic variation. New genetic variations are introduced through recombination, genetic drift, and mutations in this process, which is commonly referred to as molecular evolution. Advantageous mutations are passed down through populations over successive generations, resulting in molecular evolutionary changes. Molecular selection is evident in a number of situations, including the emergence of novel metabolic pathways and the bacterial development of antibiotic resistance. One instance of how selection pressure can drive the fixation of advantageous mutations is the quick rise in antibiotic-resistant bacterial strains. Similar to this, organisms frequently evolve particular molecular functions through positive selection in order to adapt to new environmental niches.

Phylogenetics, comparative genomics, bioinformatics, and other methods are used in the study of molecular evolution to examine the genetic alterations that underlie evolutionary processes. Through the use of comparative genomics, scientists can determine which sequences are conserved and which are divergent across various species, providing insight into the functional importance of particular genes and regulatory components. By reconstructing evolutionary relationships, phylogenetic analysis aids in identifying lineage-specific adaptations and the background of genetic alterations. Moreover, directed evolution and artificial selection in lab environments are included in the notion of molecular evolution. By using these principles, researchers are able to evolve proteins or nucleic acids that have favorable characteristics, like increased binding affinity or catalytic activity. In order to maximize molecular functions, directed evolution mimics natural evolutionary processes through repeated cycles of mutation and selection. All things considered, the processes through which genetic diversity is created, preserved, and molded over time are covered by molecular selection and evolution. The intricacy and diversity seen in the natural world are a result of these processes, which are essential to an organism's ability to adapt and survive. Researchers can learn more about the origins of genetic illnesses, the evolution of life, and the possibility for biotechnological breakthroughs by examining molecular evolution.

10.1. ORIGINS OF MOLECULAR EVOLUTION, THE MOLECULAR CLOCK, AND THE NEUTRAL THEORY

In the 1950s and 1960s, biologists began gathering data from a variety of organisms after they figured out how to determine the sequence of proteins. This is how the field of study known as molecular evolution kind of happened by accident. Hemoglobin and histones, two proteins selected for their biological significance, provided important early data. Because of their obvious functional significance and the fact that sequencing techniques for these biological molecules were developed first, the study of proteins has received special attention. Two key patterns emerged from the comparison of sequence divergence among proteins: (1) distinct proteins evolved at different rates, but (2) each protein appeared to accumulate changes at a surprisingly consistent rate, a pattern that gave rise to the idea of a "molecular clock" (Figure 10.1). These patterns were found in fibrin peptides, hemoglobin, and histones from a variety of vertebrates with well-defined fossil records.

The phrase "molecular clock" describes a mechanism as well as an instrument used in evolutionary research. As a mechanism, the observation of a nearly continuous accumulation of change suggested that the majority of the molecular change that has been observed is the result of chance rather than local adaptation. The existence of a clock that represented a molecule meant that, as a tool, if its rate could be adjusted using organisms with known ages (e.g., from the fossil record), then sequence differences that have been observed for organisms for which there is no fossil record can also be "dated."

The majority of evolutionary biologists believed that natural selection was the main factor influencing evolutionary change up until the middle of the 1960s. Genetic drift was rarely taken into consideration, and most models concentrated on selection. For instance, drift was important in Wright's shifting balance theory, but it wasn't the main force behind evolution; rather, it was a source of novel allele combinations on which selection could act. Genetic drift was thought to be significant only in small populations (Freese, 1962).

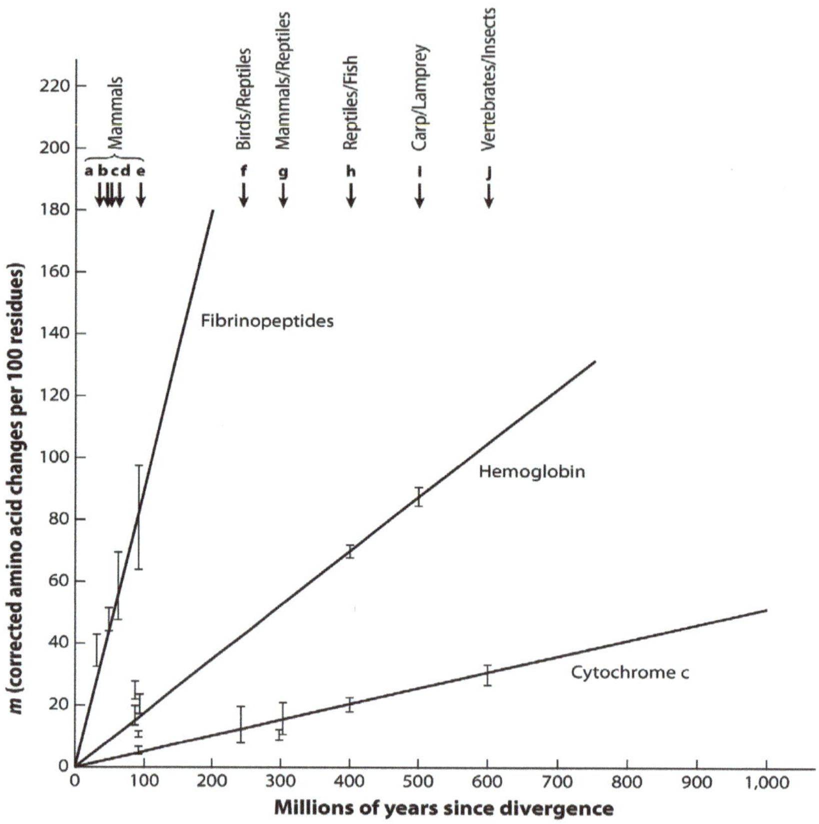

Figure 10.1. *Molecular clocks. Rates of amino acid substitution in three proteins: fibrinopeptides, hemoglobin, and cytochrome c. The number of amino acid differences (per 100 residues and corrected for multiple changes at the same residue) is plotted for comparisons between various mammals, birds and reptiles, mammals and reptiles, reptiles and fish, carp and lamprey, and vertebrates and insects, all lineages of organisms for which fossil data provided estimates of the time since divergence. Note that while some comparisons of the three proteins are from the same pair of organisms and time of divergence, the three proteins are evolving at very different rates (i.e., fibrinopeptide the fastest, and cytochrome c the slowest). In addition, the rough linearity of the rate of accumulation of molecular divergence with time illustrates the molecular clock concept.*

Source: https://edisciplinas.usp.br/pluginfile.php/4481131/mod_resource/content/1/Princeton%20Encyclopedia%20verbete%20de%20evolu%C3%A7%C3%A3o%20molecular.pdf.

Numerous findings in the middle of the 1960s and the beginning of the 1970s cast doubt on selection's primacy as the primary force behind evolution. Initially, elevated quantities of protein polymorphism (i.e., variation within populations) in humans, bacteria, and fruit flies. Second, extrapolating from available data, Motoo Kimura estimated that there were as many as two amino acid replacements per generation across the genome in mammals—could all that variation within populations really be maintained by natural selection? Third, Jack King and Thomas Jukes took into consideration the genetic code with its built-in redundancy (meaning some nucleotide changes were "silent" and didn't change the encoded protein sequence) and the conservative characteristics of many amino acid changes seen between proteins isolated from different organisms.

Again, could selection really drive that many amino acid replacements to fixation without an intolerable reduction in population fitness? Like Kimura, they contended that a large portion of the variation in proteins found within and between species was the result of genetic drift and mutation rather than alteration in function. Although Kimura acknowledged that adaptation did occur, she contended that it only affected a small percentage of the genome at any given time and that it was unlikely for natural selection to be responsible for the fixation of variation between species as well as the maintenance of the wide molecular variation seen within species. Furthermore, it was suggested that many mutations were detrimental and should be eliminated from populations in order to preserve the functionally essential portions of genes and the genome. Should selection fail to provide a plausible explanation for these observations, drift may very well play a major role.

The neutral allele theory of molecular evolution is the name given to this broad idea, which has formed a conceptual and model framework that has served as the foundation for much of the current field. The realization that all organism populations have a finite size and that genetic drift is a random process that creates a background on which all other evolutionary forces act is arguably the most significant. Although not all variation is selectively neutral, the so-called "neutral theory" can be misleading because it acknowledges that a sizable percentage of new mutations are highly harmful and should be eliminated from the population almost immediately. Yes, the theory does permit some adaptive mutations, but only a small number of selectively equivalent (neutral) mutations are allowed, and genetic drift alone determines their dynamics. Therefore, it is assumed that most of the variation we observe within and between species has no bearing on an organism's fitness.

An unstoppable transition to the study of DNA sequences in the field of molecular evolution started in the 1970s with the development of techniques for directly sequencing DNA, which were far less time-consuming than sequencing proteins. The close genetic ties between neighboring nucleotides ensure that sequences retain more of their evolutionary past in addition to allowing DNA sequence data to estimate the frequency of variation at individual nucleotides. These correlated evolutionary histories made it possible to test molecular evolution models—especially the neutral theory—using new statistical and computational techniques. The field was no longer devoid of data and rich in theory because the amount of data was now coming in at an incredible rate. The capacity to acquire DNA sequences in larger sample sizes has also improved statistical power to distinguish between models and to deduce demographics, evolutionary history, and the targets and strength of selection operating on the genome. Although the strict neutral theory was obviously oversimplified, it has given researchers a useful reminder of the significance of stochastic processes in all populations as well as a useful null hypothesis to compare data with (Dietrich et al., 1998).

10.1.1. Order From Complexity

An organism can be simply described as a system of chemicals with the ability to catalyze its own reproduction. This is the essence of life. According to this perspective, autocatalytic closure between a group of molecular species is the fundamental characteristic of life. Collectively, the system of molecules is "alive," whereas individual molecular species are "dead" when they exist alone. For the chemical biologist, this idea is highly persuasive because biocatalysis is now indispensable for the genesis of life rather than merely supporting it. It is not possible for catalysis to create life on its own. For the purpose of survival and self-reproduction, all living systems "eat," or take in matter and energy. Hence, closed autocatalytic but open (non-equilibrium) thermodynamic systems must be the analog of living systems. Kauffman argues that the remarkable order that permeates living systems and appears to be in direct opposition to the second law of thermodynamics arises from molecular diversity and complexity in an open thermodynamic system. Even if life can emerge from autocatalytic closure among a group of molecular species, how does this theory of life explain it? To put it another way, in an open thermodynamic system, molecular diversity and complexity directly lead to the emergence of self-organization., "free order."

10.1.2. Evolution From the Molecular Level

Natural selection arguments have dominated biologists' conversations regarding the beginnings of life. Self-organization appears to have been neglected more often than not. This is probably due to the fundamental difficulty in understanding how self-organization resulting from molecular diversity and complexity and the "forces" of natural selection can both govern living systems simultaneously. However, a final theory of biology must permit the commingling of self-organization and selection processes as an expression of an even deeper order if we are to comprehend the origin of life correctly. Self-organization undoubtedly occurs before natural selection, as Kauffman contends. For natural selection to be able to operate there should be two main criteria satisfied:

- The existence of compartmentalization or the means of compartmentalization;
- The existence of low energy, self-organized molecule states ('robust systems') that provide a foundation upon which to exercise the forces of natural selection to create and mold living organisms.

If for no other reason than to maintain spatial integrity and avoid dilution of the reacting molecules that make up each living system, compartmentalization is an essential step in the self-organization process. Lipids' well-established capacity to self-assemble into lipid bilayers and more intricate mesophases is widely acknowledged to play a crucial role in biological compartmentalization. Natural selection can be used to shape and create living things based on stable, low-energy, self-organized states known as "robust systems," such as DNA, RNA, and even folded proteins (including the prosthetic groups and cofactors they contain). Random structural variations, or mutations, that represent a molecular pool of diversity upon which selection pressures can act must exist in the primary structure of all such robust systems (specifically, base variations in nucleic acids and amino-acid residue sequence variations in proteins) in order for the forces of natural selection to exert their influence. Then, pressures from natural selection should favor the mutation or set of mutations from a particular robust system that are most advantageous to the survival of the self-assembled living system. The forces of "natural selection," operating under the influence of a particular set of metabolic selection pressures, should, for example, choose for an amino-acid residue mutation or group of mutations that promote the necessary enhancements in catalytic efficiency of a key enzyme or group of enzymes if enhanced catalytic efficiency is required. A type of molecular evolution is this.

Molecular evolution is believed to drive macro-evolution by a series of single point mutations that lead to increasingly robust systems within self-assembled living organisms. This process promotes the well-being and competitiveness of the organism. When combined with the concept of natural selection being preceded by self-organization, a holistic molecular hypothesis for the origin of life emerges. The origin of life can be explained as follows: self-organization arises from molecular complexity and diversity to compartmentalize a system of reacting molecules, leading to the formation of stable DNA, RNA, and folded proteins. These systems can then be shaped by natural selection through single point mutations to perform necessary functions for living organisms to develop and thrive. Reproduction is the final stage, relying on the autocatalytic replication of robust systems within a suitable chemical environment, transforming reacting molecules into sustainable organisms.

10.1.3. Chemical Self-Organization From Complexity

The existence of self-organization as a direct result of molecular diversity and complexity needs to be verified through both theoretical mathematical treatments and experimental testing. Put differently, through well-designed pre-biotic chemistry experiments, chemical biology should be able to clearly demonstrate how molecular diversity and complexity lead to life-like "order for free." The main objective of pre-biotic chemistry is to try to generate by experimental means chemical structures that accurately presage the biological macromolecules of life including their cofactors and prosthetic groups, by reaction from prebiotic 'starting materials.' In so doing, credible molecular links can be established from these pre-biotic 'starting materials' to recognizable robust systems, and hence 'living' organisms. There are three key early experiments that provide the foundations for pre-biotic chemistry, namely the Miller, Oro and Fischer experiments. In the Miller experiment, proteinogenic amino acids were generated spontaneously by electrical discharge through anaerobic mixtures containing methane, ammonia, hydrogen and water vapor. In the Oro experiment, adenine was generated through the heating of aqueous ammonium cyanide. Finally, in the Fischer experiment, formaldehyde and water generated sugars in the presence of calcium hydroxide (formose reaction).

Studies using "starting mixtures" of compounds known to be present in interstellar gas clouds or "starting mixtures" of compounds identified from among gas phase products generated in methane/nitrogen mixtures following electrical discharge have surpassed the three types of experiments, despite their importance. In both instances, the molecular compositions are strikingly similar. Water and oxygen are obviously absent, but they could be added later, when it's more suitable. Notably, there are a lot of "starting materials" with triple bonds that can store chemical energy (in an anaerobic environment) and most definitely have the chemical potential for significant forward reactions. Some of these forward reactions have actually been studied recently, and from the pool of anaerobic, interstellar/electrical discharge "starting materials," some important structures that form frequently have been identified. From the perspective of chemistry, these are referred to as elementary structures. Anaerobic equivalents of amino acids, bases, prosthetic groups, and cofactors must all be elementary structures that can form from lists of triple-bond compounds by any one of several energetically permissible pathways, according to a molecular theory of the origins of life. There seems to be increasing evidence to support the idea that this is the case.

10.1.4. Origins of Biological Macromolecules of Life

Protein prosthetic groups and cofactors have a valid reason to emerge before proteins and enzymes, as the former are frequently directly accountable for a protein's chemical reactivity, which is reinforced by the surrounding polypeptide structure. Because of this, cofactors and prosthetic groups are occasionally referred to as "molecular fossils." Uroporphyrinogen III, the primary precursor of protoporphyrin IX, which is present in myoglobin, hemoglobin, and cytochrome c proteins, is an excellent illustration of a molecular fossil. Cyano-uroporphyrinogen III, the anaerobic form of uroporphyrinogen III, can be generated from cyano-aspartate and cyano-glycine (α-amino nitriles) through various processes involving heat or acid catalysis. Hence, cyano-uroporphyrinogen III is a classic elementary structure! Cyano-uroporphyrinogen is composed of four main isomers, but despite this relatively primitive level of pre-biotic chemistry, the type III isomer still predominates over the other three and easily produces protoporphyrin IX upon oxidation and hydrolysis. Moreover, the primary purine and pyrimidine bases of DNA and RNA are elementary structures. Even more astounding is the development of anaerobic forms of each of these from such basic chemical starting points. It should be noted that potential intermediaries that developed along the way may even point to pre-biotic pathways leading to anaerobic versions of the redox active cofactors and prosthetic groups. As a result, these could also be considered basic structures.

It is not easy to map the putative pathway(s) from such anaerobic precursors to fully functional proteins or nucleic acids. Routes to DNA in particular require careful thought. Despite the appearance of catalytic RNA (ribozymes), the formose reaction results in complex product mixtures and low levels of racemic 5-carbon ribose, which may contradict Gilbert's theory of a pre-biotic "RNA world." The apparent preference for hexose over pentose sugars is one of the mysteries of pre-biotic chemistry that remains to be solved. The question of why pentose sugars, rather than hexose sugars, are used to construct DNA and RNA in light of this preference may be answered by the finding that hexose DNA, or homo-DNA, can form antiparallel helices without strictly adhering to the Watson-Crick base pair rules to the point where this essential component of helical cohesion is not applicable. Therefore, base-sequence-encoded genetic information could not be propagated through autocatalytic molecular replication of homo-DNA. On the other hand, pentose DNA strictly adheres to the Watson-Crick base pair rules, making it the most effective method of transmitting genetic information encoded in base sequences at the same time as autocatalytic molecular replication. The chemical aetiology of nucleic acid structure now flows from this discovery. Essentially, hexose nucleic acid "evolutionary intermediates" are necessary for the convergence upon pentose DNA and RNA in the molecular evolution of nucleic acids.

10.1.5. Predictions of the Neutral Theory for Variation Within and Between Species

A neutral allele's frequency in the population equals its likelihood of eventually becoming fixed. Furthermore, the "neutral" mutation rate per generation is practically equal to the substitution rate, which measures how quickly new alleles are fixed in a population. Therefore, the rate of evolution should remain constant if the neutral mutation rate does. The average time to fixation (measured in generations) for newly discovered neutral mutations that will be fixed by drift is roughly four times the size of the population over the long run. The average time to loss for mutations that are doomed to fail is relatively short. Large populations should therefore be expected to take a long time to fix, resulting in a lot of "transient" genetic variation that slowly diffuses through the population. When taken as a whole, these mechanisms imply that population size and mutation rate determine the degree of variation within a species. An equilibrium level of variation is produced for a stable population by the interaction between new mutations and loss or fixation through drift. It can be demonstrated that this degree of variation is such that the likelihood that an average nucleotide site exhibits heterozygosity in a randomly selected diploid sexually reproducing organism, or demonstrates a difference between two randomly selected chromosomes, is roughly equal to four times the long-term population size multiplied by the rate of mutation. The expected number of differences between two randomly selected chromosomes in the ancestral population plus an additional amount equal to the mutation rate times twice the time since speciation will determine the amount of divergence between two sequences sampled from two different species. Strictly neutral variation within a species should positively correlate with strictly neutral variation between species because variation within a species is merely an extension of variation between species, and both are ultimately driven by mutation (Kimura Motoo et al., 1983).

10.1.6. The Impact of Natural Selection on Molecular Variation and Evolution

Positive selection causes mutations that confer a fitness advantage to become more common in the population. When a variant reaches 100% frequency, it means that one variant has been replaced by another in the population (e.g., the ancestral G nucleotide has been replaced by a new A). Fixation rates resulting from positive selection can be orders of magnitude faster than those resulting from genetic drift alone.

Due to the strong genetic connections between neighboring nucleotides, selection affects both the advantageous mutation and the area of the genome that contains that variant. Therefore, rapid fixation can fix not only the favored variant but also the surrounding genomic segment. This produces a "selective sweep," or a genomic region where there is initially very little or no adjacent neutral variation. Variability will return to this region only with time and new mutations. The average divergence of linked neutral sites does not differ from the neutral prediction, which may come as a surprise. These patterns shed a great deal of light on the distribution and frequency of adaptive fixations across the genome. Natural selection maintains the population's rate of new mutations at the same level, but it significantly increases the fixation rate of advantageous mutations, increasing the sequence divergence between species for those particular sites under positive selection when compared to nearby neutral variants whose dynamics are solely determined by stochastic processes of genetic drift. One particularly good example of how this contrasting pattern of positively selected, negatively selected, and neutral variation can be used to deduce where and how natural selection has operated in the genome is the study of protein-coding sequences. mRNA uses a triplet code, or three nucleotides, to encode amino acids. There are only 20 common amino acids, but there are 61 possible combinations of three nucleotides that can encode amino acids (three more can encode "stop" signals for protein synthesis). More than one nucleotide triplet can encode an amino acid. The synonymous or silent variants between those triplet codons that encode the same amino acid are what set them apart. Those that alter the amino acid that is encoded are referred to as nonsynonymous or replacement variants.

The amino acid sequence of a protein plays a major role in determining its function, so nonsynonymous mutations have far more fitness consequences than synonymous mutations. Strong purifying selection is applied to nonsynonymous variants due to constraints on protein function, which keeps them from going to fixation (substitutions) or reaching significant frequencies in populations (polymorphism). Certain proteins, like the histones in Figure 10.1, evolve extremely slowly. Proteins with extremely lax restrictions on their amino acid sequence (such as fibrinopeptides, which are unnecessary "spacer peptides" that are cleaved from fibrinogen during blood clotting; see Figure 10.1) evolve at a rate that is almost neutral. Certain proteins are essential for responding to biotic or abiotic stimuli or for adapting to new enzyme substrates.

Here, new amino acid variants are preferred by positive selection, which speeds up the fixation of mutations. Certain adaptive responses necessitate frequent modifications to the protein sequence (at the antigen-binding sites of certain immunity proteins, for example), which leads to progressively faster replacements. This positive and negative selection has little to no effect on the rates of substitution of adjacent strictly neutral variants, which is somewhat counterintuitive. Therefore, estimates of the strength of purifying and positive selection for a protein-coding gene can be obtained from contrasting levels of variation and/or divergence at nonsynonymous to synonymous sites. Pseudogenes, or "dead genes" that are no longer in use, provide the best neutral benchmark because their levels of variation and divergence are typically comparable to those observed at synonymous sites. These differences between synonymous and nonsynonymous variation and divergence serve as the foundation for a number of tests that identify the functional targets of natural selection and identify its effects on genomes.

Functional and population genetics studies have shown that many mutations affect function only slightly, most in a slightly negative manner but sometimes improving function slightly. It is not true that all mutations are simply strictly neutral, lethal, or strongly favored. In these cases, population size may affect a mutant's ability to function as a neutral variant. The mutant will behave as a neutral variant even if it would have a (slight) selective advantage or disadvantage in an infinitely large population if the difference in reproductive success (fitness) is less than the reciprocal of the long-term population size, according to an analysis of the relative strengths of natural selection and genetic drift.

Research data has shown that a significant portion of DNA variation in a population is nearly neutral, with some being mildly harmful. This type of genetic variation can be affected by changes in population size, leading to fluctuations in natural selection's ability to act on these variants. This phenomenon is most noticeable in the "generation time effect" seen in certain types of variations. For instance, germ line mutations are often the result of DNA replication, meaning that short-lived mammals tend to have more mutations per year compared to long-lived species. However, short-lived mammals also tend to have larger population sizes, allowing for more efficient natural selection and potentially more nearly neutral mutations. As a result, protein evolution rates are slower in short-lived mammals on a per generation basis compared to substitution rates in long-lived mammals.

10.1.7. Biological Insights From the Study of Molecular Evolution

Comparing sequences from various organisms, especially short-generation organisms like viruses and bacteria, has produced one of the best examples of a fundamental evolutionary principle: descent with modification. For instance, when we have only sequence data from the extant end points of the evolutionary process for study, these important studies have provided real data sets with known phylogenies with which to evaluate the accuracy of statistical methods to estimate the phylogenetic relationships of organisms and their ancestors (Figure 10.2). Direct testing of evolutionary theories regarding adaptation, such as whether recurrent adaptation occurs through the same or different mechanisms, is made possible by experimental microbial and viral evolution studies. This is especially true because samples of each stage of the evolutionary process can be preserved for later research and even functional reanalysis.

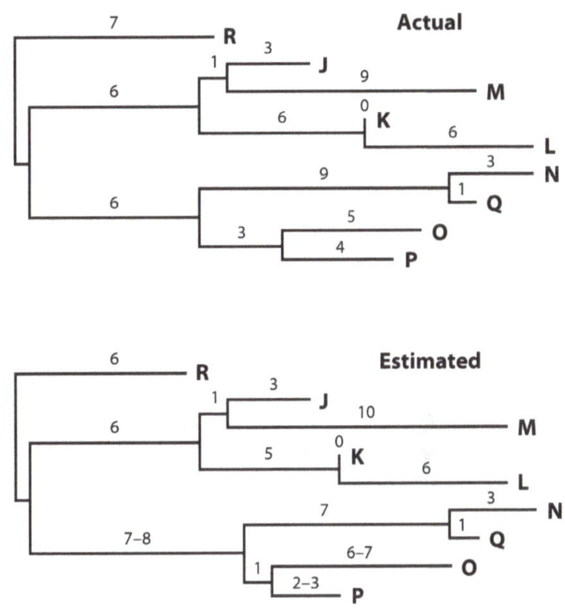

Figure 10.2. *Mutational change in virus cultures demonstrates descent with modification and molecular evolution, and the molecular clock concept. Shown is a comparison of an actual "true" phylogeny of a virus (bacteriophage T7) with an estimated phylogeny constructed using only DNA sequences of the viruses in the experimental population. This population was initiated with one virus and split in binary fashion into several sequential derived lineages, the end points being denoted as the letters on the right side of the phylogeny. Numbers above the branches indicate the actual or estimated number of nucleotide substitutions that occurred along each branch, respectively. Actual substitutions were determined by sequencing 1091 base pairs of the ancestral viruses at each branch point in the tree.*

Source: https://edisciplinas.usp.br/pluginfile.php/4481131/mod_resource/content/1/Princeton%20Encyclopedia%20verbete%20de%20evolu%C3%A7%C3%A3o%20molecular.pdf.

Molecular evolution has gone from being data limited to data overloaded, with no organism now technically unreachable, thanks to the advent of the polymerase chain reaction (PCR), which allows for the amplification of specific genomic regions from very small samples (and even ancient bones, scat, and skins), combined with wave after wave of advancements in sequencing technology and automation. The genomic study of the microbes that live in our stomachs, on our skin, in the soil, and in the oceans is a startling example. Since many of these microbes could not be grown in a lab or identified solely by morphology, science was not even aware of them. Many microbes are currently only recognized and identified by their DNA sequences, and the degree of sequence divergence from other known microbes is revealing information about these microbes' relationships. There are two ways in which molecular evolution patterns and rates can provide insight into function. First, unchanging portions of genes and genomes probably serve a crucial and constant purpose. All of biology is based on the fundamental idea of molecular evolutionary conservation, which has also served as a foundation for the study of function in the rapidly increasing number of genome sequences that have been completed for almost all kinds of plants, animals, microbes, and viruses. The shift in biology from merely focusing on the organism to using comparative functional data about an interest gene across all studied forms can be attributed to conserved regions of genes. Second, it has been discovered that the parts of molecules that undergo change are both those with little to no function and those whose rapid change is adaptive in and of itself. It can be difficult to distinguish between relaxed functional constraint and adaptive change, but over the past 20 years, a variety of statistical and computational techniques have been developed that make this distinction possible. Additionally, assessing the functional and fitness implications of adaptive fixations is a highly active field of research.

The amount of positive selection that has been detected that seems to be connected to conflict (such as "arms races" between hosts and pathogens) rather than traditional adaptation (such as adaptation to new environments or nutrients) is another aspect of molecular evolution that has emerged in the last 10 years. However, other conflicts between males and females, between host genomes and transposable elements, and even between hosts and endosymbionts have emerged as important drivers of rapid molecular evolution, including the evolution of new genes and new gene functions. It is evident that microbial and viral pathogens have driven, and continue to drive, much of the rapid molecular evolution in genomes.

In recent years, it has also become increasingly clear how important noncoding regions of the genome are for adaptation. Certain adaptive evolutionary changes in molecules and phenotype have been shown to be largely dependent on variation in regulatory sequences, which include enhancers, splicing machinery, and transcription factors. Long noncoding RNAs (lncRNAs), microRNAs, and long intergenic noncoding RNAs (lincRNAs) are examples of new and unexpected types of functional sequences that have been discovered. These sequences not only show remarkably high levels of evolutionary conservation and control over important developmental and cellular processes, but they may also be the cause of adaptive change in certain situations. There is obviously still a great deal to learn about genomes and the processes that allow molecules of life to change and adapt.

10.2. MOLECULAR BREEDING; NATURAL SELECTION ACTING ON SELF-ORGANIZATION

In essence, breeding is a natural selection process that humans have created. For millennia, humans have utilized breeding to select for desired traits in other species. Horses have been bred for speed and strength; crops for improved yield and quality of produce; yeasts for baking and fermentation; livestock for increased nutritional value and docility; and cats and dogs for company. For thousands of years, the only thing known was that crossing individual organisms with desired traits frequently resulted in the accumulation of those traits in their offspring and succeeding generations. This approach was and is wholly empirical; without a true grasp of the underlying principles, improvements in function and utility were made. We now understand the genes that "code for traits" and how breeding aims to alter or mutate genes thanks to Mendel's groundbreaking research. Additionally, we now know more about how these genes are regulated and/or controlled to code for different traits that the breeder needs to fill a biological niche, traits that are deemed desirable or that may have some selective advantage.

Although the entire breeding process cannot yet be precisely duplicated at the molecular level in the lab, parts of it that involve the alteration or mutation of genes that result in changes to proteins can. This could be considered the most basic form of molecular breeding. According to one theory of molecular natural selection, the most malleable class of biological macromolecules is protein structure, which can be altered and its function modified by successive mutations. As a result, a key component of organismal adaptation to selection pressure is protein mutation. The availability of a pool of protein mutations is therefore a fundamental prerequisite for the molecular evolutionary process. There is, however, disagreement among experts regarding the best ways to achieve improvements or novel functions. Some believe that the path to progress is always a sequential evolutionary series of protein species, each of which differs from its neighboring species by a single point mutation, while others allow for multiple mutations in each evolutionary series of protein species. Either way, it is evident how to go about molecularly breeding new or better protein functions. As long as a suitable pool of single or multiple mutations is subjected to some kind of artificial selection pressure, it should be possible to identify and isolate mutant proteins with new or improved functions, as desired by the molecular breeder.

In chemistry, the conventional method is to first comprehend structure and mechanism in order to offer insights that facilitate a logical approach to better or changed binding or reactivities. This method has been used throughout most of synthetic organic chemistry to create an amazing body of knowledge that links structure and reaction mechanism. This idea has also been applied to rational protein design; for example, increasing an enzyme's catalytic efficiency can be achieved by reasoned site-directed mutagenesis in conjunction with structure and mechanism knowledge. The opposite method is molecular breeding. Molecular breeding, in theory, makes it possible to find modifications or enhancements in the binding or biocatalytic activities of proteins, RNA, and, increasingly, DNA, without having to know much about their structures and workings beforehand. Therefore, the problems in molecular breeding are very different from the typical problems in chemistry: how do we create molecular diversity and then select for particular desired traits in an iterative process? This molecular breeding approach is showing some success in both academia and practice, even though it is not yet orthodox. Practically speaking, evolved proteins are already available for purchase. Examples include enhanced lipases and evolved modified green fluorescent protein. From an academic perspective, *in vitro* evolution sheds light on the evolutionary process itself. Research on the evolution of self-replicating systems may even shed light on the molecular roots of life.

10.3. DIRECTED EVOLUTION OF PROTEIN FUNCTION

One of the most basic strategies for molecular breeding involves starting with an enzyme gene that carries out a specific function, then using a variety of techniques, such as chemical mutagenesis, erroneous PCR, gene shuffling, or incremental truncation, to introduce variation or molecular diversity at the genetic level into the gene. Almost all fundamental molecular breeding techniques begin with a protein that, at least partially, possesses a desired trait; the objective of the *in vitro* experiment is to enhance or evolve that trait. One term for this type of molecular breeding experiment is directed evolution of protein function. Following the generation of mutant genes, protein variants are expressed and subjected to artificial selection pressure through the use of a particular assay. After determining which mutant protein variant is the most improved, the entire procedure is repeated, beginning with the gene corresponding to the improved mutant protein variant.

10.3.1. Random Mutagenesis and PCR

Using a technique of random mutagenesis by error-prone polymerase chain reaction (error-prone PCR) is a helpful way to generate molecular diversity starting from a given gene (Figure 10.3). Beneficial mutations are uncommon but harmful mutations are common when using error-prone PCR, so the mutation rate needs to be carefully adjusted. All advantageous mutations may be rendered inactive by secondary harmful mutations if the mutation rate is too high. On the other hand, too few advantageous mutations will be seen if the mutation rate is too low. Nevertheless, this approach has a major flaw in that there will typically only be one beneficial mutation for each selection event. The process of evolving a protein is comparatively slow, even with sequential random mutagenesis. Negative mutations are likely to persist while many beneficial mutations may be eliminated (perhaps only to be discovered with the help of a second mutation). Error-prone PCR can be carried out in experiments in a variety of ways. Since Taq polymerase, the standard PCR enzyme, is remarkably accurate, alternative methods—like substituting Mn^{2+} for Mg^{2+} and altering the concentration of dNTPs beyond equimolar amounts—are needed to increase the error rate. Eliminating bias from a PCR experiment that is prone to errors is a challenging task.

For example, error bias refers to the fact that certain mutations are more likely to occur than others due to the properties of the polymerase. The fact that some amino acids are coded for by a single codon while others are coded by numerous codons causes issues with the genetic code itself. This means that the probability of replacing a specific amino acid within a given codon varies from amino acid to amino acid (codon bias). The iterative nature of PCR itself is another source of bias; any mutation that arises early in the process is likely to appear in a sizable portion of the final DNA sequences (amplification bias). Chemical mutagenesis is an alternate strategy to error-prone PCR that involves introducing less biased variations into the gene coding the original protein of interest using chemical mutagens. Using mutator strains is an additional strategy. These bacterial strains have abnormalities in their DNA repair pathways, which causes a far higher rate of mutation in their DNA than is typical. The lack of control over the degree of mutation is a major issue with both of these methods; mutations will be discovered in both the chromosome and the construct that contains the desired target gene. Furthermore, the rate of mutation cannot be easily controlled when mutator strains are used. Consequently, because error-prone PCR allows for more control over mutagenesis rate and location, it has generally supplanted both of these approaches as the method of choice for causing random mutagenesis.

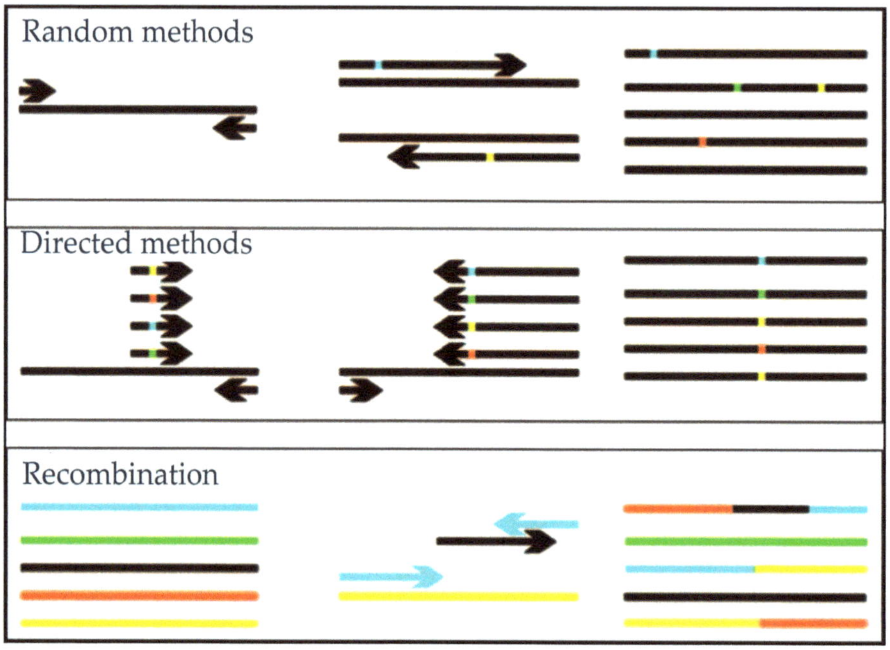

Figure 10.3. *Methods for the randomization of DNA sequences using PCR. Each line in left column represents DNA duplex and each arrow a primer; the 5'-sense primer initiates strand copying using the complementary strand as a template, and the 5'-complementary primer initiates strand copying of the complementary strand using the sense strand as a template. Random methods such as error-prone PCR introduce random changes at positions throughout the gene sequence. Directed methods will introduce changes at a specific position(s) defined by lesions in the primers themselves. Recombination methods such as family shuffling start from a library of homologous genes from related species that can be artificially fragmented by DNAse I treatment and recombined*

by PCR (without added primers) leading to chimeric genes that code for chimeric proteins. These methods confer high levels of molecular diversity upon which to exercise selection screening for novel or improved characteristics.

Source: https://www.semanticscholar.org/paper/Chemical-and-biochemical-strategies-for-the-of-DNA-Neylon/ab2233c55dcffca2de92fc67b27d9cfc1e7c7d24.

10.3.2. Mutagenesis and DNA Shuffling

The development of DNA shuffling as a combinatorial method for searching "sequence space" marked a significant advancement. DNA shuffling is the process of combining related sequences that contain mutations (Figure 10.3). The above-mentioned issue of the loss of potentially beneficial mutations when utilizing error-prone PCR is resolved by DNA shuffling, or sexual PCR. There are several steps involved in shuffling DNA. Initially, erroneous PCR is used to introduce random mutations into a sequence. The first step in screening these seemingly advantageous random mutations is to identify which ones they are. Second, a PCR reaction is conducted without the addition of primers to recombine the gene sequences that were randomly fragmented using the enzyme DNAse I after the beneficial mutants have been combined. As a result, every advantageous mutation is incorporated into a blend of rearranged genes at distinct points within the gene sequence. Furthermore, at this stage, new mutations could potentially arise. Ultimately, primers are introduced prior to the last PCR cycle in order to produce a full-length PCR result. Starting from a library of homologous genes that were originally from related species is an extension of the DNA shuffling technique. Family shuffling is the name given to this technique, which can produce intriguing chimeric genes and proteins that combine characteristics of several parent genes into a single gene that codes for a single polypeptide. DNA shuffling is, in many respects, far more similar to the natural processes that give rise to the pool of random mutations that serve as the basis for the forces of natural selection driving the evolution of better or novel protein activities than error-prone PCR.

10.3.3. Oligodeoxynucleotide Cassette Mutagenesis

Oligonucleotide cassette mutagenesis can be used to target specific sequences of the gene coding for the desired protein when more restricted randomness is desired or necessary. Primers may be selected for focused, error-prone PCR within a short, 10–30 base gene segment that is typically chosen for mutagenesis. The amplified region is then recombined back into the wild-type gene, replacing the original short region PCR. Or, using oligodeoxynucleotide mutagenesis primers (also known as "dirty oligos") that are "spiked" at each position of the primer with a small percentage of the three alternative nucleotides to the correct, complementary Watson-Crick base pair nucleotide at each position, numerous "site directed mutagenesis" experiments can be carried out. "Spiking" is intended to guarantee that, at the end of the mutagenesis experiment, each mutant gene produced will have only one, possibly two, base mutations. Therefore, the "spiking" should make sure that every potential single-point mutation in an amino acid residue can be sampled as nearly as possible in a mutant library throughout the whole range of the selected short region within the original protein of interest. This procedure is the purest form of saturation or random mutagenesis. Mutagenesis primers, on the other hand, might be made to exclusively support single point mutations of individual amino acid residues that belong to the same type, class, or identity. Actually, there are a plethora of options available when employing cassette mutagenesis to introduce molecular diversity into a constrained structural region of a particular protein of interest.

10.3.4. Screening Strategies

A molecular breeding experiment's ability to screen a large number of protein variants is frequently constrained by how quickly and easily the desired activity can be assayed. Thus, the screen needs to be well-designed to make it simple to recognize enhanced proteins when they appear. For example, the screen only needs to measure increases in protein fluorescence above a predetermined threshold if improved fluorescence qualities are needed in a molecular breeding experiment. Additional goals of a molecular breeding experiment could be to alter the environment in which biocatalysis occurs, such as in non-aqueous solvents, at different pH levels, or at higher temperatures. Under these circumstances, straightforward screens can also be developed to find better proteins. The design of efficient screens becomes more difficult when altered enzyme specificity is needed, particularly when de novo biocatalytic activity is needed. This worked example should help illustrate how a molecular breeding experiment's screening procedure works in tandem with the mutagenesis steps to identify better proteins that perform the desired functions.

A chiral catalyst's ability to catalyze an achiral substrate to a chiral product with a bias toward the formation of one of the enantiomers is known as asymmetric catalysis. In the chemical industry, there is a great need for chiral products, especially in the pharmaceutical sector where, in many cases, only one of the enantiomers of chiral drugs is pharmaceutically active and, in the case of thalidomide, the other enantiomer may even be hazardous. Even with the increased demand, asymmetric catalysis with conventional organic chemistry is still very difficult. Using directed evolution of a carefully chosen enzyme to evolve the necessary, desired enantioselectivity is one way to overcome the challenge. There are two main challenges in realizing such a molecule breeding experiment.

- Designing methods to introduce molecular diversity, which is likely to be beneficial.
- Designing a simple high-throughput screen to identifying improved proteins.

In the following case, we shall look at the molecular breeding of the lipase from the bacterium *Pseudomonas aeruginosa*. This particular molecular breeding experiment was the first example of the directed evolution of an enantioselective enzyme. The wild-type enzyme catalyzes the hydrolysis of esters to carboxylic acid and shows very little enantioselectivity – only 2% enantiomeric excess biased towards the (S) configuration.

Error-prone PCR was used in the initial experiment to introduce molecular diversity. The conditions were set up to guarantee that a single or possibly two base mutations were introduced per mutant gene, in an effort to sample every possible single point amino acid residue mutation across the entire primary structure of the original protein of interest. The numbers begin to tell the story on their own. The target lipase has 285 amino acids in it. 5415 single point amino-acid residue mutants could be created if each amino acid mutated to any one of the other 19 natural amino acids. This number increases to 15 million unique variations when double point mutants are considered, and to 52 billion variations when unique triple point mutants are considered. These figures are typical for protein mutant libraries and show how difficult it is to screen effectively and efficiently. High throughput screening is indispensable given such molecular diversity. In the case of this lipase example, the assay used was a simple UV–visible spectroscopy assay monitoring effective enzymatic hydrolysis in terms of the appearance of the p-nitrophenolate anion (detected by absorbance A_{405}). Each mutant protein was tested in a pairwise assay, testing hydrolysis of both the (S)- and the (R)-substrates. The relative rates of hydrolysis gave a measure of the enantiomeric selectivity (the enantiomeric excess, ee, as a percentage) as given by:

$$ee = \left(\frac{[R] - [S]}{[R] + [S]}\right) \times 100$$

Where; [R] and [S] are the molar concentrations of the (R)- and (S)-products respectively.

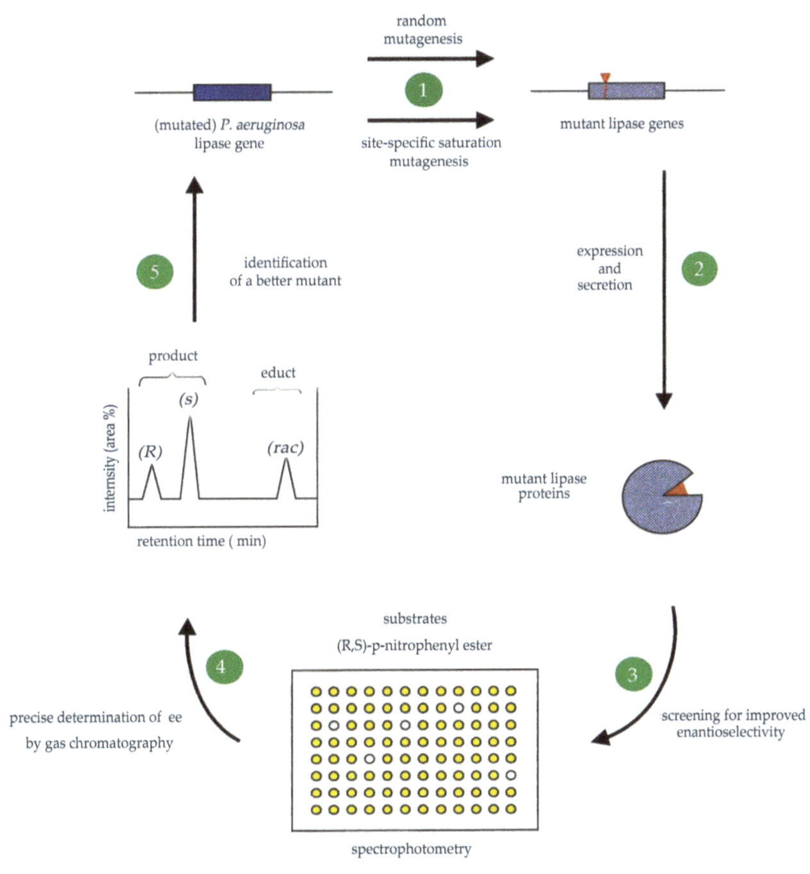

Figure 10.4. *Strategy for evolving an enantioselective lipase. Molecular diversity is introduced by random mutagenesis of the lipase gene and site-specific random or saturation mutagenesis. Mutant genes resulting are segregated and separately expressed in individual bacterial clones.*

Protein products are secreted into the extracellular medium and a spectrophotometric assay used to assay for nitrophenolate production from (R) and (S) substrates by action of active lipase enzymes. Relative quantities of (R) and (S) product are quantified by gas chromatography and where enantioselectivity is also discovered then the corresponding gene is harvested for another round of mutagenesis and selection.

Source: https://www.sciencedirect.com/science/article/pii/S1074552100000156.

During the first screening, about 1% of the 1000 mutant enzymes produced by the first and subsequent rounds of mutation demonstrated enhanced enantioselectivity, reaching up to 31% ee, which was a notable improvement over the wild-type enzyme. Following three more rounds of mutation and selection on these mutant enzymes, a mutant with enantioselectivity of up to 81% ee (corresponding to an E value of 11.3) was identified. Subsequently, local "hot spots" were identified in the primary sequence where amino-acid residue modifications were believed to be most likely to be advantageous. This allowed for the creation of a mutant enzyme capable of producing an E value of 25 through saturation mutagenesis. A new error-prone PCR step that was altered to add about three mutations per gene per generation, followed by DNA shuffles, produced a mutant protein with an E value of 32 post-selection, and further enhanced this efficacy. Finally, a much improved mutant with the capacity to produce an E value of over > 52 was identified through a modified form of DNA shuffling known as combinatorial multiple-cassette mutagenesis.

10.4. DIRECTED EVOLUTION OF NUCLEIC ACIDS

Compared to the evolution of proteins, the evolution of nucleic acids has occurred in a more direct manner. The complete cycle of transcription and translation is necessary for directed protein evolution, which usually means that certain steps of the process must be carried out using a living system to translate the protein. But nucleic acid selection and evolution can be regularly achieved through *in vitro* molecular breeding, a completely chemical *in vitro* procedure. Nucleic acids have binding strengths that are comparable to those of proteins, despite having a much smaller catalytic repertoire. We will discuss methods for the selection and evolution of DNA and RNA for binding and catalysis in this section.

10.4.1. Aptamers

Nucleic acids are not just responsible for storing the genetic material that codes for proteins. Nucleic acids with a single strand can fold into complex structures that are able to recognize molecules and even catalyze reactions. Similar to how the amino-acid residue sequence determines the three-dimensional structures of polypeptides, the primary structure, or deoxynucleotide (or nucleotide, for RNA) sequence 5' →3,' specifies three-dimensional structures. Certain ribosomal RNAs (rRNAs) have the ability to catalyze essential steps even within the protein synthesis pathways themselves, while transfer RNAs (tRNAs) use their three-dimensional shape for molecular recognition in nature.

From a technology perspective, a massive pool of nucleic acids (RNA, DNA or modified nucleotides) (that provides for the required molecular diversity) represents a huge library from which to select, amplify and identify novel binding partners (aptamers) or even novel biocatalysts (aptazymes). This type of molecular breeding is known as the systematic evolution of ligands by exponential enrichment (SELEX) process. Using the SELEX process, it is possible to evolve naive nucleic acids until they are capable of binding specifically to proteins, peptides, cells, nucleic acids and even other small molecules. Nucleic acids capable of catalyzing chemical reactions may also be evolved. SELEX may be used to select from both RNA and DNA libraries. RNA has greater potential structural diversity than DNA, hence affinities for a potential target should be generally higher overall. DNA has less structural diversity than RNA, hence affinities for a potential target should be generally lower overall, albeit with a higher degree of overall intrinsic stability. SELEX steps are outlined below.

10.4.1.1. Design and Construction of a Polydeoxynucleotide/Polynucleotide Library

Standard solid phase techniques are used to create a polydeoxynucleotide library for molecular breeding; however, randomized single-stranded DNA, or pre-template DNA, is created using a 3'-flanking region of known sequence that can be identified by a primer in a PCR reaction. To ensure that there is no deoxynucleotide bias in the final library, randomization is achieved during solid phase synthesis by using an approximately equal mix of the four phosphoramidite building blocks weighted according to differences in coupling efficiencies. The molecular diversity is equal to 4^n when a fully randomized polydeoxynucleotide is created from the four primary deoxynucleotide building blocks (dA, dG, dC, and dT), where n is the number of randomized base positions (in this case, 25, meaning a total of 10^{15} library sequences). The molecular diversity rises to a library of 10^{18} if n is increased to 30. Molecular diversity becomes uncontrollable after this point. Because the molecular diversity would approach the number of elementary particles in the universe, for example, if a stretch of 100 random positions were to be studied, 25–30 randomized positions are usually used in preference. Randomized double-stranded DNA (template DNA) is produced by synthesizing complementary strands using the Klenow fragment of DNA polymerase I and annealing a primer to the known sequence region after the pre-template DNA is synthesized.

A high affinity binding polynucleotide should likewise have a length of 25–30 nucleotide residues, as natural RNA–protein recognition sites consist of 15–25 nucleotide residues. Because T7 RNA polymerase operates from a T7 promoter element inserted during the synthesis of the template DNA, it is possible to generate an appropriate RNA library by direct transcription from template DNA. To give RNA molecules more stability, modified nucleotides are frequently used in place of natural ones. This is crucial when designing aptamers for medical applications. The most popular method involves adding 2'-F or 2'-NH2 modifications to β-d-ribofuranose rings linked to pyrimidine bases, which provide resistance against the majority of RNAases. It is also possible to change the uracil base's 5-position. Fortunately, modified bases and β-d-ribofuranose rings inserted at the 2 -position can be tolerated fairly well by the T7 RNA polymerase.

10.4.1.2. Partition, Amplification and Iteration

In order to separate DNA molecules that are able to accomplish the necessary task from those that cannot, basic screening should be the next step in the molecular breeding process of functional DNA molecules from template DNA. The target ligand could be immobilized on a column and affinity chromatography applied, for example, if DNA aptamers are desired. Nucleic acids that bind to the column would bind to the target ligand, while those that do not would pass through. In SELEX experiments, immobilizing the target on a nitrocellulose filter is a common substitute. The most crucial factor in a SELEX experiment is partitioning it must be possible to effectively reduce the initial complexity of a template DNA library's 10^{15}–10^{18} sequences to a manageable number of sequences in as few screening rounds as feasible. Preventing or reducing the likelihood of non-specific nucleic acid binding to column or filter media, which could result in false positive results, is another aspect of this procedure. Before moving on to sequence characterization, a number of screening iterations may be performed to further hone the pool of highly affinity DNA binders to the preferred target ligand.

When screening an RNA library for RNA aptamers, the process begins with isolating functional mRNA that makes it through the selection process. This pool of mRNA is then transformed into next-generation pre-template DNA via reverse transcription and next-generation template DNA through PCR (which is the same as cDNA synthesis). After that, transcription using T7 RNA polymerase is carried out, allowing RNA selection to recommence from this smaller, next-generation mRNA library of "positive hits." Once high affinity RNA binders to the preferred target ligand are identified, selection continues through additional full rounds of target ligand binding and release. The template DNA can be added to a cloning vector if necessary, and each individual clone can then be sequenced.

In cases where RNA or DNA aptamer identification is necessary, alignment of either set of sequences following multiple screening iterations ought to yield the identification of crucial consensus sequences necessary for the intended aptameric activity. Competition experiments can be conducted to determine whether these families compete for the same target, as this often leads to families of aptamers. There are several methods available for refining the chosen aptamer families. First, the shortest aptamer that can bind to the target with a high degree of affinity can be found using truncation. The alignments of the different sequences can occasionally reveal potential truncations; truncations are simple to design and test using common molecular biology methods. As an alternative, digestions against the target from either end could be tried and then separated to determine the precise length needed for high affinity binding. Aptamer complex structures with their targets have been solved for a number of them.

10.4.1.3. Applications of Aptamers

Therapeutic aptamers and therapeutic antibodies have many similarities. Aptamers outperform antibodies in the following areas, despite the fact that therapeutic antibodies may have a longer history.

- Aptamers are selected in an entirely *in vitro* process that allows for specificity and affinity to be tightly controlled, and allows selection against toxic or non-immunogenic targets.
- Aptamers generally bind more strongly to a target than an antibody, and are able to disrupt protein-protein interactions.
- Aptamers show little immunogenicity and no toxicity, and can be delivered subcutaneously.
- Aptamers are easily synthesized, readily producible in bulk and easily stored as lyophilized powders.

Aptamers can bind to almost any target with an affinity in the nano- to micromolar range, just like antibodies can. They need to bind firmly and inhibit a particular biomolecule function while exhibiting no negative side effects in order to have a therapeutic effect. One such aptamer that was examined in a therapeutic setting was a DNA aptamer that can extend the duration of a blood clot *in vitro* by competing with thrombin. Additionally, aptamers are becoming more and more significant in the field of molecular diagnostics. For instance, they can be connected to fluorescence tags and transformed into beacons, which indicate the presence of important disease markers at extremely low detection levels.

10.4.2. Selection of Catalytic RNA

A catalytic RNA evolves in two general steps. First, a screening procedure that distinguishes between active RNA with the desired trait and those without it needs to be implemented from a pool of sequence variants. Secondly, before the next round of selection, the survivors' genomes need to be duplicated and amplified using DNA intermediates. The method is iterative and mainly depends on how simple it is to copy nucleic acids using PCR amplification. There are then two ways to select: direct selection and indirect selection. The catalytic RNA undergoes a chemical transformation in a direct selection process. In an indirect selection method, an RNA aptamer approach is used, but the RNA aptamers are found and modified to bind to a transition state in the desired catalyzed reaction.

In direct selection, one needs a method to separate catalytic from non-catalytic RNA. There are three general approaches:

- differential migration by polyacrylamide gel electrophoresis
- primer-binding site tagging technique
- affinity probe tagging technique.

Since the catalytic RNA is a reaction product in differential migration, it has the potential to be selected. The target could be RNA self-cleavage or ligation of RNA that is already covalently bound to the catalytic region in another location. It should be obvious that in the event of a reaction, the outcome would differ electrophoretically from that observed in a non-reactive system. Therefore, cleanly reacting and non-reacting scenarios can be distinguished using PAGE. As an alternative, the primer-binding site tagging method works well when analyzing ligation reactions, for instance. In this instance, a stretch of RNA sequence that could serve as a PCR primer site is introduced onto the catalytic RNA itself by successful ligation. Consequently, only if the ligation reaction was successful in the first place would successful PCR amplification be feasible. As a result, PCR amplification can and will only amplify RNA that is correctly catalytic. Finally, catalytic RNA that can catalyze additional chemical reactions can be identified using affinity probe tagging. An affinity probe, like biotin, should be used in the chemical reaction of interest to cause the RNA to self-tag. Thus, using streptavidin protein affinity chromatography, catalytic RNA can be easily distinguished from non-catalytic RNA. Numerous attempts have been made to evolve catalytic RNAs, also known as ribozymes, from pools of random sequence RNA as well as from naturally occurring ribozymes. The advantage of natural ribozymes is that they can be evolved because they already have a natural active site; these methods have been somewhat successful in modifying or enhancing catalytic properties. Other ribozymes that have evolved thus far include polynucleotide kinases, ligases, and catalytic RNA nucleases. Similar approaches have also been employed to evolve ribozymes that catalyze reactions at carbon bonds, such as functional group transfers, including N-alkylation (e.g., using the 2-amino functional group of guanosine to displace iodide from iodoacetamide) and S-alkylation (e.g., using 5 -phosphorothioate to displace bromide from an N-bromacetyl peptide). Because non-nucleotide substrates require more complex conformations to bind, catalytic RNAs that catalyze reactions at carbon bonds typically have more complex structures than those that catalyze reactions at phosphodiester bonds. There is much room for more research in this field because, on the whole, natural ribozymes prove to be much more effective catalysts than evolved ribozymes.

10.4.3. DNA Aptamers and Catalytic DNAs

Although no DNA-based catalysts have yet been discovered in nature, single-stranded DNA can adopt complex tertiary structures in a manner similar to RNA, which makes catalytic DNA possible. Using similar molecular breeding techniques to those employed to evolve RNA aptamers and ribozymes, it is possible to evolve both DNA aptamers and DNA catalysts (deoxyribozymes). However, because each deoxynucleotide residue in DNA lacks a 2' -hydroxyl group in its sugar ring, DNA polydeoxynucleotide sequences are less adaptable and flexible than their RNA counterparts.

However, DNA aptamers and deoxyribozymes are more resistant to nucleases compared to unmodified RNA aptamers and ribozymes. The discovery of deoxyribozyme catalysts has sparked a new discussion on the origins of life and has brought the RNA world concept back into the spotlight for further examination. Single-stranded DNA aptamers have been found to bind thrombin with a Kd value ranging from 25 to 200 nM. These aptamers typically have a central core of 15–17 residues that form a G-quartet structure, a common motif in both DNA and RNA aptamers. Other defining structural features of DNA aptamers that bind proteins include bulges, stem-loops, and pseudoknots. Some DNA aptamers have also been engineered to bind specific small molecules, such as ATP with a 6 μM affinity. Interestingly, DNA aptamers have been successfully designed to bind a variety of porphyrin rings. It is important to note that while DNA and RNA aptamers may be functionally similar, they exhibit structural diversity and show no sequence homology when evolved to bind the same target ligands.

Going beyond DNA aptamers, catalytic DNA, or deoxyribozymes, have been evolved to cleave RNA or DNA, metalate porphyrins, cleave N-glycosidic bonds and ligate DNA. The majority of deoxyribozymes isolated catalyze RNA cleavage by phosphoester transfer, usually in the presence of a metal cofactor. The general approach for discovery of a cleaving deoxyribozyme is follows. An initial population of single-stranded DNA is synthesized that incorporates a primer target site, a target sequence, a randomized sequence with a second 3' -primer target site, and a 5' -end biotin label for simple immobilization. This multifunctional polydeoxynucleotide is first immobilized on a streptavidin protein matrix, then non-binding DNA molecules are washed away. Those immobilized DNA molecules are then incubated under suitable reaction conditions, importantly incorporating potential deoxyribozyme metal cofactors. Thereafter, successful catalytic DNA molecules that self-cleave from the column are washed clear and then amplified using PCR and a biotin labeled primer. Double-stranded DNA from PCR amplification is then rebounded to the streptavidin matrix. The process is repeated until the catalytic activity of the deoxyribozyme reaches a plateau. At this stage the evolved deoxyribozyme may then be eluted, sequenced and the different species aligned to define the catalytic motif. In order to evolve deoxyribozymes against other fissile bonds or links, the screening procedure is actually the same, although the appropriate target cleavage site should be included in the initial multifunctional polydeoxynucleotide.

Finding a deoxyribozyme that can ligate RNA is a great illustration of how catalytic DNA development works. An original multifunctional polydeoxynucleotide with two regions complementary to two random RNA substrates served as the starting point for the process. A randomized sequence that was intended to evolve and enrich ligase activity was positioned between these two arms. Otherwise, to reduce steric hindrance to DNA, a small number of RNA bases were left unpaired. Right-side RNA was annealed to the DNA and ligated using the T4 RNA ligase enzyme as the first step in the selection process. The generated DNA/RNA duplex was then separated and purified using PAGE. The left-hand side RNA with a 2,' 3' cyclic phosphate was added in the second step. It was anticipated that this second RNA would anneal to the accessible DNA binding site through Watson-Crick base pairing, creating the conditions for ligation. Potential deoxyribozyme metal cofactors were once more incorporated to facilitate any such potential ligation reactions. A shift in the ligated product's electrophoretic mobility on polyacrylamide gels relative to that of the non-ligated, unanalyzed condition served as a sign of successful ligation. Catalytic and non-catalytic DNA were progressively separated in this manner. Iteratively, this entire process was carried out over a number of cycles until the majority of the remaining DNA pool was shown to exhibit some ligase activity.

10.5. CATALYTIC ANTIBODIES

There are numerous similarities between the application of antibodies and aptamers. The primary structural distinction between aptamers and antibodies is, of course, that the former are based on polynucleotide or polydeoxynucleotide chains, while the latter are based on polypeptide chains, specifically the immunoglobulin fold. A vertebrate immune system normally possesses the ability to biosynthesize at least 10^{10} structurally unique antibodies in immune cells called B-cells, each of which has a particular affinity for foreign substances known as antigens. The humoral immune response directed against foreign agents or invasive pathogens includes the production of antibodies. A chemical ligand, or hapten, can be highly immunogenic in terms of the humoral immune response when it is coupled to a carrier protein, such as keyhole limpet hemocyanin (KLH) or bovine serum albumin (BSA). A mouse or rabbit injected with such a product develops a significant blood-borne response, which is mirrored in the massive production of B cells in their spleens. In order to create immortalized B-cell lines (hybridomas), these animals are slaughtered and their spleens removed. The B-cell populations are then fused to cancer cell lines like myeloma cells. The purpose of hybridomas is to function as factories of antibody production; each hybridoma should be able to generate a distinct antibody that is particular to a distinct antigen. When hybridomas produce anti-hapten antibodies, they are found through cell population screening, separated, and cloned using cell culture. Every hybridoma that has been cloned in this way ought to produce a single kind of antibody (monoclonal antibody). To find the population of monoclonal antibodies that is best suited for hapten binding, monoclonal antibodies usually need to go through one last screening.

The most efficient method of achieving catalysis, regardless of the quantity of substrate available for biocatalysis, is, according to the most recent theories of enzyme catalysis, the differential stabilization of reaction transition states with respect to ground states. Therefore, an antibody must have the best affinity for binding to a reaction transition state rather than a ground state molecule like an antigen in order to function as an enzyme. Thus, any monoclonal antibody that has been raised against an analog of a transition state ought to have the ability to function as a biocatalyst for the associated reaction. It takes creativity and a certain amount of chemical surrogacy to design a transition state analog. For example, before the second alcohol elimination step, phosphonate seems to be a good proxy for the rate-determining tetrahedral transition state that is created when water is added to an ester. In fact, the P-O bond (1.521 A) is intermediate in length between the C-O bond found in the actual tetrahedral intermediate and the assumed C-O bond length projected in the transition state. In addition, the phosphonate group is mono-charged. Additional illustrations of catalytic antibody catalysis and transition state analogs. Additionally, the biocatalysis of biomolecular reactions like transamidation may be impacted by catalytic antibodies. A word of caution, though: catalytic antibodies generally do not influence biocatalysis with the same efficiency as natural enzymes, just as natural ribozymes do with natural enzymes. In fact, they typically come up far short. Therefore, alternative, additional theories of catalysis, such as general acid–base, nucleophilic, and/or electrophilic catalysis, have been used to improve catalytic–antibody-mediated biocatalysis. The field of catalytic antibodies is still advancing.

10.5.1. History of Catalytic Antibodies

- In 1948, Linus Pauling an American biochemist explained the principle of enzyme catalysis in detail. He proposed that compounds resembling the middle state or transition state of an enzyme reaction will be very potent inhibitors of those enzymes and hence these mimics were called transition analogs.
- In 1969, Jenks, suggested that one could obtain antibodies with catalytic activity by generating antibodies raised against a stable analog of the transition state.
- In 1986, Peter Schultz and Richard Lerner generated abzymes which catalyze ester hydrolysis for the first time.
- In 1989 Paul et al., discovered the first natural catalytic antibodies (IgG) which hydrolyze the vasoactive intestinal peptide (VIP) in bronchial asthmatics. This discovery was followed by IgG (DNase activity) and IgG (RNase activity) in systemic lupus erythematosus (SLE) and in several autoimmune (AI), viral disorders etc.

10.5.2. Structure and Mechanism of Action of a Catalytic Antibody

CatAbs share structural similarities with the Y-shaped Ig G-type, which is made up of two identical heterodimers joined by S-S bonds. Each heterodimer is made up of two chains: one light and one heavy. Conserved domains (F_c), which are constant regions formed in between two heavy chains, make up one end of the immunoglobulin molecule. Except for the variable epitope-paratope interaction, conserved domains have been found to have similar amino acid compositions in most immunoglobulins. The variable domains (F_v), on the other hand, are in charge of selectively binding antigen. There is a single deep pocket (antigen binding site) where two V_H and V_L regions converge. Complementarity determining regions are hot spots within the variable domains that are essential for antigen specificities (CDRs). Through non-covalent interactions, the amino acids in the hot spots interact with the antigen.

The transition state analog haptens that have been used to generate them during the immunization process are known to bind extremely tightly to CatAbs. The binding forces in the antibody binding sites are necessary to stabilize the transition states and intermediates, lowering the total energy barrier of the reaction and speeding up the rate of reaction. Kinetic parameters like V_{max} and KM are used to measure the rate of reaction that a CatAb catalyzes. The low K_M of catalytic antibodies indicates that they have a higher affinity to bind a target molecule. However, due to their low V_{max} values, which suggest a slower reaction rate, CatABs are still unable to approach or even come close to the rates of reaction that biological enzymes catalyze. Furthermore, CatAbs use a binding mode known as "induced fit," which improves complementarity between the hapten and the antibody combining site. Real enzymes, on the other hand, are usually able to alter their own conformation when binding the transition state.

10.5.3. Construction and Generation of Catalytic Antibodies

"Abzyme" is the term for a synthetic catalytic antibody that is created based on demand. Usually, it's a monoclonal antibody that was produced in a lab to specifically bind and catalyze a reaction against particular synthetic haptens. The major components for Abzyme generation are:

- A transition state analog which is able to mimic the transition state of desired reaction is synthesized. These are called as haptens
- Haptens are attached to carrier molecule which is capable of eliciting an antibody response. These are called as antigens
- Specific antigen designed is injected to invivo models like mouse or rabbit
- Monoclonal antibodies are isolated and purified
- These monoclonal antibodies are further tested for catalytic activity by cat ELISA, HPLC etc

The major stages associated with the production of catalytic antibodies include designing haptens, generating monoclonal antibodies by hybridoma technology and further purification and screening for catalysis.

10.5.3.1. Designing of Haptens

Four different methods are most commonly used for designing haptens

Stable transition state analog: This most widely used method is regarded as the cornerstone of enzyme research. Because of prior knowledge about the nature of reaction mechanisms and extensive experience using phosphoryl species as stable mimics, enough work has been done to understand the hydrolytic reactions of acyl species. The majority of these acyl transfer reactions involve the nucleophile being added stepwise, followed by the leaving group—a transient, highly energetic tetrahedral intermediate—being expelled. The most progress in this direction has been made with regard to phosphorus-based transition states.

- **Bait and Switch Hapten Design:** Charge-charge complementarity is one of the most remarkable characteristics that contributes to the precise and strong binding of antibodies to their corresponding antigens. The "bait and switch" tactic has taken advantage of this feature. Antibodies with induced charges serve as general acid/base or nucleophilic catalysts because the charges on the hapten cause complementary charged residues to form in the active site. The b-elimination reaction, acyltransfer procedures, cis-trans alkene isomerizations, and dehydration reactions are all frequently carried out using this technique. Using the bait and switch technique, a hapten is used as "bait." Put differently, the hapten serves as a modified substrate by incorporating ionic functions meant to simulate the expected coulombic distribution in the transition state. The altered substrate is then replaced with the original substrate in the following stage. This "switch" therefore looks for these antibodies' catalytic ability and screens for product formation. SUGA has also produced esterolytic antibodies by employing a different "bait and switch" tactic.

- **Strain Induced Hapten Design:** A minor modification of the transition state analog approach serves as the foundation for the strain-induced hapten design that elicits CatAbs. In this method, the substrate is chemically altered to distort the hapten that results. The strain-induced hapten causes changed substrate binding in the antibody combining site during hyper-immunization, which lowers the energy of the transition state and speeds up the chemical reaction. One substrate for which a 7G12 strain-induced hapten-based antibody mimic specific to the enzyme ferrochelatase was produced is N-methylmesoprphyrin. Ferrochelatase is an enzyme which catalyzes the insertion of Fe^{2+} ions into protoporphyrin IX, which is the last step in heme biosynthesis.
- **Reactive Immunization:** The goal of this hapten design strategy is to increase CatAbs' catalytic efficiency relative to that of natural enzymes. This is accomplished by utilizing a hapten that, during the hyperimmunization process, experiences conformational changes that trap chemical reactivity at the B-cell level. During immunization, a nucleophilic residue in the antibody combining state covalently traps the reactive hapten. The analog of the transition state is the trapped monoester.

10.5.3.2. Production of Catalytic Antibodies

Currently three methods are primarily used for eliciting catalytic antibodies. Polyclonal antibodies, Monoclonal antibodies and Antibody fragment based (scF_v or F_{ab}).

10.5.3.3. Preparation and Purification of Antibodies

The designed hapten was coupled via a butanoic acid linking arm to transferrin and bovine serum albumin (carriers) in order to raise antibodies and measure their binding affinities. By immunizing a suitable strain of mice or rabbit that has been cloned using hybridoma technology, antibodies were prepared. After being separated from serum, the antibodies underwent a gel electrophoresis procedure to verify their purity. Competitive ELISA will be used to further select antibody clones that exhibit strong binding to the hapten's substrate conjugate in order to facilitate catalysis studies.

10.5.3.4. Applications

The various applications of abzymes include:

- Catalytic antibodies are extremely important to the pharmaceutical and drug industries. Enzymes are designed to aid in the detoxification of drugs such as cocaine. CatAbs, which cleave the cocaine molecule at particular bonds, have been developed to eradicate the drug's toxic effects.

- TCA-tumor cell antigens, which are distinctive surface determinants of hyperproliferating cancer cells, are also utilized in Abyzme technology to destroy these cells. Antibodies specific to these TCAs, dubbed "magic bullets" because they have two distinct paratopes—one of which is bivalent and binds to both TCA and a cytotoxic prodrug or toxin—can be used to target and destroy cancerous cells. Patients receive an intravenous injection of the antibody, which binds specifically to TCA before prodrug is given. Subsequently, the prodrug is infused into the bloodstream; however, it only becomes active when the targeted antibody is in close proximity.
- The world's first catalytic antibody to be sold commercially is 38C2. This was employed as a catalyst for the aldol addition of numerous cyclic and open-chain aliphatic ketones to different aromatic and aliphatic aldehydes. It has been possible to identify over 100 distinct substrate combinations for cross- or intramolecular-aldol reactions.
- Novel catalytic antibodies were developed for synthetic carbohydrate synthesis, insect pheromones production, and for producing derivatives of the epothilone (anti-cancer drugs).
- Catalytic antibodies against allergy, viral and bacterial infections were also developed.

CASE STUDY

INTEGRATING DEVELOPMENT WITH EVOLUTION: A CASE STUDY WITH BEETLE HORNS

Developmental mechanisms are equally relevant to small-scale evolutionary processes because they may influence the response of a population to natural or sexual selection (e.g., Alberch, 1982; Riska, 1986; Pigliucci et al., 1996). For example, if some types of phenotypic variation are difficult or impossible to produce because of the nature of the underlying developmental processes, then these biases in the available phenotypic variation may constrain the direction of evolution in natural populations (Alberch, 1982; Bonner, 1982; Raff, 1996). Similarly, if some phenotypes are especially easy to produce (e.g., if many different genetic mutations all result in the same phenotype), then independent evolutionary lineages may be unusually likely to converge on the same phenotype (Nijhout 1991). Studies of patterning mechanisms in butterfly wings (Nijhout 1991, 1994a) and avian plumage (Price & Pavelka, 1996) reveal biases inherent in the types of color patterns that can be produced, so that some wing patterns are more easily generated than others. Similarly, studies of the development of tetrapod limbs reveal limits to the range of possible morphologies (Alberch & Gale 1985; Oster et al., 1988; Hinchliffe, 1994). In each of these examples, studies of mechanism have illuminated boundaries to the range of likely phenotypic variants, and this developmental perspective helps explain patterns of natural variation within and among related species. In principle, understanding how development affects the expression of morphological traits should help explain the evolution of those traits. However, empirical studies demonstrating an immediate relevance of development for understanding evolution in natural populations are rare because most population biologists do not study the developmental mechanisms regulating the expression of their traits of interest.

One trait in which to examine this question is the horns of beetles. The behavior associated with horns, the evolution of horns, and the development of horns have been explored for the same two species; consequently, it is now possible to integrate the results from these studies and to explore how knowledge regarding the mechanism of horn development influences our understanding of beetle horn evolution. In this article, I show how combining this information has enhanced our understanding of the evolutionary significance of beetle horns and illustrate how the integration of developmental studies with evolutionary studies may be used to identify unexpected avenues for future research.

- Sexual Selection and the Evolution of Beetle Horns

Beetle horns are rigid extensions of the exoskeleton that are as dramatic as the antlers of elk or moose (although, unlike antlers, beetle horns are not shed; beetles must bear their horns for their entire adult life). Horns have arisen repeatedly during the evolution of beetles, and many thousands of extant species have horns (e.g., Arrow, 1951; Enrodi, 1985). This rich evolutionary history is reflected in the tremendous variety of horn morphologies: long cylindrical rods, curved flat blades, distended mandibles, and sundry spines, knobs, or tubercles. Even in closely related beetle taxa, horn sizes, shapes, and locations can vary extensively.

Despite this diversity, all beetle horns are used for basically the same thing. Beetle horns generally are expressed only in males; in most species, females either have reduced horns or no horns. In all of the species studied so far, horns are used by males in combat with other males for access to resources used by females (e.g., Eberhard, 1978, 1982; Palmer, 1978; Siva-Jothy, 1987; Conner, 1988; Rasmussen, 1994; Emlen, 1997a). For example, stag beetles (*Dendrobias mandibularis*), harlequin beetles (*Acrocinus longimanus*), and some rhinoceros beetles (e.g., *Allomyrina dichotoma*) battle for small wounds or sap flows on trees that are visited by females (Goldsmith, 1987; Siva Jothy, 1987; Zeh & Zeh, 1992). Other rhinoceros beetles battle for emergent bamboo shoots or branches on shrubs that also are visited by females (e.g., *Golofa porteri*; Eber hard, 1978) or for burrows carved into plant stems, in which females feed and lay eggs (e.g., *Podischnus agenor*; Eberhard 1982). Male dung beetles use their horns in battles for access to subterranean tunnels containing females (Palmer, 1978; Otronen, 1988; Cook, 1990; Rasmussen, 1994; Emlen, 1997a). Although the resources in question vary considerably from species to species, they share one key characteristic: they all tend to occur in discrete, readily defendable patches. Males with enlarged weapons are able to gain disproportionate access to these contested sites and, consequently, gain disproportionate access to females (Eberhard, 1982; Goldsmith, 1987; Siva-Jothy, 1987; Zeh & Zeh, 1992; Rasmussen, 1994; Emlen, 1997a; Moczek & Emlen, 2000). But beetle horns do not come without costs. Horns are likely to constitute a considerable investment by the bearer. Beetle horns comprising more than 10% of the animal's total body mass are not uncommon, and such large structures are energetically expensive to produce. Horn growth prolongs beetle development and can increase the risk of larval mortality (e.g., from soil nematodes that kill larvae; Hunt & Simmons, 1997), and nutrients allocated to horn growth deplete resources that would have been available for growth of other tissues (Nijhout & Emlen, 1998). In addition, horns may impede the mobility or performance of individuals in ways that increase the risk of predation or starvation (as in Magnhagen, 1991). For example, the increased wing loading associated with horns may make flight more energetically expensive, and horns may physically impede movement in some substrates. As a result, dispersal, foraging, and escape may be more difficult for beetles with well-developed horns. Nevertheless, although beetle horns are likely to be expensive to generate and difficult to bear, they are useful for gaining reproductive access to females, and the reproductive benefits gained by having large horns presumably outweigh the costs of their production and maintenance. The intense sexual selection generated by contests over patchy, localized resources may have favored extravagant investment by males into the weapons that aid them in securing access to females, leading to the evolution of some of the world's most dramatic sexually selected traits. One genus of dung beetle that is characterized by prolific evolution of horns is the genus Onthophagus. This genus contains over 2000 described species, making it a contender for the largest genus of plant or animal in the world. In most of the species, males bear some type of horn, although the shapes and locations of these horns vary tremendously. Onthophagine beetles occur on six continents and span most climate and habitat extremes, from tropical rain forests to temperate plains, African savanna, and even the Australian outback and the deserts of the southwestern United States. They feed on dung from antelope, buffalo, elephants, rabbits, kangaroos, emus, tapirs, deer, horses, monkeys, marmots, and even packrats.

Despite this wide array of habitat types and food resources, all of these beetles appear to behave in basically the same way: females dig tunnels in the soil beneath the dung and bury the dung below ground to use as larval provisioning, sometimes with the assistance of males (Halffter & Edmonds, 1982; Cook, 1990; Rasmussen, 1994; Emlen, 1997a; Hunt & Simmons, 1998; Moczek, 1998). Males fight with other males for possession of these tunnels, and males successful at guarding tunnels gain reproductive access to the females inside (Cook, 1990; Rasmussen, 1994; Emlen, 1997a; Hunt & Simmons 1997; Simmons et al., 1999; Moczek & Emlen, 2000).

In two species, *Onthophagus taurus* and *Onthophagus acuminatus*, the reproductive behavior occurring inside tunnels has been studied in detail. Males of both species produce a pair of horns that extend from the base of the head. The length of these horns varies considerably among individuals, and this variation scales with body size: large males produce long horns, whereas small males produce much shorter horns. The scaling relationship between horn length and body size is sigmoidal in both species. The shape of this scaling relationship means that the transition between minimal and complete horn expression occurs abruptly, at approximately the body size that is associated with the inflection of the sigmoid curve (3.35 mm;). Males larger than this "threshold" size produce fully developed horns, whereas males smaller than this size produce only minimal horns or no horns at all. As a result, the distribution of horn sizes in both of these species is bimodal: populations contain large numbers of males with fully developed horns, large numbers of males with no or small horns, and relatively few males with horns of intermediate length (inset). Consequently, not only are these species sexually dimorphic with respect to horn expression (no females express horns), but a dimorphism exists within males. As it turns out, the two types of males use entirely different behavioral tactics to encounter and mate with females.

Females of both species dig tunnels into the soil beneath pieces of dung. They reside within these tunnels, where they sequester pieces of dung for themselves and for larval provisioning. Females may spend several days inside a single tunnel, pulling pieces of dung down to the blind ends of tunnel branches and packing the pieces into dense oval masses (brood balls) that will eventually serve as food for developing larvae (Bornemisszza, 1970; Halffter & Edmonds, 1982). In both *O. acuminatus* and *O. taurus*, females lay a single egg in a chamber at the top of each brood ball, and larvae complete their development in isolation within the confines of a single brood ball (Main, 1922; Emlen, 1994). Females mate repeatedly with males during the days preceding and including egg laying. Because females are inside the tunnels during this period, entry into a tunnel is a prerequisite for mating. Male reproductive behavior revolves around methods of securing access to the females inside these tunnels; in both species, hornless males accomplish this task in a very different way from horned males.

Large, horned males guard entrances to tunnels that contain females, and these males fight with all other males that try to enter the tunnel. Large males win significantly more fights than small males; moreover, when the effects of body size are experimentally controlled for, males with longer horns win significantly more fights than same-sized rival males with shorter horns (Emlen, 1997a; Moczek & Emlen, 2000; see Rasmussen 1994 for similar results in *Phanaeus difformis*). Consequently, the largest horned males are the most likely to successfully defend access to tunnels containing females, generating strong sexual selection for long horns in males. By contrast, small, hornless males generally are not effective at guarding entrances to tunnels, and these males are evicted from tunnels whenever they encounter a larger, horned male. Hornless males therefore employ an alternative tactic to encounter and mate with females. These males attempt to enter guarded tunnels by circumventing the guarding male, either by slipping undetected past him (Emlen, 1997a; Moczek & Emlen, 2000) or by digging a side tunnel that intercepts the guarded tunnel below ground (Emlen, 1997a). Occasionally, hornless males manage to sneak into guarded tunnels, mate with the female, and leave without being caught by the guarding male (Emlen, 1997a; Moczek & Emlen, 2000). If such "sneaker" males get caught by the resident male, not only do they get chased out of the tunnel, but the resident male returns to the female and mates with her immediately thereafter. This immediate remating may either dilute or displace sperm from a prior mating with the sneaker male (see Simmons et al. 1999 for discussion of sperm competition in *O. taurus*). In principle, horns can influence the performance of males using either behavioral tactic, but the effect of the horns will differ dramatically depending on the behavioral tactic used. Long horns are clearly beneficial to males that guard. They improve male fighting ability significantly (Emlen 1997a; Moczek & Emlen, 2000), probably because they provide leverage during contests and because they help block the tunnel entrance, making it more difficult for intruding males to push past the guarding male and into the tunnel. In contrast, horns appear detrimental to sneaker males, for the simple reason that they get in the way. Success at sneaking depends on rapid, undetected entry into tunnels. Horns impede movement below ground by scraping against tunnel walls as beetles run, which can slow males down (Moczek & Emlen, 2000) and generate vibrations that alert resident males to the presence of an intruder (Douglas J. Emlen, unpublished observations). Consequently, the alternative methods for encountering females may have selected for opposing horn morphologies in large and small males. Males larger than most of their competitors can effectively guard tunnels, and these males guard tunnels more successfully if they have long horns. Small males enter guarded tunnels on the sly, and these males may enter and exit tunnels faster, and with less risk of detection, if they do not have horns. The observed patterns of sexual selection on horns are consistent with the morphological variation present in natural populations, where large males produce the horns and smaller males do not. How is this adaptive association between horn expression and body size achieved? Experiments conducted on both of these species make it possible to begin to address this question.

- The Development of Beetle Horns

Male horn expression in beetles is facultative. Despite the importance of horns to sexual selection and male reproductive success, horn production shows insignificant levels of heritable genetic variation in natural populations (e.g., $h^2 = 0.006$, $P = 0.98$ in *O. acuminatus*, Emlen 1994; $h^2 = -0.162$, $P = 0.59$ in *O. taurus*, Moczek & Emlen, 1999). Instead, nutritional conditions encountered by each larva as it develops determine both the final adult body size of that animal and the length of its horns (Emlen 1994, 1996, 1997b, Hunt & Simmons, 1997, in press, Moczek, 1998; Moczek & Emlen, 1999). Males who encounter favorable larval conditions (e.g., large amounts of food) grow large and produce horns, whereas related (even sibling) males who encounter poor larval conditions remain small and do not produce horns. The developmental switch between hornlessness and horn growth occurs abruptly around a threshold body size: Individual males whose body sizes are below this threshold do not produce horns, and those whose body sizes are larger do produce horns. But what exactly are beetle horns, when do they grow, and how is their phenotypic expression coupled with growth in body size?

- *Horns Grow During the Larval Period*

Onthophagine beetles develop in isolation inside balls of dung that have been buried below ground (Main 1922; Halffter & Edmonds, 1982). After hatching, beetles pass through three larval stages, undergoing exponential growth for most of this period. At the end of the third larval stage, the animals stop feeding and begin to purge their gut of all contents in preparation for metamorphosis. During this prepupal period they synthesize most of their adult structures, including their horns. Cells in localized regions of the epidermis that have detached from the larval exoskeleton (called imaginal discs) divide prolifically to produce the adult eyes, legs, wings, genitalia, antennae, and mouthparts (for review, see Cohen, 1993). Consequently, a late third-instar larva (i.e., a prepupa) already has many adult morphological traits. But these structures are not visible on the outside—the larva still looks like a larva. Instead, these structures are folded up and tucked away on the inside of the animal. They unfold to take their adult form during molting, when beetles emerge from their old larval exoskeleton and expand to unwrinkle and fill out their new pupal exoskeleton. Beetle horns develop at the same time and in the same manner as these other adult structures—that is, they begin as regional invaginations of the larval epidermis (Emlen & Nijhout, 1999). In *O. taurus*, two small pockets of epidermis detach from the cuticle at the base of the head— one for each of the two horns. These detached regions of epidermis undergo rapid cell divisions and dramatically increase in surface area, producing long, hollow tubes. However, because the outer larval cuticle is still in place, the horns cannot extend as they grow larger, so they remain trapped inside the larval exoskeleton (in *O. taurus*, the growing horns are actually trapped inside the rigid larval head capsule). Consequently, these tubes of growing cells invaginate repeatedly to produce a series of dense, concentric folds. Externally, these animals still do not have visible horns, but large male larvae now have bulging pockets of folded epidermis trapped beneath the head cuticle. When the animals shed their larval cuticle, the folded sacs expand rapidly to form the fully extended horns that are visible in pupae and adults. In small males (below the threshold size), this same region of larval epidermis grows minimally, and the males emerge with only rudimentary horns or with no horns (Emlen & Nijhout, 1999).

With this knowledge of the details of horn development, we can refine the question of how horn growth is regulated. Factors influencing the rate or amount of growth of these locally detached regions of larval epidermis ultimately determine whether a male produces horns. It is the localized growth of these pockets of larval epidermal cells that appears to be regulated by the body size threshold: males growing larger than a critical body size show pronounced cell proliferation in the regions where the horns will form, whereas males that have not attained this critical body size show minimal cell proliferation. Somehow, then, developing animals are able to incorporate information about their body size into the "decision" of whether or not to grow horns.

- *Horn Growth Is Mediated by Juvenile Hormone*

This developmental decision appears to be mediated by levels of a circulating hormone in the blood of larvae (Emlen & Nijhout, 1999). Hormones serve as messengers between tissues and may initiate changes in patterns of gene expression within cells expressing appropriate hormone receptors (e.g., Nijhout, 1994b). In *O. taurus*, male horn growth is mediated by juvenile hormone (Emlen & Nijhout, 1999). At the end of the third larval stage, after larvae have completed all of their growth in body size and have begun to purge their guts of the remaining contents, the epidermal cells that will develop into horns become sensitive to the presence and level of juvenile hormone (JH). During a brief (30-hour) period, the hemolymph level of JH determines the amount of proliferation in these cells. High levels of JH appear to initiate rapid cell proliferation, resulting in future horned adults, whereas reduced levels of JH initiate only minimal cell proliferation, resulting in future hornless adults (Emlen & Nijhout, 1999). But what determines which individuals will have high levels of JH and which will not?

- *Levels of JH Are Affected by Each Individual's Body Size*

Hormone release is often triggered by the sensory system, and such release is one way in which insects regulate developmental events based on stimuli perceived from their environment (Nijhout 1994b, 1999). Sensory stimuli can include abiotic aspects of the physical environment, such as photoperiod or temperature—for example, hormonal changes in response to photoperiod initiate the onset of overwintering diapause (e.g., Tauber et al., 1986). Stimuli can also include aspects of the social or nutritional environment, as when hormonal changes initiate wing production by insects developing under crowded or deteriorated conditions (e.g., Pener, 1985; Zera & Denno, 1997). Finally, sensory stimuli can pertain to the growth or size of the animal itself. Indeed, a number of developmental events are initiated only after animals attain a certain critical body size (e.g., the onset of metamorphosis in tobacco hornworms is body size dependent; Nijhout & Williams 1974. Nijhout, 1975, 1979, 1994b, Jones et al., 1981). In this final situation, conditions related to the body size of the animal (e.g., stretch in the softer regions of the larval cuticle; Beckel & Friend, 1964; Anwyl, 1972; Nijhout, 1981, 1984; Chiang & Davey, 1988) influence cellular events within the developing insect by affecting the levels of circulating hormones (Nijhout, 1994b). One hormone that may regulate body size–dependent expression of traits is JH (Wheeler & Nijhout, 1984; for additional possibilities, see Stern & Emlen 1999). In a variety of insects, larval nutritional conditions (e.g., the amount or quality of food) and larval growth (e.g., attainment of a critical body size) directly affect circulating levels of JH (Johansson, 1958; Asencot & Lensky, 1976; Lenz, 1976; Velthius, 1976; Dogra et al., 1977, de Wilde & Beetsma, 1982; Rembold, 1987; Rachinsky & Hartfelder, 1990). For example, JH levels are thought to correlate with larval body size in the ant *Pheidole bicarinata*, and JH regulates the body size–dependent switch between worker and soldier development in that species (Wheeler & Nijhout, 1983, 1984). Similar mechanisms may operate in *O. taurus*. Although the exact levels of JH in *O. taurus* larvae have not yet been measured, it appears that larger animals have higher concentrations of JH during this critical period than smaller individuals (see Emlen & Nijhout, 1999). With this mechanism, cells initiating horn growth in response to high concentrations of JH also end up initiating horn growth only in the largest individuals.

The most convincing evidence that JH links horn growth with body size in beetles comes from an experiment that perturbed JH levels during the 30-hour sensitive period for horn growth. By experimentally augmenting levels of JH, it was possible to induce tiny *O. taurus* males to produce horns (the morphology of large males was unaffected by this manipulation; Emlen & Nijhout, 1999). This finding suggests that amplifying the levels of JH in small males caused the cells that grew to become horns to "misread" the body sizes of those individuals. This ability of extra JH to "trick" small males into producing morphological traits that typically are expressed only at large body sizes implicates JH as the link between larval body size and horn growth.

It is also noteworthy that horn length is not determined until very late in larval development. The cells that produce the horns respond to perturbations of JH only during a brief period that occurs after larvae have stopped feeding and have begun to purge their guts (Emlen & Nijhout, 1999). It is at the end of the feeding stage that larvae can fully assess their nutritional state. At this time, all increases in body size have ceased, and the size of a larva should accurately predict the final body size of that animal as an adult (as in insects generally, beetles do not grow as adults). If horn lengths match the size of a larva during this prepupal period, they will also match the size of the adult that emerges.

- *Synthesis of Horn Development*

The findings discussed in this section show that beetle horns result from localized regions of larval epidermis that undergo a very rapid period of growth immediately before pupation. The amount of proliferation in these localized regions of epidermis ultimately determines the length of the adult horns. In *O. acuminatus* and *O. taurus*, horn growth is affected by how large each individual becomes, so that large male larvae show extensive growth in these epidermal regions and small male larvae show only minimal growth. The developmental switch between minimal and complete horn growth occurs abruptly and is associated with attaining a specific threshold body size. Body size, in turn, is affected by the nutritional environment. Larvae encountering favorable conditions grow large, whereas larvae encountering poor conditions remain small. Consequently, effective matches between horn length and body size require a mechanism that translates the actual growth conditions encountered by each larva into levels of horn growth. In *O. taurus*, this translation is mediated by circulating levels of JH during a brief period just before pupation (and after all growth in body size has been completed). During this period, large males appear to have higher concentrations of JH than small males, and high levels of JH initiate growth of the cells that produce the horns. The threshold body size for producing horns thus appears to be a threshold of sensitivity to JH: cells encountering a sufficient concentration of JH initiate a period of exponential growth and produce long horns, whereas cells not encountering sufficient JH show only minimal horn growth. Because horn production takes place very late in the larval period, after all increases in body size have ceased, the amount of horn growth is matched with the actual individual variations in body size, even though body size is influenced by external environmental conditions encountered by larvae as they develop. The result is a facultative mechanism of horn expression, in which horns are produced consistently in only the largest males.

A Developmental Perspective Adds to the Study of Beetle Horn Evolution

In addition to being interesting in its own right, an understanding of horn development has important implications for studies of beetle horn evolution. Information pertaining to this proximate regulatory mechanism enhances our perception and interpretation of sexual selection in beetles, illuminating how horns evolve, identifying constraints on horn evolution, and revealing hidden costs of horn production.

- **How Horns Evolve**

Even a preliminary grasp of the mechanisms that regulate the expression of morphological traits can offer insights into how specific evolutionary changes arise. The low heritabilities for beetle horn length might be interpreted as indicating that these structures have limited potential for evolutionary modification. In fact, this conclusion is not true, as evidenced by the tremendous diversity of horn types present among even closely related taxa. By studying the mechanisms regulating horn growth, we can begin to appreciate how horn evolution might have occurred.

- **Evolution of the Threshold**

For natural populations of *O. acuminatus*, there exists an average threshold body size at which the switch from minimal to complete horn growth occurs (3.35 mm). However, individuals within these populations vary slightly in the precise body size at which they initiate horn growth, and in *O. acuminatus* these differences are heritable (Emlen, 1996). Evolutionary shifts in this threshold body size were generated through artificial selection experiments (Emlen, 1996). After seven generations of selection, one population initiated horn growth at a significantly smaller body size than the starting population, and another population that had been selected in the opposite direction initiated horn growth at a significantly larger body size. Microevolutionary changes in the threshold size for horn production illustrate how genetic variation in components of a developmental regulatory mechanism can lead to evolutionary changes in the expression of morphological traits. Although shifts in the *O. acuminatus* threshold were generated in the laboratory, similar evolutionary changes have occurred in natural populations. Indeed, a comparison of several onthophagine species reveals that one major avenue of evolutionary change in this group has involved shifts in the threshold body size regulating horn growth. But what is a threshold trait, and how can it be modified by selection? By combining what is known about endocrine mechanisms that regulate the expression of threshold traits in other insects with results from experiments exploring horn development, I can now suggest a plausible scenario for how this threshold trait might evolve. Although several parts of this model are still untested (and several alternative possibilities exist), it allows us to begin to visualize how evolutionary changes in horn morphology may have arisen.

- **How Might This Threshold Evolve?**

Growth in horn length is regulated by a hormone. Hormones regulate a diverse array of developmental processes in insects, yet insects have surprisingly few hormones (Nijhout, 1994b). One of the ways a limited number of hormones can regulate so many processes is that not all cells are sensitive to a particular hormone at the same time. Often, specific tissues respond to hormones only during brief critical periods, or windows, of sensitivity (Nijhout, 1994b). In this way, a single hormone may serve different functions at different times. What determines when cells will respond to the presence of a hormone? Whether specific cells are sensitive to a hormone at any particular time depends on the receptors present in the cells: cells become sensitized to a hormone when they synthesize receptors for that hormone (e.g., Nijhout, 1994b, 1999; Fujiwara et al., 1995; Jindra et al., 1996). The duration of time that receptors are present determines the length of the critical period when cells are sensitive, and the density of receptors presumably determines how sensitive those cells are to that hormone at that time. Cells expressing high densities of receptors should be more sensitive to the hormone (and should therefore respond to lower concentrations of the hormone) than similar cells expressing fewer receptors. In *O. taurus*, the cells that will become the horns are sensitive to JH for a brief period at the end of larval development. Cells that are exposed to sufficiently large doses of JH during this period undergo rapid cell division and produce long horns. If these same cells do not encounter sufficient levels of JH, they do not produce horns. The threshold body size for horn production appears to correspond to a level of JH that is just high enough to bind to sufficient receptors to initiate horn growth. In this mechanism, whether horns are produced depends on two factors: first, how large each animal grows to be (which is reflected in the circulating levels of JH during the sensitive period) and, second, how sensitive that animal is to the level of JH (as determined by the concentration of JH receptors expressed during the sensitive period). Levels of both JH and JH receptors can vary among individuals, and genetic variation affecting the levels of either could lead to an evolutionary shift in the threshold body size at which horn growth is initiated. Mutations affecting how much JH is secreted, when JH is secreted, and how quickly JH is removed or inactivated would all change the level of JH that is present during the sensitive period, and any of these mutations could result in a shift in the threshold body size for horn growth. However, mutations affecting circulating levels of a hormone are likely to affect the expression of other morphological traits as well. All other tissues that share the same sensitive period as horns would be affected by a change in hormone level (and most of the adult structures grow at the same time as horns). In contrast, the second possible avenue for horn evolution involves mutations whose effects are tissue specific. Mutations affecting the sensitivity of particular cells (e.g., the cells that produce the horns) to a circulating hormone will affect the expression of the structure produced by those cells but should not interfere with the endocrine regulation of other traits. For this reason, changes in the sensitivity of the target cells may be a more likely avenue for horn evolution than changes in levels of circulating hormones. Consequently, for the remainder of this article, I focus on the potential implications of evolutionary changes in the sensitivity of target cells. If the cells destined to produce horns become sensitized to JH by synthesizing the appropriate receptors during the critical period, then it is possible that beetle genotypes vary in their sensitivity to JH by varying in the expression of JH receptors. Genotypes overexpressing receptors in horn imaginal cells would be more sensitive to JH and would initiate horn growth at a lower threshold concentration of hormone, and thus at a proportionately smaller body size. In this case, quantitative differences in the critical size for horn production could result from something as simple as variation in the rate of receptor production, and it is tempting to speculate that the interspecific evolutionary changes in the relationship between male horn length and body size were brought about in this way.

- *Evolution of Horn Location, Horn Size, and Sexual Dimorphism*

Visualizing the mechanism regulating horn growth may inform us about other types of horn evolution as well. The genus Onthophagus is characterized by extensive variation in horn morphology. Species differ in the location of horns, the relative sizes of horns, and which sex expresses horns. For the first time, we can begin to envision how these evolutionary changes may have been brought about because each avenue of morphological diversification can now be interpreted in light of the basic regulatory mechanism described above. Extant species of Onthophagus bear horns on several different parts of the body. Many species have horns that extend from the base of the head, as in *O. acuminatus* and *O. taurus*. Other species have horns that extend from the front of the head, or clypeus. Still others have horns that extend from the thorax. In addition, some species exhibit multiple horns, with all possible combinations of horn locations, and it is clear from preliminary phylogenetic analyzes that horn locations have changed multiple times within the history of the genus (Douglas J. Emlen, Clifford Cunningham, Leigh W. Simmons, Jen Marangelo, unpublished data). If horn growth is initiated by localized clusters of epidermal cells expressing receptors to JH during the critical period, then changing which cells express the receptors may also change which regions of epidermis produce the horns. In this case, evolutionary changes in the location of horns simply may require shifts in the specific clusters of epidermal cells that express receptors to JH during the horn-specific critical period. Consequently, it is possible to begin to visualize major (interspecific) morphological transformations that have occurred within this genus as having resulted from simple, graded alterations in JH receptor expression. A second way that onthophagine horns differ is in their relative sizes. Horns range from enormous extensions that constitute more than 10% of total body mass in some species to tiny knobs in other species. Some species do not express horns at all. Again, we can now begin to understand how evolutionary changes in the relative sizes of horns may have been brought about. One possibility is that there have been changes in the number of cells that are sensitive to JH during the critical period for horn growth. Because these cells grow exponentially to produce the horns, changing the starting conditions (i.e., the size of the patch of epidermis that grows in response to JH) can lead to changes in the final horn size. Animals with many epidermal cells expressing JH receptors at the sensitive time for horn development would grow longer horns than animals with fewer dividing cells (i.e., fewer initial cells expressing receptors for JH). Finally, sex-specific expression of these same receptors may underlie the evolution of sexual dimorphism in horn production. Most species of Onthophagus are sexually dimorphic for horns. Female larvae encounter the same range of environmental conditions as do males, and female body size varies over the same range as males. Yet in virtually all species, females do not produce horns. The simplest way to achieve sexual dimorphism, given this regulatory mechanism, would be for females to not express JH receptors in the cells that would otherwise become horns. Females in these species would be insensitive to circulating JH (in terms of horn production), and consequently would not produce horns. This same mechanism could also account for the few exceptional species in which females do express horns. By expressing JH receptors, these females may be able to initiate patterns of growth typically expressed only in males. All of these hypotheses about beetle horn evolution generate explicit predictions that should be testable in the near future. For example, the major differences between species are predicted to arise from changes in one or more of the following: which sex expresses receptors to JH at the appropriate time, which cells express receptors for JH at the appropriate time, how many cells express these receptors, and how many receptors are produced by each of these cells. One way to assess if, and how, any of these factors differ among species uses fluorescently labeled antibodies to characterize the cellular distribution of JH receptors. Once antibodies have been generated (JH receptors are only just being described; Wyatt & Davey, 1996; Jones & Sharp, 1997; Ashok et al., 1998), it should be feasible to rapidly assess where and when receptors are expressed (by assaying which regions of the larval epidermis bind the labeled antibody) and how many receptors are expressed (by quantifying how much binding occurs) in related species that differ in specific aspects of their horn morphology. Although much of this analysis is preliminary and many alternative possibilities exist to explain the evolution of beetle horns (e.g., horn evolution could have resulted from changes in hormone levels or from evolution of downstream patterning and growth genes that respond to hormone signals, instead of from changes in the expression of hormone receptors), this approach illustrates how even a rudimentary understanding of developmental mechanisms can help fill in the "black box" connecting genotype with phenotype. By learning how horn expression is regulated, we can begin to think about how horn expression might be modified. Specifically, we can explore how perturbations to the mechanism of horn expression would affect the resulting phenotype, and, using these results, we can begin to think about what types of modifications are likely to have occurred in the past to generate the diversity of forms that exists today.

- *Constraints on Horn Evolution*

Developmental processes translate genotype to phenotype, and modifications to developmental mechanisms generate morphological variation within populations. Once generated, variant individuals act and interact within their social and physical environments, and those individuals that are most successful at surviving and reproducing contribute disproportionately to subsequent generations. When specific traits or combinations of traits contribute to the relative success of individuals within populations, and when variation in the expression of these traits is heritable, then populations may evolve toward increased expression of the relevant traits. However, natural and sexual selection can act only on available phenotypic variation. Consequently, evolutionary responses to selection may be limited by the variation that the developmental mechanisms are capable of generating (Alberch 1982; Oster et al., 1988; Arnold, 1992). If developmental processes were infinitely labile, then any phenotypic variant should be possible, and, in theory, such a population should be capable of evolving in any direction. Developmental phenomena often are not infinitely flexible, however, and certain morphologies may be mechanistically much easier to produce than others. In fact, in some cases, specific morphologies (or combinations of traits) are extremely difficult—or even impossible—to produce, given a particular developmental mechanism. In these cases, the translation of genotype to phenotype leads to a biased production of morphologies, so that some forms are produced far more readily than others. By biasing the generation of phenotypes within a population, developmental processes can influence the likelihood of various directions of evolution, thus biasing or constraining the evolution of those lineages (Alberch 1982; Arnold, 1992). Our developmental studies (e.g., Nijhout & Emlen, 1998) concerning the regulation of expression of beetle horns suggest that constraints operate to limit the range of possible beetle morphologies.

- *Beetle Horns Versus Beetle Eyes*

The phenotypic expression of horns is not independent of all other morphological traits. In natural populations of both *O. acuminatus* and *O. taurus*, horns are negatively correlated with eyes; males with unusually long horns also have relatively small eyes, and males with small or no horns have relatively large eyes (Nijhout & Emlen, 1998). In addition, males in general have smaller eyes than females, which do not produce horns (Douglas J. Emlen, unpublished observations). Such negative phenotypic correlations can result from a number of causal factors (e.g., Falconer, 1989). For example, this pattern could result from horns and eyes exhibiting inverse but independent responses to the larval environment. In this case, evolutionary changes in one of the traits (e.g., horn enlargement) need not affect the expression of the other trait (eyes), and it would be inaccurate to consider the present association among horns and eyes as indication of a constraint on the evolution of either trait. An alternative possibility is that this phenotypic correlation results from a limitation in the capacity of the underlying regulatory mechanism to generate morphological variants. For example, if a resource allocation tradeoff mandates a negative correlation between horn length and eye size, then any evolutionary modification to the relative size of one of the traits will necessarily influence the expression of the other. Such a finding would indicate a bias in the generation of phenotypic variation because individuals with large horns and large eyes would be difficult to produce. It would also constitute a true source of constraint on the evolution of beetle shape because enlargement of one trait would occur only at the expense of the other.

To examine whether the negative phenotypic correlation between horns and eyes reflects a limitation in the range of phenotypes that can be generated by this regulatory mechanism, my collaborator and I used three separate experimental methods (JH application, diet manipulation, and artificial selection) to manipulate male allocation to horns during development (Emlen, 1996; Nijhout & Emlen, 1998). Each experiment generated populations of males that differed in the relative lengths of their horns. In all three cases, males induced to produce relatively long horns had smaller eyes, and vice versa. The first two of these methods (JH application and diet manipulation) operated within a single generation of beetles. Topical applications of methoprene (a JH mimic) to feeding-stage larvae and manipulations of larval diet quality each affected the relative length of horns produced by developing males (Emlen, 1997b; Nijhout & Emlen, 1998). By altering male investment in horns during growth, these experiments revealed a developmental linkage between these two traits. Horn expression was inversely coupled with eye expression, and perturbing the expression of one trait influenced expression of the other. The third experiment altered male investment in horns across generations by artificially selecting on the relative length of the horns. This evolutionary change in male investment in horns also brought about an inversely correlated response in the eyes: males selected for unusually long horns also had small eyes, whereas males selected for unusually short horns had large eyes. This result indicated that the negative correlation between horns and eyes can bias the morphological evolution of these traits.

The observation of a negative genetic correlation between male horns and male eyes indicates that there is a limit to the range of phenotypes that can be produced; males with long horns and small eyes are quite easy to produce, as are males with small horns and large eyes, but males with long horns and large eyes are very difficult to generate. This paucity of certain shapes in beetles may explain why evolution of relative horn length also has brought about a correlated (although inverse) change in relative eye size. It also suggests that there may be an upper limit to enlargement of male horns and illustrates one way that a developmental process may bias or constrain evolution.

CLASS ACTIVITY

1. What conclusions can you draw about the relationship between the way in which the present-day theory of evolution developed and the credibility of the theory? Explain your thinking.
   a. When the theory of evolution was first proposed, it met with a lot of criticism and disbelief, but it is widely supported today. Theories that have withstood a larger amount of criticism are more credible than those that are accepted easily
   b. The theory of evolution has its foundation in both biological and geological observations, making it a more credible theory because it can explain more about the world
   c. The theory of evolution relies on the heritability of traits, but the mechanism of this inheritance was not understood when the theory was developed. This reduces the credibility of the theory because the people who created it did not understand how it worked
   d. It is meaningful that two naturalists working independently from each other offered the same explanation for the same set of phenomena. When two people independently look at the same evidence and come to the same conclusion, this reinforces the credibility of that conclusion
2. Describe how an adaptation, such as better running speed, relates to natural selection
   a. If the adaptation of a better running speed is controlled by genetics and is beneficial to survival and reproduction, evolution will favor it.
   b. If the adaptation of better running speed is favorable in a given environment, it will be controlled by genetics and evolution will increase it.
   c. If the adaptation of better running speed is beneficial to survival and reproduction, random motion will naturally select it.
   d. If the adaptation of better running speed is favored by natural selection, the organism will naturally select an environment where it can run the farthest.

SUMMARY

- Molecular evolution describes how inherited DNA and/or RNA change over evolutionary time, and the consequences of this for proteins and other components of cells and organisms. Molecular evolution is the basis of phylogenetic approaches to describing the tree of life.
- Mutations are permanent, transmissible changes to the genetic material (DNA or RNA) of a cell or virus. Mutations result from errors in DNA replication during cell division and by exposure to radiation, chemicals, other environmental stressors, viruses, or transposable elements.
- Molecular selection and evolution are central themes in understanding the dynamics of genetic variation and the mechanisms driving the diversity of life. Natural selection operates on genetic variation, favoring alleles that confer adaptive advantages in a given environment.
- Molecular evolution is the putative engine of macro-evolution and is thought to involve primarily successions of single (point) mutations leading to ever more enhanced robust systems that in turn promote the well-being and competitiveness of the self-assembled living system.
- Asymmetric catalysis is the ability of a chiral catalyst to catalyze an achiral substrate to a chiral product with a bias for the formation of one of the enantiomers.
- Nucleic acids can play roles far beyond merely harboring the coding information for proteins. Single-stranded nucleic acids can fold into intricate structures capable of molecular recognition and even catalysis.
- A polydeoxynucleotide library is synthesized for molecular breeding by standard solid phase methodologies except that randomized single-stranded DNA (pre-template DNA) is synthesized with a 3' -flanking region of known sequence that can be recognized by a primer in a PCR reaction.

REVIEW QUESTIONS

1. How does natural selection operate at the molecular level?
2. Explain the relationship between self-organization and the emergence of life.
3. What are the primary hypotheses regarding the origin of biological macromolecules?
4. Describe an example of a molecular adaptation driven by natural selection.
5. Define molecular breeding and its purpose.
6. What is random mutagenesis and how is it performed using PCR?
7. What are the benefits of using DNA shuffling in directed evolution experiments?
8. Discuss the importance of sensitivity and specificity in screening assays.
9. Define DNA aptamers and their applications.

10. Explain the mechanism by which a catalytic antibody catalyzes a chemical reaction.

REFERENCES

1. Bofkin, L., & Goldman, N. (2006). Variation in evolutionary processes at different codon positions. *Molecular Biology and Evolution, 24*(2), 513–521.
2. Dietrich, M. R. (1998). Paradox and persuasion: Negotiating the place of molecular evolution within evolutionary biology. *Journal of the History of Biology, 31*(1), 85–111.
3. Duret, L., & Galtier, N. (2009). Biased gene conversion and the evolution of mammalian genomic landscapes. *Annual Review of Genomics and Human Genetics, 10*, 285–311.
4. Fay, J. C., Wyckoff, G. J., & Wu, C. I. (2002). Testing the neutral theory of molecular evolution with genomic data from Drosophila. *Nature, 415*(6875), 1024–1026.
5. Freese, E. (1962). On the evolution of the base composition of DNA. *Journal of Theoretical Biology, 3*(1), 82–101.
6. Graur, D., & Li, W.-H. (2000). *Fundamentals of Molecular Evolution* (2nd ed.). Sinauer.
7. Kimura, M. (1983). *The Neutral Theory of Molecular Evolution*. Cambridge University Press.
8. Kumar, S. (2005). Molecular clocks: Four decades of evolution. *Nature Reviews Genetics, 6*, 654–662.
9. Lynch, M. (2007). *The Origins of Genome Architecture*. Sinauer.
10. Nei, M., & Kumar, S. (2000). *Molecular Evolution and Phylogenetics*. Oxford University Press. Retrieved from: http://www.megasoftware.net/ (accessed on 12 July 2024).
11. Nielsen, R. (2005). Molecular signatures of natural selection. *Annual Review of Genetics, 39*, 197–218.
12. Yang, Z. (2006). *Computational Molecular Evolution*. Oxford University Press.

Index

C

D

E

FSC
www.fsc.org
MIX
Papier | Fördert gute Waldnutzung
FSC® C083411

Kolibri 360 GmbH
Ferdinand-Jühlke-Straße 7
99095 Erfurt, Deutschland
produktsicherheit@kolibri360.de